# Recent Advances in Sedimentology

# Recent Advances in Sedimentology

Guest Editors

**George Kontakiotis**
**Angelos G. Maravelis**
**Avraam Zelilidis**

Basel • Beijing • Wuhan • Barcelona • Belgrade • Novi Sad • Cluj • Manchester

*Guest Editors*

George Kontakiotis
National and Kapodistrian
University of Athens
Athens
Greece

Angelos G. Maravelis
Aristotle University of
Thessaloniki
Thessaloníki
Greece

Avraam Zelilidis
University of Patras
Patras
Greece

*Editorial Office*
MDPI AG
Grosspeteranlage 5
4052 Basel, Switzerland

This is a reprint of the Special Issue, published open access by the journal *Journal of Marine Science and Engineering* (ISSN 2077-1312), freely accessible at: https://www.mdpi.com/journal/jmse/special_issues/cl_gsg2022_sedimentology.

For citation purposes, cite each article independently as indicated on the article page online and as indicated below:

Lastname, A.A.; Lastname, B.B. Article Title. *Journal Name* **Year**, *Volume Number*, Page Range.

**ISBN** 978-3-7258-2745-9 (Hbk)
**ISBN** 978-3-7258-2746-6 (PDF)
https://doi.org/10.3390/books978-3-7258-2746-6

© 2024 by the authors. Articles in this book are Open Access and distributed under the Creative Commons Attribution (CC BY) license. The book as a whole is distributed by MDPI under the terms and conditions of the Creative Commons Attribution-NonCommercial-NoDerivs (CC BY-NC-ND) license (https://creativecommons.org/licenses/by-nc-nd/4.0/).

# Contents

About the Editors . . . . . . . . . . . . . . . . . . . . . . . . . . . . . . . . . . . . . . . . . vii

Preface . . . . . . . . . . . . . . . . . . . . . . . . . . . . . . . . . . . . . . . . . . . . . . . ix

George Kontakiotis, Angelos G. Maravelis and Avraam Zelilidis
Recent Advances in Sedimentology
Reprinted from: *J. Mar. Sci. Eng.* **2024**, *12*, 1935, https://doi.org/10.3390/jmse12111935 . . . . . 1

Ahmer Bilal, Renchao Yang, Muhammad Saleem Mughal, Hammad Tariq Janjuhah,
Muhammad Zaheer and George Kontakiotis
Sedimentology and Diagenesis of the Early–Middle Eocene Carbonate Deposits of the
Ceno-Tethys Ocean
Reprinted from: *J. Mar. Sci. Eng.* **2022**, *10*, 1794, https://doi.org/10.3390/jmse10111794 . . . . . 6

Christos Kanellopoulos, Vasiliki Lamprinou, Artemis Politi, Panagiotis Voudouris, Ioannis
Iliopoulos, Maria Kokkaliari, et al.
Speleothems and Biomineralization Processes in Hot Spring Environment: The Case of Aedipsos
(Edipsos), Euboea (Evia) Island, Greece
Reprinted from: *J. Mar. Sci. Eng.* **2022**, *10*, 1909, https://doi.org/10.3390/jmse10121909 . . . . . 34

Leonidas Moforis, George Kontakiotis, Hammad Tariq Janjuhah, Alexandra Zambetakis-Lekkas,
Dimitrios Galanakis, Panagiotis Paschos, et al.
Sedimentary and Diagenetic Controls across the Cretaceous—Paleogene Transition: New
Paleoenvironmental Insights of the External Ionian Zone from the Pelagic Carbonates of the
Gardiki Section (Epirus, Western Greece)
Reprinted from: *J. Mar. Sci. Eng.* **2022**, *10*, 1948, https://doi.org/10.3390/jmse10121948 . . . . . 47

Isabella Lapietra, Stefania Lisco, Luigi Capozzoli, Francesco De Giosa, Giuseppe Mastronuzzi,
Daniela Mele, et al.
A Potential Beach Monitoring Based on Integrated Methods
Reprinted from: *J. Mar. Sci. Eng.* **2022**, *10*, 1949, https://doi.org/10.3390/jmse10121949 . . . . . 70

Qian Zhang, Mingming Tang, Shuangfang Lu, Xueping Liu and Sichen Xiong
Effects of Mud Supply and Hydrodynamic Conditions on the Sedimentary Distribution of
Estuaries: Insights from Sediment Dynamic Numerical Simulation
Reprinted from: *J. Mar. Sci. Eng.* **2023**, *11*, 174, https://doi.org/10.3390/jmse11010174 . . . . . . 94

Avraam Zelilidis, Nicolina Bourli, Konstantinos Andriopoulos, Eleftherios Georgoulas,
Savvas Peridis, Dimitrios Asimakopoulos and Angelos G. Maravelis
Unraveling the Origin of the Messinian? Evaporites in Zakynthos Island, Ionian Sea: Implications
for the Sealing Capacity in the Mediterranean Sea
Reprinted from: *J. Mar. Sci. Eng.* **2023**, *11*, 271, https://doi.org/10.3390/jmse11020271 . . . . . . 112

Syed Kamran Ali, Rafiq Ahmad Lashari, Ali Ghulam Sahito, George Kontakiotis, Hammad
Tariq Janjuhah, Muhammad Saleem Mughal, et al.
Sedimentological and Petrographical Characterization of the Cambrian Abbottabad Formation
in Kamsar Section, Muzaffarabad Area: Implications for Proto-Tethys Ocean Evolution
Reprinted from: *J. Mar. Sci. Eng.* **2023**, *11*, 526, https://doi.org/10.3390/jmse11030526 . . . . . . 126

**Mubashir Mehmood, Abbas Ali Naseem, Maryam Saleem, Junaid ur Rehman, George Kontakiotis, Hammad Tariq Janjuhah, et al.**
Sedimentary Facies, Architectural Elements, and Depositional Environments of the Maastrichtian Pab Formation in the Rakhi Gorge, Eastern Sulaiman Ranges, Pakistan
Reprinted from: *J. Mar. Sci. Eng.* **2023**, *11*, 726, https://doi.org/10.3390/jmse11040726 . . . . . . **150**

**Kamil A. Qureshi, Mohamad Arif, Abdul Basit, Sajjad Ahmad, Hammad Tariq Janjuhah and George Kontakiotis**
Sedimentological Controls on the Reservoir Characteristics of the Mid-Triassic Tredian Formation in the Salt and Trans-Indus Surghar Ranges, Pakistan: Integration of Outcrop, Petrographic, and SEM Analyses
Reprinted from: *J. Mar. Sci. Eng.* **2023**, *11*, 1019, https://doi.org/10.3390/jmse11051019 . . . . . **174**

**Osip Kokin, Irina Usyagina, Nikita Meshcheriakov, Roman Ananiev, Vasiliy Arkhipov, Aino Kirillova, et al.**
Pb-210 Dating of Ice Scour in the Kara Sea
Reprinted from: *J. Mar. Sci. Eng.* **2023**, *11*, 1404, https://doi.org/10.3390/jmse11071404 . . . . . **198**

# About the Editors

**George Kontakiotis**

George Kontakiotis graduated with a Ph.D. in Paleoceanography from the University of Athens in 2012, where he later worked as a Laboratory and Teaching Staff in the fields of Marine Geology and Sedimentology and, recently, as an Associate Professor in Sedimentology–Paleoceanography–Chemostratigraphy. His major research contributions include developing novel approaches on the distribution and pathways of diagenesis in Mg/Ca paleothermometry. He has also worked on sedimentological and paleoceanographic reconstructions at different time scales by means of marine cores and land sections. His main research topics are summarized as follows: environmental sedimentology; marine petroleum systems; exploitation of natural energy resources; integrated bio-cyclo-tephro-stratigraphy; carbonate reservoirs; calibration–validation–application of geochemical proxies for sea surface temperature (SST) and salinity (SSS); applied environmental micropaleontology as a bio-monitoring tool; sea-level variations; and ocean/climate changes. He has published over 100 peer-reviewed articles in international journals.

**Angelos G. Maravelis**

Angelos G. Maravelis graduated with a Ph.D. in sedimentology from the Department of Geology, University of Patras, Greece, in July 2009. His research initially focused on classic sedimentology and the elucidation of clastic environments and sub-environments of deposition. His research activity became progressively enriched in aspects such as geochemistry (organic and inorganic) and petrography of siliciclastic sediments. Angelos Maravelis is strongly interested in high frequency sequence stratigraphy using high-resolution stratigraphic cross-sections and detailed biostratigraphic data. Currently, his research is focused on the global-scale subduction and collision processes related to the development of the accretionary complex and fold and thrust belts, along with their relationships with the evolution of the past climate of the Earth and impact on the accumulation of organic matter. Angelos Maravelis has spent the last few years in the exploration for natural resources in Greece, the Eastern Mediterranean, and Australia, and he has a strong publication record in this field. He has participated in several projects on hydrocarbon evaluation for oil companies using subsurface and surface data, and he has also carried out many projects with field courses for oil companies.

**Avraam Zelilidis**

Avraam Zelilidis was born in 1960 in Naousa Imathias of Macedonia in Greece. He received his B.Sc degree in Geology in 1984 and Ph.D. in Sedimentology in 1988 from the University of Patras, Greece. He has worked as a Lecturer since 1993 and as a Full Professor, in the Department of Geology at the University of Patras, since 2009. He was the Dean of the Natural Science School at the University of Patras from 2006 to 2010. He was the Head of the Department of Geology from 2017 to 2020. Since 2020, he has been the Director of the Hydrocarbon Institute at the University of Patras. His expertise is in sedimentary basin analysis, sequence stratigraphy, and the petroleum geology. He has carried out several projects on hydrocarbon evaluation for oil companies, using subsurface and surface data, and he has carried out many projects with field courses for oil companies. He has published more than 130 papers in international journals, most of which were focused on hydrocarbon prospectivity in Greece, and he has presented many of these papers at international meetings around the world (Japan, Europe, Africa, and America). He was a supervisor for twelve Ph.D. theses related to petroleum geology. He has worked in the Gulf of Suez and has published collaborative research between Greece and Egypt, with international interest.

# Preface

In the current Special Issue, we present a collection of articles that delve into the sedimentary environments, including depositional and diagenetic controls, geochemistry, and economic potentiality of sedimentary deposits worldwide. The published papers of this Special Issue also fill in some of the knowledge gaps of sedimentology, such as ice scour depositional settings and speleothem biomineralization processes, from very ancient carbonate to modern clastic sedimentary distribution patterns in seashore environments. As we reflect on the wealth of insights presented in this Special Issue, it is evident that the study of sedimentary geology remains a dynamic and multifaceted field ripe for future exploration. By continuing to unravel the complexities of such environments and addressing key knowledge gaps, we can better understand past environmental changes, decipher geological evolution, and inform future research endeavors.

**George Kontakiotis, Angelos G. Maravelis, and Avraam Zelilidis**
*Guest Editors*

*Editorial*

# Recent Advances in Sedimentology

**George Kontakiotis [1,*], Angelos G. Maravelis [2] and Avraam Zelilidis [3]**

[1] Department of Historical Geology and Palaeontology, Faculty of Geology and Geoenvironment, National and Kapodistrian University of Athens, Panepistimiopolis, 15784 Athens, Greece
[2] Department of Geology, Aristotle University of Thessaloniki, 54124 Thessaloniki, Greece; angmar@geo.auth.gr
[3] Department of Geology, University of Patras, 26504 Patras, Greece; a.zelilidis@upatras.gr
* Correspondence: gkontak@geol.uoa.gr

## 1. Introduction

Sedimentary rocks represent a vital component of the Earth's geological framework, playing a significant role in the Earth's surface morphology, as well as in paleoenvironmental reconstructions. Over the last few decades, apart from their geological significance, the diverse and dynamic nature of sedimentary deposits has intrigued academic researchers and industrial companies striving to obtain sustainable resource management applications, particularly in construction [1] and hydrocarbon exploration and production [2], highlighting their economic potential. Despite considerable progress in sedimentary geology, several knowledge gaps persist, necessitating a more comprehensive examination of both carbonate and clastic sedimentary systems. Through an interdisciplinary approach, drawing upon cutting-edge research and the latest methodological advancements and insights in this dynamic field, this Special Issue addresses gaps in the research and fosters interdisciplinary collaboration between the scientific and industrial communities. By focusing on the paleoenvironmental conditions and clarifying the interplay between the depositional processes, diagenetic alterations, and inherent geochemical signatures, researchers can unravel the complex history of sedimentary successions and discern the driving factors behind their evolution. Depositional characteristics provide invaluable insights into the environmental conditions and sedimentary processes that govern the formation of sedimentary rocks [3], particularly the decipherment of sedimentary facies and structures, bedding patterns, and fossil assemblages in carbonate and/or clastic formations, which can serve as archives of past climates, tectonic events, and biological evolution [4–8]. Geochemical indicators such as isotopes and trace and major elements are extremely sensitive to paleoenvironmental reconstructions [9–12], and therefore, their application is essential in the context of geochemical and mineralogical investigations. Diagenetic processes such as dolomitization, compaction, cementation, and dissolution can modify the original mineralogy, texture, and porosity of sediments, thereby influencing their reservoir properties and hydrocarbon potential [13–18]. In this regard, the addition of the geological time dimension through the integration of sedimentological and sequence stratigraphic data and related correlations offers crucial information for potential reservoir characterization and exploration strategies [19,20].

## 2. An Overview of Published Articles

The Special Issue "Recent Advances in Sedimentology" of the *Journal of Marine Science and Engineering* comprises a selection of ten peer-reviewed research articles, half of which are presented in Session S1, "Evolving techniques in the study of sediments" from the 16th International Congress of the Geological Society of Greece that was held in Patras, Greece, from 17 to 19 October 2022. This session was organized by the Hellenic Committee of Sedimentology of the Geological Society of Greece and focused on the presentation of recent results of sedimentological research in Greece. The other five scientific articles published in the current Special Issue represent important sedimentary geology-related topics beyond Greece, such as different sedimentary basins within the Kara Sea, the Mediterranean Sea, and the Proto-Tethys Ocean. Although each of them focuses on a particular topic and often a particular region, they mark the recent advances in this discipline and, taken together, are valuable for realizing the complexity of the conceptual framework of sedimentology. Overall, based on their fundamental principles to advance cutting-edge methodologies, all these contributions improve the state of the art of their corresponding sub-disciplines by facilitating comprehension and fostering a deeper appreciation of the complexities inherent within described sedimentary sequences, offering further diverse perspectives and challenges.

Bilal et al. (Contribution 1) offer an in-depth exploration of the sedimentological and diagenetic processes of the Eocene carbonate deposits of the Ceno-Tethys Ocean employing a multi-proxy approach, which involves field observations, sedimentological microfacies characterization, paleontological, petrographic, and scanning electron microscopy analyses. This study provides valuable insights into the complex history of Early–Middle Eocene Margalla Hill Limestone and Chogali Formation across the eastern margin of the Upper Indus Basin in Pakistan, shedding light on the depositional diagenetic stages and highlighting the reservoir characteristics of these carbonates for possible hydrocarbon exploration in the future.

The research by Kanellopoulos et al. (Contribution 2) examines the biodiversity and biomineralization processes in caves. The authors are particularly focused on those recorded in the hot spring environment of the Aedipsos area (NW Euboea Island, Greece). This study presents a comprehensive analysis through the mineralogical composition of speleothems and the cyanobacteria biomineralization processes, highlighting the importance of the geo-micro-biological study of such deposits, particularly in the extreme environments of hot springs.

Moving forward, Moforis et al. (Contribution 3) present a comprehensive, integrated sedimentological analysis of the pelagic carbonates of the Ionian zone (Epirus, western Greece), including lithostratigraphic determination, microfacies analysis, depositional environments, and the diagenetic history of Senonian and Microbreccious Limestone Formations in the Gardiki section. This study also provides valuable biostratigraphic and paleoecological insights across the Late Cretaceous–Early Paleocene Transition, highlighting any gradual or sudden foraminiferal extinctions and the related evolutionary and/or ecological crises that occurred globally during that time. Moreover, it provides a basis for the further evaluation of the hydrocarbon potential in western continental Greece, which contains proven reserves and is of crucial economic and strategic importance.

The fourth contribution by Lapietra et al. deals with the geomorphological, sedimentological, and geophysical characterization of two Apulian sandy beaches (Torre Guaceto and Le Dune beach), which are representative of the coastal dynamics of a large sector of the central/northern Mediterranean Sea involving the southern Adriatic Sea and the northern Ionian Sea. The authors propose a potential procedure for monitoring the morphosedimentary processes of sandy beaches by analyzing the textural and compositional characteristics of sand and quantifying the volumes involved in the coastal dynamics.

Zhang et al. (Contribution 5) investigate the effects of mud supply and hydrodynamic conditions on the sedimentary distribution of estuaries. In this study, the effects of mud concentration, mud transport properties, fluvial discharge, and tidal amplitude on the sedimentary characteristics of an estuary are systematically analyzed using sedimentary dynamic numerical simulation. The results indicate that the sedimentary dynamic numerical simulation can provide insights into an efficient quantitative method for analyzing the effects of mud components on the sediment processes of estuaries.

Zelilidis et al. (Contribution 6) provide insights into the depositional conditions of the Messinian evaporites on Zakynthos Island, highlighting the significance of their origin, thickness, and distribution for the sealing capacity in hydrocarbon exploration targets within the Mediterranean Sea.

Ali et al. (Contribution 7) offer an in-depth exploration of the Proto-Tethys Ocean during the Cambrian period on the northern margin of the Indian Plate, employing a multi-proxy approach involving sedimentological and petrographic characteristics of the Abbottabad Formation in the Muzaffarabad area, Upper Indus Basin, Pakistan. The detailed litho- and petrofacies characterization, along with the elemental analysis and sedimentological hierarchy of the dolomitic facies, offer valuable contributions to our understanding of carbonate systems worldwide.

Mehmood et al. (Contribution 8) performed a study on the Maastrichtian Pab Formation in the Rakhi Gorge, Eastern Sulaiman Ranges, Pakistan, which represents part of the eastern Tethys. The paleoenvironment and sequence stratigraphy were studied in the Cretaceous fluvio-deltaic sandstone succession through lithofacies analysis and associated architectural elements.

Qureshi et al. (Contribution 9) investigate the sedimentary processes, depositional architecture, and reservoir rock potential of the Mid-Triassic mixed siliciclastic and carbonate succession of the Tredian Formation in the Salt and Trans-Indus Ranges. The depositional environment was reconstructed based on lithofacies characterization, while the paragenetic sequence was determined based on petrographic and SEM observations.

The paper by Kokin et al. (Contribution 10) presents an application of the $^{210}$Pb dating of the largest ice scour in the Baydaratskaya Bay area (Kara Sea). This method of dating using the study of ice-scouring processes has become very important in recent times due to climate change. In this regard, the authors consider that the studied ice scour was formed no later than the end of the Little Ice Age. Such findings on the sedimentation chronology in ice scours help to establish the periods of active ice scouring on the glaciated continental margins and supplement knowledge about sedimentation on the Arctic shelf.

**Conflicts of Interest:** The authors declare no conflicts of interest.

**List of Contributions**

1. Bilal, A.; Yang, R.; Mughal, M.S.; Janjuhah, H.T.; Zaheer, M.; Kontakiotis, G. Sedimentology and Diagenesis of the Early–Middle Eocene Carbonate Deposits of the Ceno-Tethys Ocean. *J. Mar. Sci. Eng.* **2022**, *10*, 1794. https://doi.org/10.3390/jmse10111794
2. Kanellopoulos, C.; Lamprinou, V.; Politi, A.; Voudouris, P.; Iliopoulos, I.; Kokkaliari, M.; Moforis, L.; Economou-Amilli, A. Speleothems and Biomineralization Processes in Hot Spring Environment: The Case of Aedipsos (Edipsos), Euboea (Evia) Island, Greece. *J. Mar. Sci. Eng.* **2022**, *10*, 1909. https://doi.org/10.3390/jmse10121909
3. Moforis, L.; Kontakiotis, G.; Janjuhah, H.T.; Zambetakis-Lekkas, A.; Galanakis, D.; Paschos, P.; Kanellopoulos, C.; Sboras, S.; Besiou, E.; Karakitsios, V.; et al. Sedimentary and Diagenetic Controls across the Cretaceous—Paleogene Transition: New Paleoenvironmental Insights of the External Ionian Zone from the Pelagic Carbonates of the Gardiki Section (Epirus, Western Greece). *J. Mar. Sci. Eng.* **2022**, *10*, 1948. https://doi.org/10.3390/jmse10121948
4. Lapietra, I.; Lisco, S.; Capozzoli, L.; De Giosa, F.; Mastronuzzi, G.; Mele, D.; Milli, S.; Romano, G.; Sabatier, F.; Scardino, G.; et al. A Potential Beach Monitoring Based on Integrated Methods. *J. Mar. Sci. Eng.* **2022**, *10*, 1949. https://doi.org/10.3390/jmse10121949
5. Zhang, Q.; Tang, M.; Lu, S.; Liu, X.; Xiong, S. Effects of Mud Supply and Hydrodynamic Conditions on the Sedimentary Distribution of Estuaries: Insights from Sediment Dynamic Numerical Simulation. *J. Mar. Sci. Eng.* **2023**, *11*, 174. https://doi.org/10.3390/jmse11010174
6. Zelilidis, A.; Bourli, N.; Andriopoulos, K.; Georgoulas, E.; Peridis, S.; Asimakopoulos, D.; Maravelis, A.G. Unraveling the Origin of the Messinian? Evaporites in Zakynthos Island, Ionian Sea: Implications for the Sealing Capacity in the Mediterranean Sea. *J. Mar. Sci. Eng.* **2023**, *11*, 271. https://doi.org/10.3390/jmse11020271

7. Ali, S.K.; Lashari, R.A.; Sahito, A.G.; Kontakiotis, G.; Janjuhah, H.T.; Mughal, M.S.; Bilal, A.; Mehmood, T.; Majeed, K.U. Sedimentological and Petrographical Characterization of the Cambrian Abbottabad Formation in Kamsar Section, Muzaffarabad Area: Implications for Proto-Tethys Ocean Evolution. *J. Mar. Sci. Eng.* **2023**, *11*, 526. https://doi.org/10.3390/jmse11030526
8. Mehmood, M.; Naseem, A.A.; Saleem, M.; Rehman, J.u.; Kontakiotis, G.; Janjuhah, H.T.; Khan, E.U.; Antonarakou, A.; Khan, I.; Rehman, A.u.; et al. Sedimentary Facies, Architectural Elements, and Depositional Environments of the Maastrichtian Pab Formation in the Rakhi Gorge, Eastern Sulaiman Ranges, Pakistan. *J. Mar. Sci. Eng.* **2023**, *11*, 726. https://doi.org/10.3390/jmse11040726
9. Qureshi, K.A.; Arif, M.; Basit, A.; Ahmad, S.; Janjuhah, H.T.; Kontakiotis, G. Sedimentological Controls on the Reservoir Characteristics of the Mid-Triassic Tredian Formation in the Salt and Trans-Indus Surghar Ranges, Pakistan: Integration of Outcrop, Petrographic, and SEM Analyses. *J. Mar. Sci. Eng.* **2023**, *11*, 1019. https://doi.org/10.3390/jmse11051019
10. Kokin, O.; Usyagina, I.; Meshcheriakov, N.; Ananiev, R.; Arkhipov, V.; Kirillova, A.; Maznev, S.; Nikiforov, S.; Sorokhtin, N. Pb-210 Dating of Ice Scour in the Kara Sea. *J. Mar. Sci. Eng.* **2023**, *11*, 1404. https://doi.org/10.3390/jmse11071404

# References

1. Kamran, A.; Ali, L.; Ahmed, W.; Zoreen, S.; Jehan, S.; Janjuhah, H.T.; Vasilatos, C.; Kontakiotis, G. Aggregate Evaluation and Geochemical Investigation of Limestone for Construction Industries in Pakistan: An Approach for Sustainable Economic Development. *Sustainability* **2022**, *14*, 10812. [CrossRef]
2. Burchette Trevor, P. Carbonate rocks and petroleum reservoirs: A geological perspective from the industry. *Geol. Soc. Lond. Spec. Publ.* **2012**, *370*, 17–37. [CrossRef]
3. Dunham, R.J. Classification of Carbonate Rocks According to Depositional Texture. *Classif. Carbonate Rocks—A Symp.* **1962**, *1*, 108–121.
4. Ghazi, S.; Mountney, N.P. Facies and architectural element analysis of a meandering fluvial succession: The Permian Warchha Sandstone, Salt Range, Pakistan. *Sediment. Geol.* **2009**, *221*, 99–126. [CrossRef]
5. Ahmad, S.; Khan, S. Improved diagnosis of the carbonate reservoirs: A case study from the Potwar Basin, northwest Pakistan. *Carbonates Evaporites* **2024**, *39*, 44. [CrossRef]
6. Flügel, E.; Munnecke, A. *Microfacies of Carbonate Rocks: Analysis, Interpretation and Application*; Springer: Berlin/Heidelberg, Germany, 2010; Volume 976.
7. El-Kahtany, K.; Farouk, S.; Ahmad, F.; Tanner, L.; Mohammed, I.Q. Facies associations and stratigraphic sequence of the Dhruma Formation (Middle Jurassic) at its type locality, Khashm adh Dhibi, Saudi Arabia. *Carbonates Evaporites* **2024**, *39*, 9. [CrossRef]
8. Chen, Z.-Q.; Hu, X.; Montañez, I.P.; Ogg, J.G. Sedimentology as a Key to Understanding Earth and Life Processes. *Earth-Sci. Rev.* **2019**, *189*, 1–5. [CrossRef]
9. Rehman, S.U.; Munawar, M.J.; Shah, M.M.; Ahsan, N.; Kashif, M.; Janjuhah, H.T.; Lianou, V.; Kontakiotis, G. Diagenetic Evolution of Upper Cretaceous Kawagarh Carbonates from Attock Hazara Fold and Thrust Belt, Pakistan. *Minerals* **2023**, *13*, 1438. [CrossRef]
10. Cao, J.; Wu, M.; Chen, Y.; Hu, K.; Bian, L.; Wang, L.; Zhang, Y. Trace and rare earth element geochemistry of Jurassic mudstones in the northern Qaidam Basin, northwest China. *Geochemistry* **2012**, *72*, 245–252. [CrossRef]
11. Li, Q.; Wu, S.; Xia, D.; You, X.; Zhang, H.; Lu, H. Major and trace element geochemistry of the lacustrine organic-rich shales from the Upper Triassic Chang 7 Member in the southwestern Ordos Basin, China: Implications for paleoenvironment and organic matter accumulation. *Mar. Pet. Geol.* **2020**, *111*, 852–867. [CrossRef]
12. Guo, W.; Mosenfelder, J.L.; Goddard, W.A.; Eiler, J.M. Isotopic fractionations associated with phosphoric acid digestion of carbonate minerals: Insights from first-principles theoretical modeling and clumped isotope measurements. *Geochim. Cosmochim. Acta* **2009**, *73*, 7203–7225. [CrossRef]
13. Gomez-Rivas, E.; Martín-Martín, J.D.; Bons, P.D.; Koehn, D.; Griera, A.; Travé, A.; Llorens, M.-G.; Humphrey, E.; Neilson, J. Stylolites and stylolite networks as primary controls on the geometry and distribution of carbonate diagenetic alterations. *Mar. Pet. Geol.* **2022**, *136*, 105444. [CrossRef]
14. Janjuhah, H.T.; Alansari, A.; Santha, P.R. Interrelationship Between Facies Association, Diagenetic Alteration and Reservoir Properties Evolution in the Middle Miocene Carbonate Build Up, Central Luconia, Offshore Sarawak, Malaysia. *Arab. J. Sci. Eng.* **2019**, *44*, 341–356. [CrossRef]
15. Janjuhah, H.T.; Kontakiotis, G.; Wahid, A.; Khan, D.M.; Zarkogiannis, S.D.; Antonarakou, A. Integrated Porosity Classification and Quantification Scheme for Enhanced Carbonate Reservoir Quality: Implications from the Miocene Malaysian Carbonates. *J. Mar. Sci. Eng.* **2021**, *9*, 1410. [CrossRef]

16. Morad, S.; Al-Aasm, I.S.; Nader, F.H.; Ceriani, A.; Gasparrini, M.; Mansurbeg, H. Impact of diagenesis on the spatial and temporal distribution of reservoir quality in the Jurassic Arab D and C members, offshore Abu Dhabi oilfield, United Arab Emirates. *GeoArabia* **2012**, *17*, 17–56. [CrossRef]
17. Umar, M.; Friis, H.; Khan, A.S.; Kassi, A.M.; Kasi, A.K. The effects of diagenesis on the reservoir characters in sandstones of the Late Cretaceous Pab Formation, Kirthar Fold Belt, southern Pakistan. *J. Asian Earth Sci.* **2011**, *40*, 622–635. [CrossRef]
18. Janjuhah, H.T.; Alansari, A.; Gámez Vintaned, J.A. Quantification of microporosity and its effect on permeability and acoustic velocity in Miocene carbonates, Central Luconia, offshore Sarawak, Malaysia. *J. Pet. Sci. Eng.* **2019**, *175*, 108–119. [CrossRef]
19. Wadood, B.; Khan, S.; Li, H.; Liu, Y.; Ahmad, S.; Jiao, X. Sequence Stratigraphic Framework of the Jurassic Samana Suk Carbonate Formation, North Pakistan: Implications for Reservoir Potential. *Arab. J. Sci. Eng.* **2021**, *46*, 525–542. [CrossRef]
20. Wadood, B.; Aziz, M.; Ali, J.; Khan, N.; Wadood, J.; Khan, A.; Shafiq, M.; Ullah, M. Depositional, diagenetic, and sequence stratigraphic constrains on reservoir characterization: A case study of middle Jurassic Samana Suk Formation, western Salt Range Pakistan. *J. Sediment. Environ.* **2021**, *6*, 131–147. [CrossRef]

**Disclaimer/Publisher's Note:** The statements, opinions and data contained in all publications are solely those of the individual author(s) and contributor(s) and not of MDPI and/or the editor(s). MDPI and/or the editor(s) disclaim responsibility for any injury to people or property resulting from any ideas, methods, instructions or products referred to in the content.

Article

# Sedimentology and Diagenesis of the Early–Middle Eocene Carbonate Deposits of the Ceno-Tethys Ocean

Ahmer Bilal [1], Renchao Yang [1,2,*], Muhammad Saleem Mughal [3], Hammad Tariq Janjuhah [4,*], Muhammad Zaheer [3] and George Kontakiotis [5]

1. Shandong Provincial Key Laboratory of Depositional Mineralization & Sedimentary Minerals, Shandong University of Science and Technology, Qingdao 266590, China
2. Laboratory for Marine Mineral Resources, Qingdao National Laboratory for Marine Science and Technology, Qingdao 266071, China
3. Institute of Geology, University of Azad Jammu and Kashmir, Muzaffarabad 13100, Pakistan
4. Department of Geology, Shaheed Benazir Bhutto University, Sheringal 18050, Pakistan
5. Department of Historical Geology-Paleontology, Faculty of Geology and Geoenvironment, School of Earth Sciences, National and Kapodistrian University of Athens, Panepistimiopolis, Zografou, 15784 Athens, Greece
* Correspondence: r.yang@sdust.edu.cn (R.Y.); hammad@sbbu.edu.pk (H.T.J.)

**Abstract:** An integrated study based on field observation, petrography, and scanning electron microscopy (SEM) on the Early–Middle Eocene carbonate rocks has been carried out, which were deposited in the Ceno-Tethys Ocean. The study area of the Yadgaar Section lies on the eastern margin of the Upper Indus Basin, Pakistan. The Early–Middle Eocene Margalla Hill Limestone and Chorgali Formation act as reservoir rocks in other parts of the basin and are also present in the Yadgaar Section. The lack of comprehensive study in this area makes these reservoir rocks highly attractive for sedimentological evaluations and future exploration of hydrocarbons. The Early–Middle Eocene carbonate rocks are divided into nine microfacies: dolomicritic foraminiferal mudstone–wackestone microfacies (EMI); green algae dominated, mixed foraminiferal wackestone–packstone microfacies (EMII); ostracod, green algae and gypsum dominating mudstone–wackestone microfacies (EMIII); algae and mixed foraminiferal wackestone–packstone microfacies (EMIV); *Nummulites* dominating mudstone–wackestone microfacies (EMV); algal limestone mudstone microfacies (EMVI); *Assilina* bed wackestone–packstone microfacies (EMVII); micritized larger benthic foraminiferal wackestone–packstone microfacies (EMVIII); and algal limestone, mudstone microfacies (EMIX). The transgressive-regressive environment in the Ceno-Tethys Ocean leads to the deposition of these microfacies in the platform interior, open marine platform, platform edge, platform margin reef, toe of the slope apron, arid–humid platform interior, platform edge, open marine platform interior, and restricted marine platform interior, respectively. Initial post-depositional diagenetic stages are identified from the base to the top of the strata by their respective cement types, i.e., the base–lower middle part of the strata demonstrates an eogenetic sub-stage with the appearance of drusy cement, the middle section indicates a mesogenetic sub-stage by the appearance of blocky cement, while the top portion again reveals an eogenetic sub-stage of diagenesis by the presence of drusy and blocky types of cement. The ascending–descending hierarchy of cement generations is directly proportional to the grade of diagenesis from the base to the top of the carbonate strata. Variable diagenetic effects on the various microfacies also increase the secondary porosity range and enhance the reservoir characteristics of the Formations. The presence of foraminifera microfossils determined that these carbonate formations date from the Early–Middle Eocene.

**Keywords:** cement generation; diagenesis; gypsum; hydrocarbons; microfacies; source rock

## 1. Introduction

In the Eocene age, carbonate strata were deposited on the shallow carbonate platform due to the complex lagoonal conditions of the Ceno-Tethys Ocean. Green and coralline algae, along with benthic foraminifera from Tethyan deposits, are worldwide proven to be excellent indicators of paleoenvironmental settings [1]. A high abundance of benthic foraminiferas in shallow marine Tethyan rocks indicates high temperature after the Paleocene Eocene thermal maximum (PETM) or middle Eocene Climatic Optimum [2,3]. *Assilina* beds and *Alveolina*-rich microfacies in early Lutetian are an indicator of a hyperthermal event in shallow marine platform deposits of this time of the Tethys Ocean [1,4]. Micritic carbonate strata represent these deposits with shallow shelf carbonate ramp and inner platform settings in the Tethys Ocean [5].

The rocks of the Upper Indus Basin's (UIB) geology range in age from Precambrian to recent [6,7]. However, in the Yadgaar Section, only the Cambrian to Miocene rock sequence is exposed [8]. Sandstone and carbonate strata of Paleocene–Eocene rocks are unconformably overlain by Cambrian-aged dolomite of the Abbottabad Formation. The sandstone sequence is represented by the early Paleocene Hangu Formation, while the late Paleocene carbonate rocks are represented by the Lockhart Limestone and Patala Formation [9]. The Patala Formation and Lockhart Limestone's upper and lower contacts, respectively, are marked as unconformable. Many researchers worked on Eocene Tethyan deposits regionally [1–3,10–12]. The Early–Middle Eocene nodular to fused nodular carbonate deposits of Pakistan are represented by the Margalla Hill Limestone and the Chorgali Formation, respectively. The Kohat, Potwar, Abbottabad, and Murree regions are further UIB locations where these deposits were observed. This comprehensive petrological investigation demonstrates a variance in microfauna and depositional settings between those zones and the targeted area, despite the limited aspects of research in those areas [13–17]. These rocks are present in the study area of the Yadgaar section of the UIB. However, only a few researchers in the UIB and surrounding areas have focused on these rocks and their coeval strata. These previous researchers targeted only individual Formations and their limited aspects [13–16,18–21]. These deposits are unconformably underlain by sandstone rocks of the Eocene–Oligocene-aged Kuldana Formation. The Kuldana Formation is finally overlain unconformably by the Murree Formation in the study area [8,17].

A comprehensive petrological, paleontological, and sedimentological study of Early–Middle Eocene rocks in the Yadgaar Section of the UIB is still missing to unfold microfacies variations, depositional settings, and post-depositional diagenetic stages and their effect on their reservoir characteristics for possible hydrocarbon exploration in the future (Figure 1). The goal of the current study is to carry out a combined and comprehensive evaluation of Early–Middle Eocene rocks to determine their age, depositional environment, diagenetic stages and tectonic settings of the area during and after the deposition of these rocks by detailed petrographical, sedimentological, paleontological, and mineralogical instigations.

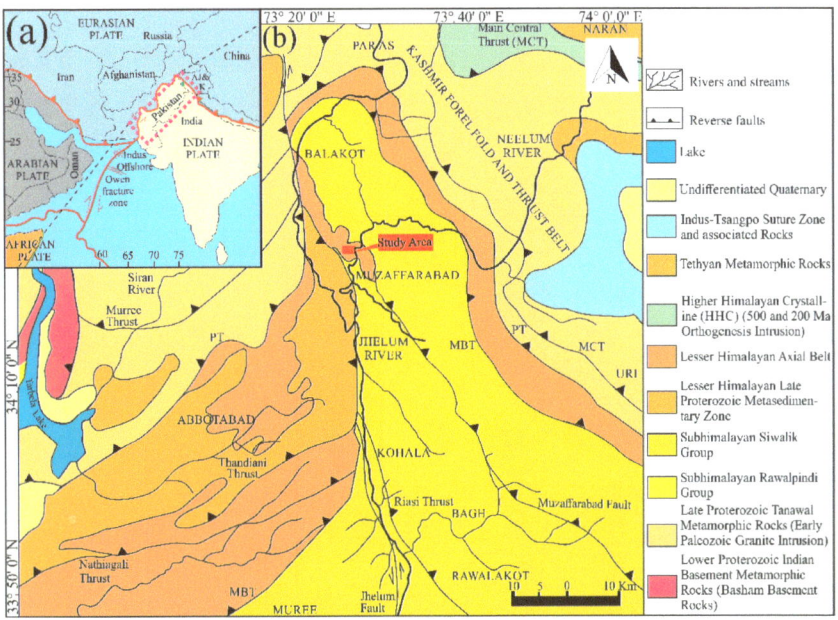

**Figure 1.** (**a**) Inset shows the regional location of Indus Basin in Pakistan with the dotted purple line; (**b**) study area of the Yadgaar Section in the UIB.

## 2. Material and Methods

### 2.1. Fieldwork

Fieldwork involved identification of Early–Middle Eocene Margalla Hill Limestone and the Chorgali Formation's upper and lower contacts and recognition of outcrop lithofacies (Figure 1). For the petrographic evaluation, 44 representative fresh rock samples were collected from the outcrop of Margalla Hill Limestone (Table 1) and 32 samples from the Chorgali Formation (Table 2). The Jacob's Staff technique was used to measure the thickness of the section, which was used to help determine the lithology of the section. In the lab of the Institute of Geology at the University of Azad Jammu and Kashmir, thin sections were prepared.

**Table 1.** Modal mineralogical composition of the Early–Middle Eocene carbonate rocks of Yadgaar Section. (CM-carbonaceous material).

| S. No. | Spar (%) | Micrite (%) | Bioclasts (%) | CM (%) | Pyrite (%) | Hematite (%) | Gypsum (%) | Dolomite (%) | Calcite Vein (%) | Dunham Classification |
|---|---|---|---|---|---|---|---|---|---|---|
| YMG1 | 2 | 10 | 10 | 2 | 4 | 7 | - | 65 | - | Mudstone–wackestone |
| YMG2 | 1 | 13 | 8 | 2 | 4 | 5 | - | 67 | - | |
| YMG3 | 3 | 10 | 11 | 3 | 2 | 6 | - | 65 | - | |
| YMG4 | 2 | 9 | 12 | 2 | 6 | 6 | - | 63 | - | |
| YMG5 | 7 | 28 | 40 | 2 | 2 | 7 | 1 | 5 | 8 | Wackestone |
| YMG6 | 5 | 31 | 40 | 2 | 2 | 7 | 1 | 4 | 8 | |
| YMG7 | 7 | 33 | 38 | 3 | 2 | 5 | 1 | 4 | 7 | |
| YMG8 | 6 | 26 | 42 | 2 | 3 | 5 | 1 | 5 | 10 | |

Table 1. Cont.

| S. No. | Spar (%) | Micrite (%) | Bioclasts (%) | CM (%) | Pyrite (%) | Hematite (%) | Gypsum (%) | Dolomite (%) | Calcite Vein (%) | Dunham Classification |
|---|---|---|---|---|---|---|---|---|---|---|
| YMG9 | 7 | 32 | 50 | 1 | 4 | 1 | 1 | 4 | - | Wacke–packstone |
| YMG10 | 7 | 35 | 48 | 1 | 3 | 1 | 1 | 4 | - | |
| YMG11 | 5 | 32 | 51 | 1 | 3 | 2 | 1 | 5 | - | |
| YMG12 | 7 | 30 | 52 | 1 | 2 | 1 | 1 | 6 | - | |
| YMG13 | 14 | 10 | 53 | 1 | 1 | 2 | 1 | 15 | 2 | |
| YMG14 | 18 | 10 | 54 | 1 | 1 | 2 | 1 | 10 | 3 | |
| YMG15 | 21 | 8 | 55 | 1 | 1 | 1 | 1 | 9 | 3 | |
| YMG16 | 18 | 9 | 53 | 2 | 2 | 2 | 2 | 10 | 2 | |
| YMG17 | 5 | 26 | 50 | 2 | 2 | 5 | 2 | 5 | 3 | |
| YMG18 | 5 | 24 | 52 | 2 | 3 | 5 | 2 | 3 | 4 | |
| YMG19 | 5 | 29 | 47 | 1 | 2 | 6 | 2 | 5 | 3 | |
| YMG20 | 4 | 35 | 45 | 1 | 4 | 3 | 3 | 3 | 2 | |
| YMG21 | 3 | 62 | 15 | 2 | 3 | 6 | 1 | 7 | 1 | Mudstone-wackestone |
| YMG22 | 2 | 66 | 14 | 2 | 2 | 5 | 2 | 6 | 1 | |
| YMG23 | 4 | 58 | 14 | 3 | 3 | 7 | 2 | 8 | 1 | |
| YMG24 | 4 | 59 | 16 | 2 | 1 | 7 | 1 | 9 | 1 | |
| YMG25 | 5 | 56 | 10 | 2 | 1 | 8 | 2 | 15 | - | |
| YMG26 | 6 | 55 | 10 | 2 | 1 | 8 | 2 | 16 | - | |
| YMG27 | 3 | 52 | 12 | 3 | 1 | 9 | 2 | 18 | - | |
| YMG28 | 5 | 50 | 14 | 3 | 1 | 10 | 2 | 15 | - | |
| YMG29 | 3 | 57 | 30 | 2 | 1 | 6 | - | - | 1 | Wackestone |
| YMG30 | 5 | 57 | 28 | 3 | 1 | 5 | - | - | 1 | |
| YMG31 | 2 | 56 | 33 | 2 | 1 | 5 | - | - | 1 | |
| YMG32 | 3 | 57 | 30 | 4 | 1 | 4 | - | - | 1 | |
| YMG33 | 10 | 22 | 60 | 2 | 2 | 4 | 2 | 2 | 1 | Packstone |
| YMG34 | 10 | 24 | 51 | 2 | 1 | 6 | 1 | 2 | 1 | |
| YMG35 | 7 | 20 | 58 | 2 | 2 | 5 | 1 | 2 | 1 | |
| YMG36 | 8 | 22 | 56 | 1 | 3 | 5 | 1 | 3 | 1 | |
| YMG37 | 5 | 30 | 50 | 2 | 4 | 6 | - | 3 | 1 | Wacke–packstone |
| YMG38 | 5 | 30 | 51 | 2 | 4 | 4 | - | 3 | 1 | |
| YMG39 | 3 | 32 | 49 | 3 | 5 | 4 | - | 3 | 1 | |
| YMG40 | 4 | 28 | 50 | 2 | 6 | 5 | - | 4 | 1 | |
| YMG41 | - | 60 | 12 | 2 | 4 | 5 | 1 | 15 | 1 | Mudstone-wackestone |
| YMG42 | - | 62 | 12 | 1 | 4 | 6 | 1 | 13 | 1 | |
| YMG43 | - | 64 | 10 | 1 | 3 | 3 | 1 | 16 | 1 | |
| YMG44 | - | 64 | 10 | 4 | 3 | 2 | 1 | 15 | 1 | |

Table 2. Modal mineralogical composition of the Early–Middle Eocene carbonate rocks of Yadgaar Section.

| | Spar (%) | Micrite (%) | Bioclasts (%) | CM (%) | Pyrite (%) | Hematite (%) | Chert/ Chalcedony (%) | Gypsum (%) | Dolomite (%) | Calcite Vein (%) | Dunham Classification |
|---|---|---|---|---|---|---|---|---|---|---|---|
| YCF1 | - | 20 | 2 | 20 | 2 | 8 | 1 | 2 | 37 | 8 | Mudstone |
| YCF2 | - | 22 | 2 | 20 | 1 | 6 | 1 | 2 | 38 | 8 | |
| YCF3 | - | 21 | 3 | 18 | 2 | 7 | 1 | 2 | 39 | 7 | |
| YCF4 | - | 20 | 4 | 20 | 1 | 8 | 1 | 2 | 39 | 5 | |

Table 2. Cont.

| | Spar (%) | Micrite (%) | Bioclasts (%) | CM (%) | Pyrite (%) | Hematite (%) | Chert/ Chalcedony (%) | Gypsum (%) | Dolomite (%) | Calcite Vein (%) | Dunham Classification |
|---|---|---|---|---|---|---|---|---|---|---|---|
| YCF5 | 3 | 5 | 50 | 20 | 1 | 3 | 1 | 2 | 15 | - | Wacke-packstone |
| YCF6 | 3 | 5 | 59 | 15 | 1 | 3 | 1 | 2 | 11 | - | |
| YCF7 | 4 | 8 | 48 | 17 | 1 | 3 | 1 | 3 | 15 | - | |
| YCF8 | 3 | 9 | 55 | 13 | 1 | 3 | 1 | 2 | 14 | - | |
| YCF9 | 4 | 5 | 54 | 15 | 3 | 3 | - | 2 | 14 | - | |
| YCF10 | 4 | 7 | 53 | 16 | 3 | 2 | - | 2 | 13 | - | |
| YCF11 | 6 | 9 | 48 | 13 | 4 | 3 | - | 3 | 14 | - | |
| YCF12 | 4 | 8 | 56 | 11 | 3 | 2 | - | 2 | 14 | - | |
| YCF13 | 10 | 7 | 39 | 10 | 2 | - | 2 | 2 | 25 | 3 | Wackestone |
| YCF14 | 10 | 10 | 40 | 12 | 2 | - | 1 | 2 | 20 | 3 | |
| YCF15 | 9 | 10 | 39 | 10 | 3 | - | 1 | 2 | 23 | 3 | |
| YCF16 | 14 | 8 | 33 | 14 | 1 | - | 1 | 3 | 22 | 4 | |
| YCF17 | 10 | 6 | 44 | 12 | 2 | - | - | 3 | 20 | 3 | |
| YCF18 | 13 | 17 | 29 | 16 | 1 | - | - | 3 | 18 | 3 | |
| YCF19 | 12 | 13 | 28 | 19 | 1 | - | - | 2 | 23 | 2 | |
| YCF20 | 12 | 15 | 30 | 18 | - | - | - | 2 | 20 | 3 | |
| YCF21 | 5 | 32 | 25 | 13 | 3 | 2 | 1 | 7 | 10 | 2 | |
| YCF22 | 3 | 31 | 28 | 12 | 3 | 2 | 1 | 6 | 12 | 2 | |
| YCF23 | 1 | 38 | 23 | 13 | 2 | 1 | 1 | 7 | 10 | 4 | |
| YCF24 | - | 39 | 25 | 13 | 2 | 1 | 1 | 7 | 9 | 3 | |
| YCF25 | - | 20 | 4 | 23 | - | - | 2 | 1 | 40 | 10 | Mudstone |
| YCF26 | - | 18 | 5 | 27 | - | - | 1 | 1 | 38 | 10 | |
| YCF27 | - | 17 | 4 | 28 | - | - | 2 | 1 | 36 | 12 | |
| YCF28 | - | 19 | 4 | 27 | - | - | 2 | 1 | 37 | 10 | |
| YCF29 | - | 39 | 2 | 16 | 2 | 1 | 5 | 1 | 4 | 30 | |
| YCF30 | - | 66 | 1 | 11 | 2 | - | 6 | 1 | 2 | 11 | |
| YCF31 | - | 66 | 2 | 9 | 2 | 1 | 5 | 1 | 3 | 11 | |
| YCF32 | - | 64 | 1 | 11 | 2 | - | 5 | 1 | 3 | 13 | |

### 2.2. Petrography

A petrographic study of limestone samples collected from the Early–Middle Eocene rocks of the Yadgaar Section (Figure 2; Table 1) was carried out in the mineralogy/petrology laboratory of the University of Azad Jammu and Kashmir. A LEICA-DM 750P polarized microscope and a LEICA-S6D stereo zoom microscope were used to find out the modal mineralogy, bioclast identification, and microfacies division from these thin sections. A LEICA-EC3 camera was used to acquire photomicrographs of fossils and other microscopic features.

### 2.3. Scanning Electron Microscopy (SEM)

Limestone and shale samples were studied under a scanning electron microscope (SEM) manufactured by JEOL Japan at the Hi-Tech Laboratory of the University of Azad Jammu and Kashmir. Rock samples were placed on a stud for sputter coating with platinum. The JFC-1600 auto fine coater (JEOL, Japan) was used to make the coating, which makes the image results of the non-conducting material the clearest to study. SEM analysis was carried out to examine the cement types and generations, which was further used to interpret the grade of diagenesis. This helped us figure out more about how the Margalla Hill Limestone and Chorgali Formation in the study area could be used as a reservoir.

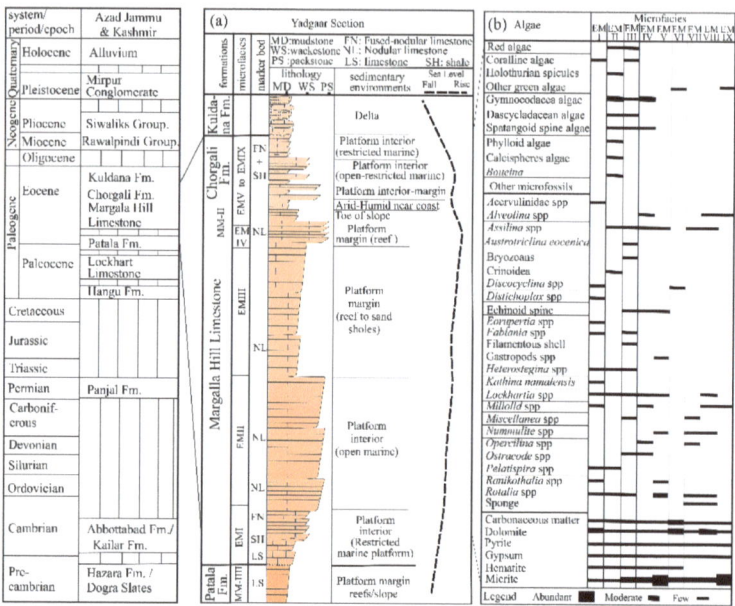

**Figure 2.** (**a**) Litholog of the Early–Middle Eocene carbonate rocks from the Yadgaar Section; (**b**) microfossils identified from the studied section. MD, Mudstone; FN, Fused nodular limestone; WS, Wackestone; NL, Nodular limestone; PS, Packstone; LS, Limestone; SH, Shales.

## 3. Results and Discussion

A shallow carbonate shelf existed in the Ceno-Tethys Ocean's margins during the Early–Middle Eocene. This led to the deposition of micritic and well-developed nodular to fused nodular limestone rocks in the area (Figures 2 and 3). The resultant Margalla Hill Limestone and the Chorgali Formation were economically highly important as they are proven reservoirs in the other areas of the UIB.

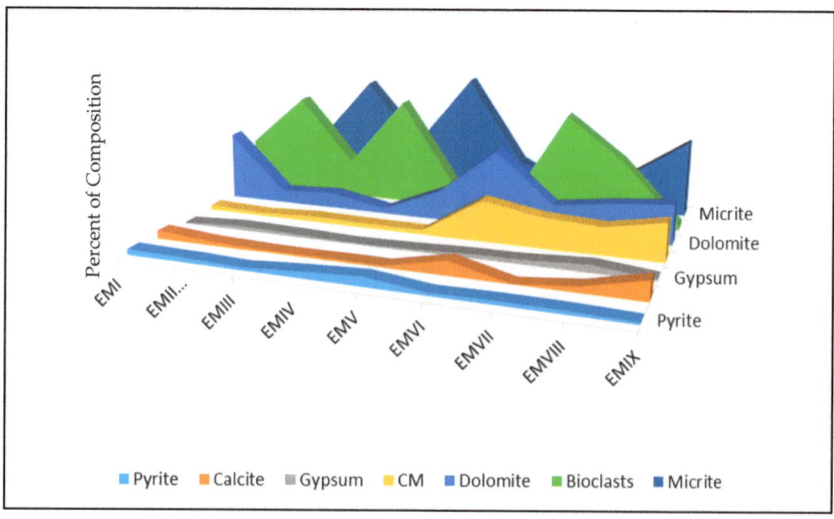

**Figure 3.** Graphic representation of Table 1. CM, carbonaceous material.

## 3.1. Lithofacies of the Early–Middle Eocene Carbonate Rocks

In the Yadgaar Section of the UIB, Early–Middle Eocene carbonate sequences have been divided into two stratigraphic units, i.e., the Margalla Hill Limestone and the Chorgali Formation. These Formations are dominantly comprised of nodular limestone in the lower part and limestone with shale intercalations in the upper part. Limestone interbedded with shale in the upper part ranges from 0.07 to 0.3 m in thickness (Figure 2). Dark brown–black colored shale exhibits carbonaceous material in its upper part. Based on the lithological variations in the field, nine lithofacies were identified.

An erosional unconformable lower contact of the Margalla Hill Limestone has been marked with the underlying Patala Formation. The first lithofacies is comprised of a 3-m-thick limestone bed and a 2-m-thick limestone bed that has intercalation of shale (Figure 4a). Light grey–bluish colored limestone is strongly compacted and hard. Shales were observed with limestone beds, and they are thin, ranging from 0.01 to 0.03 m. This lithofacies is 5 m in thickness. The second nodular limestone lithofacies is comprised of four beds of equally thick nodular limestone (Figure 4b). Each bed is 1.5 m in thickness; however, the fourth bed is a little deformed in the type section. Color variation is not marked and is found to be similar to the first lithofacies. The total thickness of this lithofacies is 4.5 m. The third marine nodular limestone lithofacies is light blue–dark dull colored and thick-bedded. The size of the nodules gradually increases from 0.07 to 0.15 m. Calcite veins (Figure 4c) are frequently observed in the outcrops. A high amount of carbonaceous material is clearly observed and distinguished in outcrops by a dull black color (Figure 4d). The thickness of this lithofacies is 30 m in the Yadgaar Section.

**Figure 4.** Photomicrographs showing (**a**) lower erosional unconformable contact of Patala Formation (PF) and Margalla Hill Limestone (MHL); (**b**) second nodular lithofacies; (**c**) calcite veins in third lithofacies; (**d**) carbonaceous material in third lithofacies; (**e**) massive nodular lithofacies; (**f**) nodular limestone; (**g**) contact between fifth and sixth lithofacies; (**h**) limestone and shale alteration lithofacies; and (**I**) upper erosional contact between Chorgali Formation (CF) and Kuldana Formation (KF).

The fourth lithofacies is a light brown color, nodular, and thick-bedded. The distinguishing feature of this microfacies is the decrease in carbonaceous content and the appearance of a light brown rusty colored limestone. Nodule sizes have also started to decrease in outcrops. The thickness of this lithofacies is also 30 m. The fifth lithofacies is a dark–light bluish grey colored, thick-bedded nodular lithofacies (Figure 4e). The rusting color starts to decrease, while the light–dark blue color starts to increase towards the top of the lithofacies. Carbonaceous content also starts to increase in this lithofacies. The distinguishing feature of this microfacies is the increasing concentration and maximum size of the nodules. The individual nodule size is $0.12 \times 0.20$ m in width and height, respectively. This lithofacies is 20 m in thickness. The sixth lithofacies is also nodular on top but medium–small bedded (Figure 4f,g). Carbonaceous content is found to decrease and the thickness of this lithofacies is 20 m. The upper contact is marked planar with the Chorgali Formation in the study area.

The seventh lithofacies is dark grey in color and medium bedded in the study area. Fossils were not found in any outcrops of this lithofacies. In the study area, thin shales are found interbedded with limestone. This lithofacies is 5 m inches in thickness. The eighth lithofacies is a dark grey colored limestone with interbedded shale (Figure 4h). This lithofacies is highly fossiliferous and fossils are exposed on surfaces and observable with a hand lens. This lithofacies is 15 m in thickness. Above this, there is another difference between these middle Eocene rocks and the deltaic limestone facies of the late Eocene-Oligocene Kuldana Formation (Figure 4I).

## 3.2. Microfacies Characterization

Early–Middle Eocene rocks are comprised of nodular limestone beds and shales in the Yadgaar section. A gradual transition from light bluish grey to dark grey is observed while moving from the base towards the top. Limestone beds are nodular and medium–thick bedded from base to center, while medium–thin bedded and fused nodules are found in the upper most part of the section. Large benthic foraminifers are frequently observed throughout the samples. Shales can be observed in a minor amount at the base; however, a considerable amount of shale with interbedded limestone is present in its upper most part. As compared to the light grey shale at the base, the color of the shale is much darker at the top due to the presence of carbonaceous material.

A petrographic study of the Early–Middle Eocene carbonate rocks revealed that micrite and bioclasts are the dominant constituents. However, dolomite, carbonaceous material, pyrite, spar, and hematite are unequally distributed throughout the Formation. Gypsum is also found throughout the section in an almost equal but minor amount. Furthermore, chert is noted only in the upper stratigraphic unit (Chorgali Formation) in minor amounts. Dunham's classification is used to categorize the limestone samples studied under thin sections. This reveals different rock types of Early–Middle Eocene carbonate strata from base to top as: mudstone–wackestone, wackestone–packstone, packstone–wackestone, mudstone–wackestone, mudstone, wackestone–packstone, wackestone, and mudstone, respectively. Detailed petrographic studies indicate different grain sizes, textural compositions, and fossil types. The average grain size in the Formation is 0.02 mm, while the bioclast sizes range from 0.1 to 30 mm in length in these strata.

Different types of rocks are categorized on the basis of the Dunham [22] classification scheme. Nine different microfacies were categorized on the basis of biofacies (differences in fossil types) and these different rock types: (1) dolomicritic foraminiferal mudstone-wackestone microfacies (EMI); (2) green algae dominated, mixed foraminiferal wackestone-packstone microfacies (EMII); (3) ostracod, green algae and gypsum dominating mudstone-wackestone microfacies (EMIII); (4) algae and mixed foraminiferal wacke–packstone microfacies (EMIV); (5) *Nummulites* dominating mudstone–wackestone microfacies (EMV); (6) algal limestone, mudstone microfacies (EMVI); (7) *Assilina* bed wackestone–packstone microfacies (EMVII); (8) micritized large benthic foraminiferal wackestone microfacies (EMVIII); and (9) algal limestone, mudstone microfacies (EMIX).

### 3.2.1. Dolomicritic Foraminiferal Mudstone–Wackstone Microfacies (EMI)

On the basis of detailed petrographic evaluations, this microfacies is subdivided into two subfacies, as follows.

Dolomicritic Mudstone–Wackstone Subfacies

The first subfacies (Dolomicritic mudstone–wackestone subfacies) is dominantly comprised of dolomicritic limestone rock bodies with an average grain size of 0.05 mm. Micrite is partly to completely dolomitized in this microfacies. Bioclasts (with an average of 10%) are partly to completely micritized. The few identified fossils are *Discocyclina ranikotenesis* (Figure 5a), *Elazigina dienii, Fabiania* spp. (Figure 5a), *Lockhartia* spp. (Figure 5b), *Pelatispira* and *Rotalia trochidiformis* (Figure 5c). Deep water agglutinated arenaceous foraminifera, *Kathina namalensis*, are also present, as well as a rare amount of *Nummulite* spp. A considerable amount of hematite (with an average of 6%) and pyrite (with an average of 4%) can also be observed in this subfacies. Spar (with an average of 3%) and carbonaceous material (with an average of 2%) are found in a minor amount. Four samples from the bottom of the Margalla Hill Limestone represent this subfacies (YMG1–YMG4).

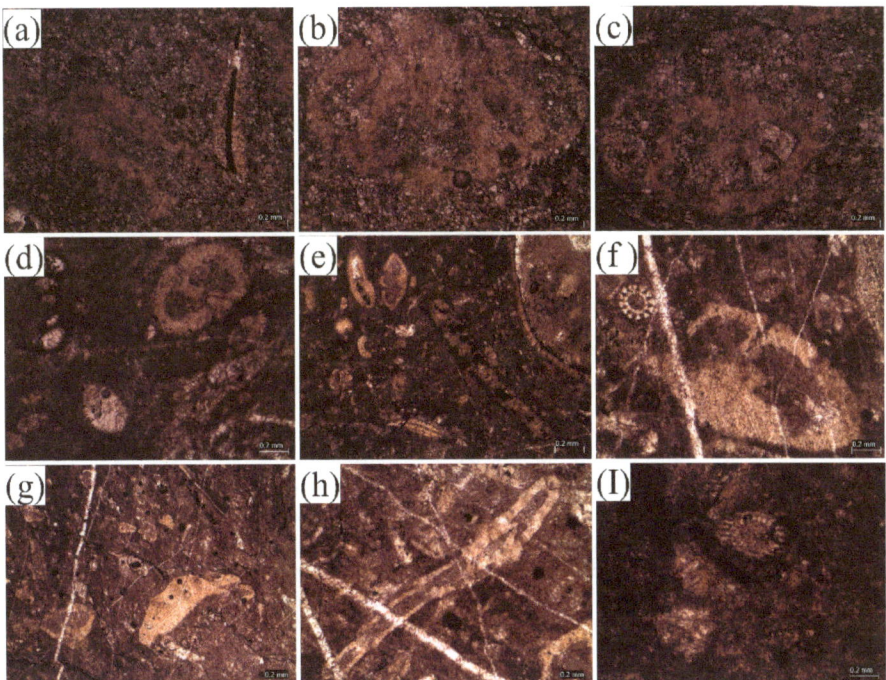

**Figure 5.** Photomicrographs showing (**a**) *Discocyclina* spp. on the right and *Fabiania* spp. on the left; (**b**) *Lockhartia* spp.; (**c**) *Rotalia trochidiformis*; (**d**) *Acervulinidae* spp. on the left and *Rotalia trochodiformis* on the right; (**e**) *Distichoplax biserialis* (coralline algae); (**f**) echinoid spine (*Echinothrix*); (**g**) *Quinqueloculina granulocostata* and calcified algae (*Vermiporella*); (**h**) *Ranikothalia nuttalli*; and (**I**) spatangoid spine algae.

Mixed Foraminiferal Wackestone Subfacies

This micritic wackestone subfacies is dominantly comprised of bioclasts (with an average of 40%). Bioclasts are found partly micritized in thin sections. Fossils identified are: *Assilina* spp., *Acervulinidae* spp. (Eocene) (Figure 5d), coralline algae, *Discocyclina* spp., *Distichoplax biserialis* (Figure 5e), echinoid spine *(Echinothrix)* (Figure 5f), *Eorupertia* spp., green algae, *Heterostegina, Lockhartia ramanae ten dam, Lockhartia daviesi, Lockhartia* tipper, *Operculina* spp., *Orbitolites* spp., Milliolidae *(Quincoloculina)*, *Quinqueloculina carinata* d'Orbigny, *Quinqueloculina granulocostata* (Figure 5g), calcified algae (*Vermiporella*) (Figure 5g), *Ranikothalia nuttalli* (Figure 5h), *Rotalia trochidiformis* and spatangoid spine algae (Figure 5I). Micrite is found as the second dominant constituent (with an average of 29%). Multiple cross-cutting and parallel calcite veins can also be seen with an average of 8% in this subfacies. Spar and hematite are observed with an average of 6% each, while dolomite and carbonaceous material are present with an average of 5% and 2%, respectively. Gypsum starts to appear in a minor amount (with an average of 1%) in this subfacies. This microfacies is represented by four thin sections (YMG5–YMG8).

**Interpretation:** This dolomicritic limestone is dominantly comprised of unimodal-shaped crystals. Dolomite crystals range in shape from subhedral to anhedral. Crystal boundaries have planar contact with each other. These shapes and textures revealed the rock Formation was subjected to low temperature (between 50–100 °C). Under low temperatures and constant Mg-rich fluid supply conditions, medium-grained subhedral-anhedral dolomite crystals are developed [23–26]. Medium–coarse crystals of dolomite can be observed sometimes in bioclasts. This resulted from the recrystallization of already formed dolomite, which eventually caused damage to the original depositional texture [27]. *Rotalia* spp. indicate a middle-Cuisian (early Ypresian) age [10]. An echinoid spine (*Echinothrix*) is identified in this microfacies. The echinoids largely influenced the coral reef's ecology [28–30]. After diagenesis of the shells of echinoderms, the high magnesium calcite ratio may be altered into low magnesium calcite. Echinoderms indicate deposition on a carbonate platform in a warm shelf setting where they are present with coral reefs, eventually facilitating carbonate mud production [31]. *Discocyclina* spp. and *Pellatispira* spp. suggest a depositional site on a deeper portion of the carbonate platform [32]. Descycladacean algae, such as *Vermiporella*, have not previously been reported in Eocene shelf carbonates [33]. The presence of *Vermiporella* and other calcified green algae indicates a photic and oxygenated zone in a shallow marine environment. Coralline red algae and other foraminifera, such as *Ranikothalia* spp., *Assilina* spp., Milliolidae, *Lockhartia* spp., *Orbitolites* spp., and *Rotalia* spp., indicate a relatively deep position of deposition on a carbonate platform [10]. This microfacies is found to be equivalent to FZ8 (restricted marine platform) of Flügel and Munnecke [34].

3.2.2. Green Algae Dominated, Mixed Foraminiferal Wackestone–Packstone Microfacies (EMII)

The second microfacies (green algae dominated, mixed foraminiferal wackestone–packstone microfacies (EMII) is recognized by the frequent abundance of calcispheres (*Vermiporella*) (Figure 6a), *Cocoarota orali* İnan, Dascycladacean green algae (Figure 6a), phylloid algae (Figure 6a), red algae (Figure 6b), *Salpingoporella melitae* Radoičić, Spatangoid spine, and a high amount of *Gymnocodiacea* green algae (Figure 6c). On the top portion of this microfacies, red algae (holothurian ossicles) start to appear in a minor amount. Other foraminifers are frequently observed and identified as: *Assilina leymeriei*, *Heterostegina* spp., *Holothurian* spicules (ossicles) (Figure 6d), *Idalina grelaudae* (Middle Eocene milliolidae) (Figure 6d), *Lockhartia haimei*, *Lockhartia prehaimei* (Figure 6e), *Lockhartia retiata*, ostracod (Figure 6f), Phylloid (Figure 6g), *Gymnocodiacea* (Figure 6g), *Pellatispira* spp. (Figure 6h), *Rotalia trochidiformis* (Leutitian age), *Saccammina grzybowskii* (Eocene foram) (Figure 6I) and *Textularia*. Crinoidea and echinoid spine are also observed. These bioclasts formed a predominant portion (with an average 50%) of this microfacies. Micrite, spar, and dolomite can be found with an average of 23%, 9%, and 6%, respectively. Hematite, carbonaceous material, pyrite, and gypsum gradually increase from base towards the top of this microfacies with averages of 2–3%, 1–2%, 2%, and 1–2%, respectively. Calcite is absent in the base and appears from the middle to the top of the microfacies with an average of 2% (Table 3). This microfacies is represented by four thin section samples (YMG9–YMG20).

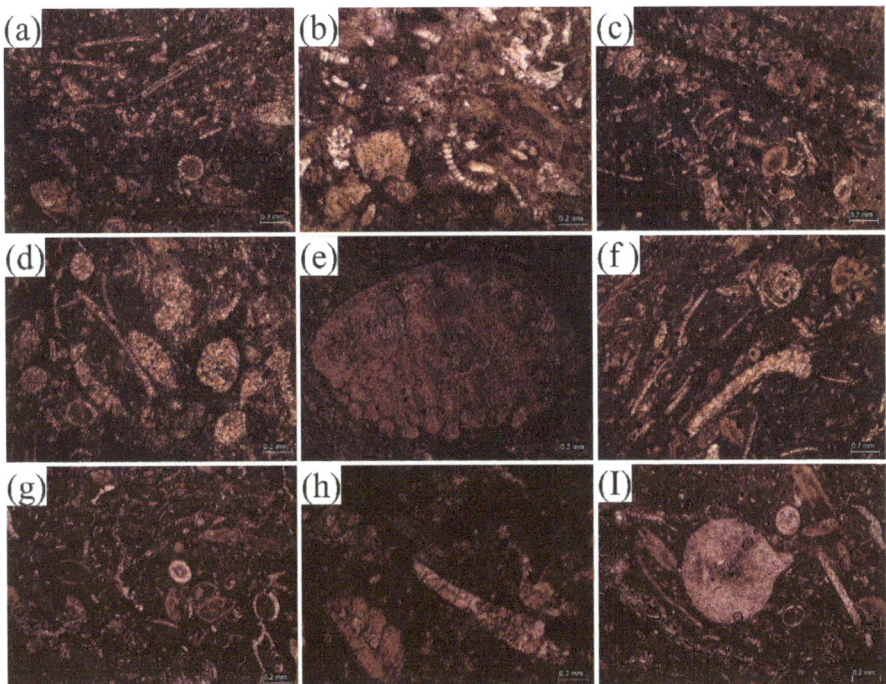

**Figure 6.** Photomicrographs showing (**a**) phylloid, dascycladacean and calcispheres algae; (**b**) mixed red and green algae; (**c**) Gymnocodiacea; (**d**) holothurian spicules (ossicles) and *Idalina grelaudae* (Middle Eocene milliolid); (**e**) *Lockhartia praehaimei*; (**f**) ostracod and green algae; (**g**) phylloid and gymnocodiacea green algae; (**h**) *Pellatispira* spp. and phylloid algae; and (**I**) *Saccammina grzybowskii* (Eocene foram).

**Table 3.** An average mineralogical composition of the Early–Middle Eocene carbonate microfacies of Yadgaar Section.

| Microfacies | Spar (%) | Micrite (%) | Bioclasts (%) | CM (%) | Pyrite (%) | Hematite (%) | Chert (%) | Gypsum (%) | Dolomite (%) | Calcite vein (%) | Deposition Site on Platform by Flügel and Munnecke |
|---|---|---|---|---|---|---|---|---|---|---|---|
| EMI | 4 | 20 | 25 | 2 | 3 | 6 | - | 0.5 | 35 | 4 | FZ8 |
| EMII | 11.5 | 25 | 55 | 1 | 2.5 | 3 | - | 2 | 7 | 2 | FZ7 |
| EMIII | 4 | 57 | 19 | 2.5 | 1 | 6.5 | - | 1 | 8 | 1 | FZ6–FZ5 |
| EMIV | 6.5 | 26 | 53 | 2 | 3 | 5 | - | 0.5 | 3 | 1 | FZ5 |
| EMV | - | 62 | 11 | 2 | 3.5 | 4 | - | 1 | 15 | 1 | FZ3 |
| EMVI | - | 21 | 3 | 19 | 4 | 7.2 | 1 | 2 | 38 | 7 | FZ9A–FZ9B |
| EMVII | 4 | 7 | 53 | 15 | 2 | 3 | 0.5 | 2 | 14 | - | FZ6–FZ7 |
| EMVIII | 8 | 19 | 33 | 22 | 3 | 0.5 | 1 | 6 | 17 | 3 | FZ7–FZ8 |
| EMIX | - | 38 | 3 | 19 | 1 | 0 | 3.5 | 1 | 20 | 107 | FZ8 |

**Interpretation:** This microfacies displays a wide range of mixed bioclasts, from green algae species to larger foraminifers. Green algae (calcispheres, Dascycladacean, echinoid spine, Gymnocodiacea, and phylloidal green algae) are identified in this microfacies. These green algae indicate a very shallow depth, normally 5–15 m on a carbonate platform. The wide range of these algae indicates a highly oxygenated environment and a photic zone [8,35–37]. Some larger foraminiferans and broken ostracod shells, *Lockartia* spp. milliolid, and *Rotalia* spp., may be reworked from adjacent microfacies due to high energy wave action and deposited in this microfacies as a mixture of bioclasts. This microfacies is equivalent to FZ7 (open marine platform) [34].

3.2.3. Ostracod, Green Algae, and Gypsum Dominating Mudstone–Wackestone Microfacies (EMIII)

This microfacies is subdivided into three subfacies, listed as follows.

Ostracod, Coralline Algae, and Green Algae Dominated Mudstone–Wackestone Subfacies

This subfacies is identified by a frequent abundance of ostracods and green algae. However, fossils are partly or completely micritized and often replaced by hematite. Micrite is the dominant constituent of this subfacies (with an average of 61%). The abundance of bioclasts appeared comparatively low (with an average 15%). Microfossils identified are *Boueina marondei*, bryozoans, coralline algae (*Sporolithon* spp.) (Figure 7a), *Idalina grelaudae* (Middle Eocene milliolid) (Figure 6a), *Discocyclina* spp., echinoid spine, green algae (Dascycladacean, Salpingoporellamelitae Radoičić), *Fabiania* spp., filamentous shells of bivalves (Figure 7b), *Heterostegina* spp., *Lockhartia prehaimei*, *Lockhartia haimei*, nummulites, *Ocoarota orali* İnan, ostracod (Figure 7c), *Assilina* spp. (Figure 7d), *Quinqueloculina* and *Rotalia* spp. Dolomite starts to increase in this subfacies (with an average 7%). Hematite, spar, carbonaceous matter, pyrite, gypsum (Figure 7e), and calcite can be observed in minor amounts with an average of 1–6%. Four samples represent this subfacies (YMG21–YMG24).

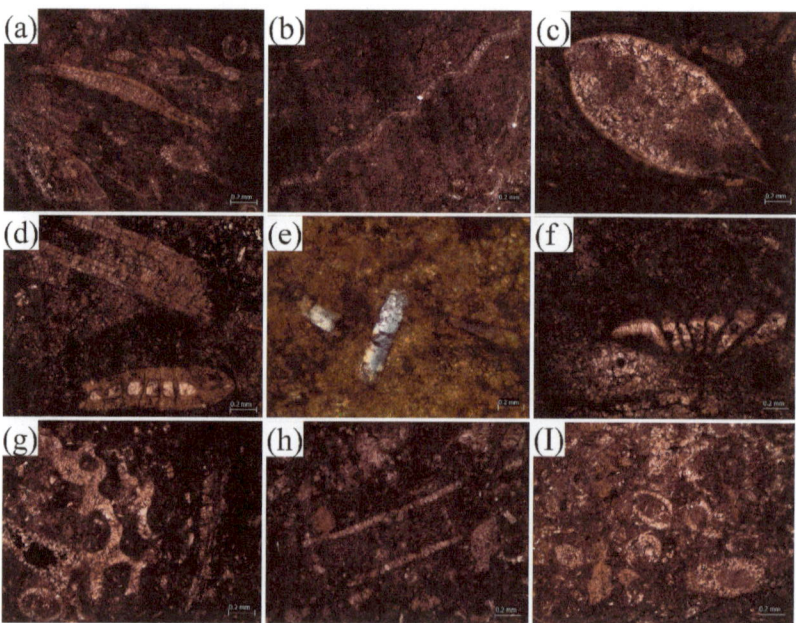

**Figure 7.** Photomicrographs showing (**a**) coralline algae and *Idalina grelaudae* (Middle Eocene milliolid); (**b**) filamentous shells of bivalves; (**c**) an ostracod; (**d**) *Assilina* spp.; (**e**) gypsum; (**f**) unidentified algal fossil; (**g**) algae; (**h**) Calcispheres green algae and Spatangoid spine algae; and (**I**) Dasycladacean green alga (*Cymopolia elongata*).

Gypsum Dominating Low Fossiliferous Mudstone–Wackestone Subfacies

This subfacies is identified on the basis of well-developed gypsum crystals (with an average of 2%). Micrite is found frequently, with an average of 53%. Small-sized fossils are micritized and are observed at an average of 12%. *Assilina* spp., *Austrotrillina eocenica*, broken red algae, *Calcarina* spp., *Entogonia formosa*, *Idalina grelaudae* (Middle Eocene milliolid), *Lockhartia praehaimei*, *Lockhartia* spp., unidentified algal fossil (Figure 7f), *Ocoarota orali* İnan, *Pellatispira madaraszi*, and *Ranikothalia* spp. Dolomite crystals start to appear considerably, with an average of 16%. Hematite is identified in the highest amounts, with an average of 9% in this subfacies. Spar, carbonaceous material, and pyrite can be observed in 5%, 2–3%, and 1%, respectively (Table 3). This subfacies is represented by four samples (YMG25–YMG28).

Green Algae Dominated Wackstone Subfacies

This subfacies is mainly comprised of green algae (Figure 7g), such as calcified algae (*Vermiporella*), spatangoid spine algae, calcisphere green algae (Figure 7h), and dascycladacean green algae (*Cymopolia elongate*) (Figure 7I). Other fossils are *Fabiania* spp., *Lockhartia* spp., *Lockhartia* spp., *Rotalia trochomoferous*, ostracods, and spatangoid spines. Fossils are highly micritized and species identification is hard. Micrite and bioclasts formed the dominant portions of the thin sections with an average of 57% and 30%, respectively. Hematite and spar are found in 5% and 3%, respectively. Carbonaceous material, pyrite, and calcite are observed in 3%, 1%, and 1%, respectively. This subfacies is represented by four thin sections (YMG29–YMG32).

**Interpretation**: Selective dolomite patches can be observed in the thin sections of this microfacies. Most bioclasts are observed to be prone to selective dolomitization. This type of dolomitization is associated with fine-grained, micritized allochem particles and their hard parts [23,38]. This may cause a lowering of the porosity values [39,40]. Coralline red algae's stratigraphic range is widely distributed and is mostly found in shallow water [41]. The co-existence of coralline algae and large benthic foraminiferans may indicate a 40–45 m (medium) water depth on a carbonate platform [42–44]. Corals are found in shallower areas as compared to sponges. They are found in association with marl or silty sandy environments with an abundance of benthic foraminiferans [45]. Corals and sponges are found in closely related environments adjacent to each other but with different environments ranging from shallow to moderately deep environments. Deep sea settings with high sea level rise may be indicated by the presence of these red coralline algae [46]. The presence of *Pellatispira*, *Heterostegina* spp., and *Discocyclina* spp. suggests a deeper depositional environment in a carbonate platform [32]. This microfacies is equivalent to FZ6 (platform margin) and FZ5 (upper slope reefs and platform-margin reefs) [34].

3.2.4. Algae and Mixed Foraminiferal Wackstone–Packstone Microfacies (EMIV)

This microfacies is subdivided into two subfacies, as follows.

Mixed Foraminiferal Micritized Packstone Microfacies

This subfacies is comprised of highly micritized bioclasts. However, the size and abundance of fossils are relatively larger than the adjacent subfacies. The bioclasts encountered in this subfacies were *Alveolina* spp. (Figure 8a)., *Assilina* spp., *Austrotrillina eocenica* (Figure 8b), *Operculina subsalsa* (Figure 8b), crinoid ossicles, echinoid spines, green algae (Gymnocodiacea, phylloid algae, and spatangoid spines) (Figure 8c), Eocene sponge (Figure 8c), *Idalina grelaudae* (middle Eocene milliolid) (Figure 8d), *Lockhartia haimei*, *Lockhartia prehamei* (Figure 8e), *Lockhartia tipper* (Figure 8e), the milliolid (*Quinqueloculina*), ostracods, *Rotalia trochidiformis*, and spatangoid spines. Bioclasts are dominant in the thin sections and their average distribution is about 56%. Micrite, spar, and hematite are observed as 22%, 9%, and 5%, respectively. Dolomite, carbonaceous material, and calcite are found in 2%, 2%, and 1%, respectively. Four samples represent this subfacies (YMG33–YMG36).

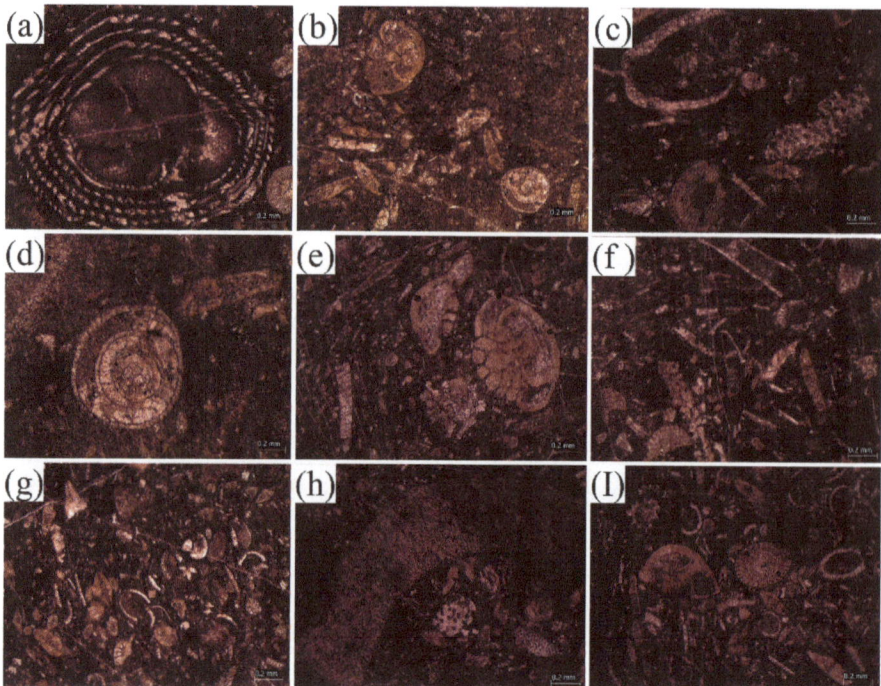

**Figure 8.** Photomicrographs showing (**a**) *Alveolina* spp.; (**b**) *Austrotrillina eocenica* on the right and *Operculina subsalsa* on the left; (**c**) Gymnocodacea in the bottom, ostracods in the top left, and sponge in the middle right; (**d**) *Idalina grelaudae* (Middle Eocene milliolid); (**e**) *Lockhartia prehaimei* on the right and *Lockhartia tipperi* on the left; (**f**) Dascycladacean green alga and *Cymopolia elongata* on top; (**g**) high concentration of green algae; (**h**) *Fabiania* spp.; and (**I**) *Rotalia trochidiformis* and echinoid spine.

Mixed Green Algae and Broken Bioclastic Wackstone–Packstone Subfacies

This subfacies is identified on the basis of the special appearance of a mixture of fossils. This mixture is dominantly comprised of green algae and broken clasts of other large fossils (with an average 50%). Fossils identified are green algae (Dascycladacean, *Cymopolia elongate*, Gymnocodiacea, echinoid spines) (Figure 8f,g), *Assilina* spp., Eocene sponge, *Fabiania* spp. (Figure 8h), *Rotalia trochidiformis* (Figure 8I), *Idalina grelaudae* (Middle Eocene milliolid), unidentified algal fossils, *Lockhartia prehaimei*, *Lockhartia tipper*, *Lockhartia retiata*, *Orbitolites* spp., ostracods, and sponges (Figure 8I). Micrite is found as the second-most abundant behind fossil fragments, with an average 30% in this subfacies. Pyrite, hematite, spar, dolomite, carbonaceous material, and calcite are 5%, 4%, 3%, 2%, and 1%, respectively. Four samples represent this subfacies (YMG37–YMG40).

**Interpretation**: Selective dolomitization in packstone–grainstone may facilitate dissolution, which can enhance porosity and reservoir characteristics [23,47]. Selective dolomitization can be observed in this microfacies. Autochthonous sponges from the Eocene sequence of the Yadgaar section have been reported for the first time. Most of these Eocene sponges are in broken fragments. The skeletons of sponges seem to be micritized, which indicates a diagenetic effect (neomorphism) on them. The association of sponges with coral indicates a relatively deep shelf area. However, the absence of coral and the appearance of sponges are indicative of shallower water depth [45]. Furthermore, broken sponge pieces indicate a moderate–high energy upper slope outer platform environment [48]. This also indicates a deep environment; however, the depth was not enough to provide a fully reduced paleogeographic constraint [45]. Ostracods characterize cold marine and deep environments [49]. This microfacies represent a wide facies belt and is equivalent to FZ5 (upper slope reefs and platform-margin reefs) of Flügel and Munnecke [34].

### 3.2.5. Nummulites Dominating Mudstone–Wackestone Microfacies (EMV)

This mudstone-wackestone microfacies is dominantly comprised of micrite material (with an average 62%). Bioclasts are partly to completely micritized (with an average 11%). Fossils identified are *Assilina* spp. (Figure 9a), gastropods and geopetal infillings (Figure 9b), *Lockhartia* spp., *Nummulites* spp. (Figure 9c,d), *Operculina* spp. (Figure 9e), *Ranikothalia* spp., and *Rotalia trochidiformis* (Figure 9f). Dolomitization is often found in the form of dolomite patches (Figure 9g) in the thin sections (with an average of 15%). Alteration of blocky cement into dolomite has been observed (Figure 9h). The changing of micrite into dolomite is obviously seen in this microfacies (Figure 9I). Carbonaceous material (with an average of 2%) and hematite (with an average of 4%) have commonly replaced fossils. Pyrite, gypsum, and calcite can also be seen with averages of 4%, 1%, and 1%, respectively (Table 3). This microfacies is represented by four samples (YMG41–YMG44).

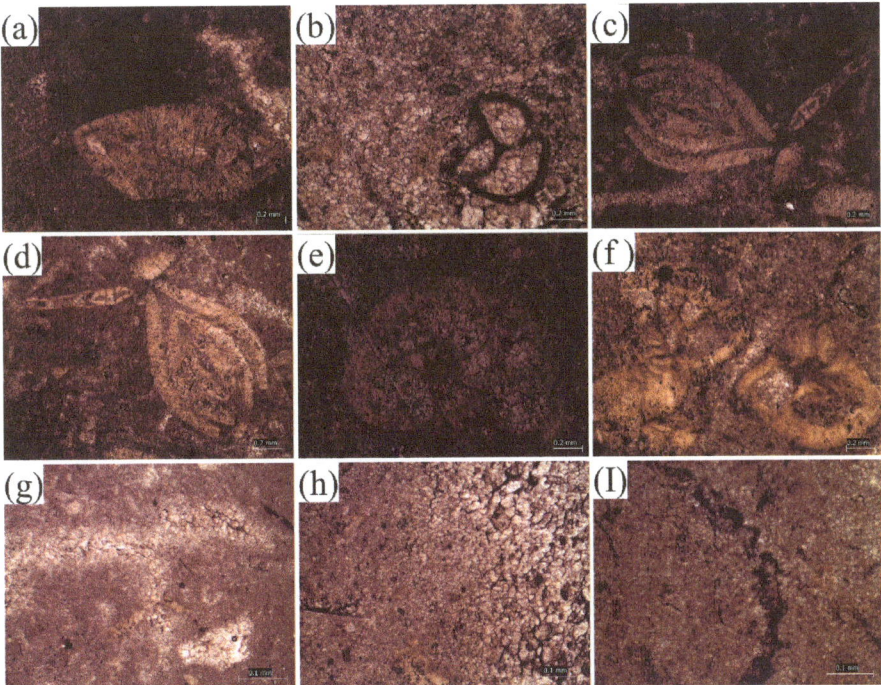

**Figure 9.** Photomicrographs showing (**a**) *Assilina* spp.; (**b**) gastropods and geopetal infillings; (**c,d**) *Nummulites* spp.; (**e**) *Operculina* spp.; (**f**) *Rotalia trochodeformis* on the left and *Lockhartia tipperi* on the right; (**g**) a dolomite patch; (**h**) blocky cement changing into dolomite; and (**I**) micrite changing into dolomite.

**Interpretation:** Gastropods indicate normal salinity and a shallow-marine shelf environment. The presence of *Nummulites* spp. characterize a subtidal environment and depth ranges from 20 to 130 m with the precipitation of lime mud [50]. *Operculina* spp. indicated the 6–15 m depth but typically dominates in the zone of 30–54 m. This also signifies the presence of a depositional site in low energy and medium light intensity in a reef slope [51]. Sometimes these types of deposits resulted from turbidites. *Operculina* spp. are identified in this microfacies. Benthic foraminifera, such as *Assilina* spp., *Operculina* spp., and *Ranikothalia* spp., indicate a depositional site in a lagoon, reef setting, or inner ramp environment in the Tethys ocean [52,53]. These Early–Middle Eocene microfossils found in this microfacies indicate close association with reefs in a lagoon environment. This microfacies is equivalent to FZ3 (toe-of-slope apron deep shelf margin) of Flügel and Munnecke [34].

3.2.6. Algal Limestone and Mudstone Microfacies (EMVI)

This microfacies is uniquely identifiable by the appearance of algal limestone. Algae and other bioclasts are completely micritized and unidentifiable. Micrite is further dolomitized (Figure 10a) in this microfacies. Dolomite is observed at an average of 38%, while micrite forms 21% of the thin sections. Carbonaceous material is also found at an average of 20% in this microfacies. Cross-cutting calcite veins (7%) and pyrite cubes (1%) are also present. Hematite, bioclasts, gypsum (Figure 10b), pyrite (Figure 10c), and chert (Figure 10d) can be observed with averages of 7%, 3%, 2%, and 1%, respectively (Table 3). This microfacies is represented by four samples (YCF1–YCF4).

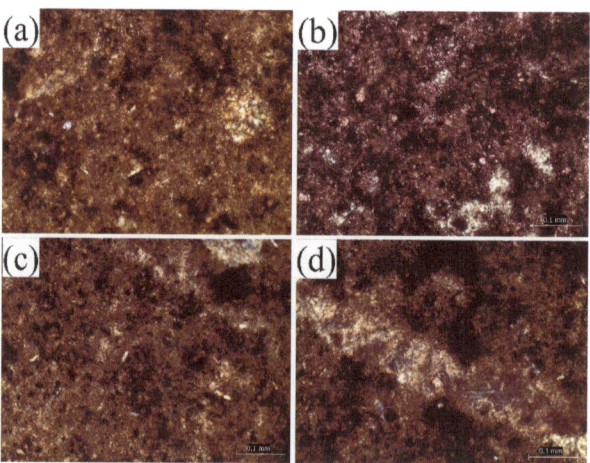

**Figure 10.** Photomicrographs showing (**a**) dolomitized bioclast and gypsum; (**b**) gypsum, hematite and micrite; (**c**) gypsum and pyrite; and (**d**) chert.

**Interpretation:** Gypsum (an evaporitic sulfate mineral) is often found in lagoon conditions associated with methane hydrate settings [54,55]. Fibrous gypsum is comprised of two main forms, a) satin spar and b) selenite [56], and is found in shallow marine limestone and shale. In evaporitic conditions where carbonate rocks react with acid sulfate, they will precipitate gypsum [57]. Gypsum is also formed by the oxidation of iron sulfide (pyrite). Recent studies have revealed a direct relationship between methane hydrate sediments, gypsum, and pyrite. The sulfate–methane transition zone often facilitates the precipitation of authigenic gypsum [56]. The coexistence of both gypsum and pyrite is tricky to explain, depending on whether the conditions are anoxic or oxic. By the activity of sulfide oxidizers, oxidized pyrite is converted into sulfate in oxic conditions, resulting in the formation of gypsum. In a sedimentary environment, these settings are called marine methane hydrate settings [58]. The connection between methane-bearing sediments and mineralization is still unclear. Gypsum is largely found to be associated with methane hydrate settings in the southwest African Margin, the Bay of Bangal, the South China Sea, and the eastern North Pacific Ocean [55]. However, in anoxic conditions, the process is reversed where sulfate-reducing bacteria convert the gypsum into pyrite [59]. Gypsum precipitation likely largely occurred after considerable and rapid evaporation, and this was presumably linked with a reduction in the size of the lake. However, the lake's bottom was still anoxic, which facilitated pyrite precipitation [54]. Sulfide/sulfate anions and Ca/Fe cations in the solution can precipitate gypsum and pyrite [56]. The pyrite and gypsum phases are tightly associated. In fresh and shallow waters, larger benthic foraminifera are absent. Their presence occurs only in relatively deep waters [60]. This microfacies is equivalent to FZ9A (arid near-coast evaporitic platforms) to FZ9B (humid near-coast brackish regions) of Flügel and Munnecke [34].

### 3.2.7. Assilina Bed Wackstone–Packstone Microfacies (EMVII)

In this microfacies, *Assilina* spp. are more abundant than the other microfossils. The size and abundance of fossils is highest among all the microfacies (with an average of 53%). This microfacies is also comprised of a high amount of carbonaceous material, with an average of 15%. Micritization and dolomitization effects are also seen in thin sections. Dolomite can be found with an average of 14%, while micrite can be observed with an average of 7%. Stylolites are frequently observed in parallel to the bioclasts. Fossils are mostly *Assilina* spp. and nummulites (Figure 11a–c), while some coralline algae (Figure 11d), unidentified algal fragments., *Discocyclina ranikotenesis* (Figure 11e), *Discocyclina zindaperensis* (Figure 11f,g), coralline algae, and *Rotalia trochidiformis* (Figure 11h,I) can be identified. Hematite, spar, gypsum, pyrite, and chert are found with averages of 3%, 4%, 2%, 2%, and <1%, respectively (Table 3). This microfacies is represented by eight samples (YCF5–YCF12).

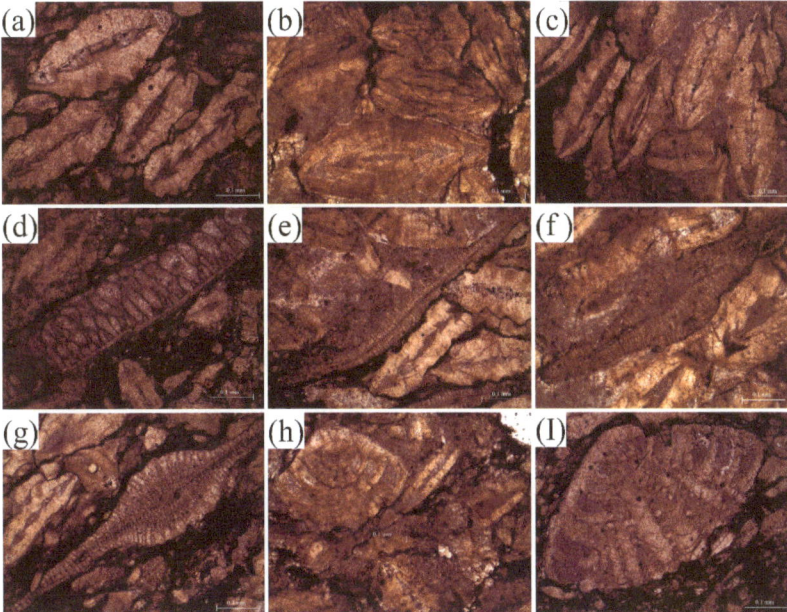

**Figure 11.** Photomicrographs showing (**a**) *Assilina* in the center to the right and *Nummulites* spp. on top; (**b**,**c**) *Assilina* spp.; (**d**) Coralline algae; (**e**) *Discocyclina ranokotenesis* in the center and *Assilina* spp. on the right; (**f**,**g**) *Discocyclina zindaperensis* in the center; and (**h**,**I**) *Rotalia trochodiformis*.

**Interpretation:** A high abundance of *Assilina* spp. (preferably called the *Assilina* bed) may indicate a highly favorable salinity range and environmental depth for the continuation of species reproduction [61]. This can cause a high abundance of *Assilina* species in the microfacies. Larger foraminifera and *Orthophragminids*, especially *Discocyclina* spp., along with *Nummulites* spp. and *Assilina* spp., indicate deposition in the photic zone of carbonate ramp settings [61,62]. This microfacies is equivalent to FZ6 (platform-edge and platform sand shoals) to FZ7 (platform interior–normal marine (open marine)) of Flügel and Munnecke [34].

### 3.2.8. Micritized Large Benthic Foraminiferal Wackstone Microfacies (EMVIII)

This is a highly micritized microfacies in which large bioclasts are almost completely micritized. Micritization is intense enough to make the species unidentifiable (Figure 12a). However, very few bioclasts can be recognized in the middle–top portion as *Alveolina* spp. (Figure 12b–d), *Assilina* spp., *Nummulites* spp. and *Assilina* spp. (Figure 12e,f), *Azzarolina daviesi*, *Lockhartia* spp., milliolid, *Operculina* spp., ostracods, and *Rotalia* spp. (Figure 12g). Bioclasts are abundantly found with an average of 32% in this microfacies. Dolomite and carbonaceous material are also found in large quantities, with averages of 17% and 13%, respectively. Micrite, spar, gypsum, calcite, pyrite, and chert (Figure 12h) were observed in 19%, 8%, 4%, 3%, 2%, and 1%, respectively (Table 3). Bioclasts are found to be replaced by carbonaceous material (Figure 12I). This microfacies is represented by four samples (YCF13–YCF24).

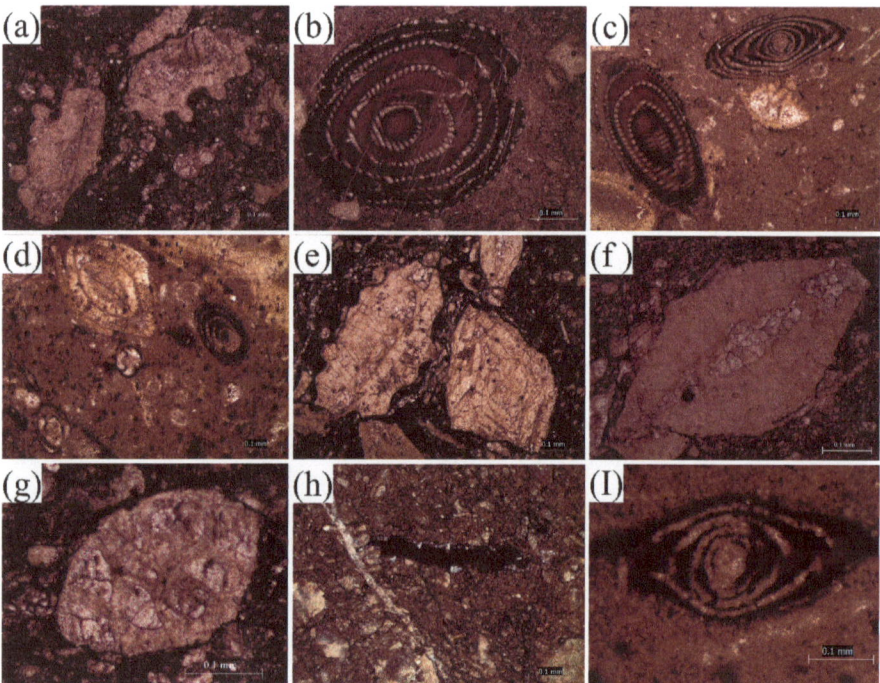

**Figure 12.** Photomicrographs showing (**a**) micritized bioclasts; (**b**,**c**) *Alveolina* spp., (**d**) Nummulites on the left and *Alveolina* spp. on the right; (**e**) *Assilina* spp. on the left and *Nummulites* spp. on the right; (**f**) micritized *Assilina* spp.; (**g**) *Rotalia trochodiformis*; (**h**) carbonaceous material and chert replacing bioclast; and (**I**) Nummulites is replaced by carbonaceous material.

**Interpretation:** The presence of *Alveolina* spp., *Nummulites* spp., and *Assilina* spp. with moderate energy conditions, indicates inner ramp settings but with a slight increase in depth relative to where the previous microfacies were likely deposited [61]. *Nummulites* spp. are normally found in relatively deep settings [60]. This association of microfauna is found in the distal inner ramp area. This is a high energy, current-dominated environment close to fair weather wave base (FWWB). *Operculina* spp. along with *Nummulites* spp. are also indicative of an increase in water depth [61]. The large quantity of *Nummulites* indicates distal inner ramp to mid-ramp environments. This indicates a high-energy environment close to FWWB. This microfacies is equivalent to FZ7 (platform interior–normal marine–open marine) to FZ8 (platform interior–restricted) of Flügel and Munnecke [34].

### 3.2.9. Algal Limestone and Mudstone Microfacies (EMIX)

The top microfacies of the Early–Middle Eocene carbonate sequence of the Yadgaar section is comprised of algal limestone (Figure 13a). Micrite is the dominant constituent with an average of 38%. Fossils are partly to completely micritized as we move towards the upper contact of the microfacies (with an average of 19%). Only very few *Alveolina* spp. (Figure 13a), green algae (Figure 13b), *Lockhartia* spp., *Nummulites* spp. (Figure 13c), milliolids and ostracods (Figure 13d) can be distinguished in this thin section study. The majority of fossils are completely micritized and unidentifiable. Dolomite crystals are abundantly found in this microfacies with an average of 20%. Calcite, bioclasts, chert (Figure 13e–h), gypsum, pyrite (Figure 13e–h), and hematite (Figure 13I) were observed with averages of 13%, 3%, 4%, 1%, 1% and <1%, respectively (Table 3). This microfacies is represented by four samples (YCF25–YCF32).

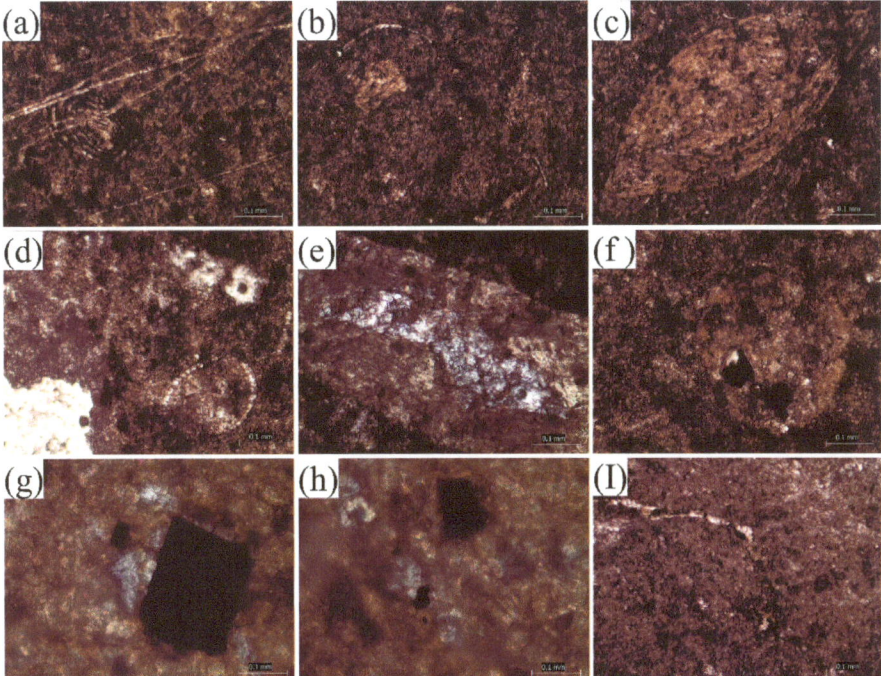

**Figure 13.** Photomicrographs showing (**a**) algal limestone and *Alveolina* spp.; (**b**) broken clasts of green algae; (**c**) *Nummulites* spp.; (**d**) an ostracod; (**e**) chert in calcite vein; (**f**–**h**) pyrite cubes in black; and (**I**) a hematite vein.

**Interpretation:** *Alveolina* spp. is compatible with a variegated salinity-temperature range and indicates inner ramp settings [63]. On a carbonate platform, it is distributed over a wide shallow water platform area <35 m [64]. This area is specifically just below the fair weather wave base in a high energy environment [65]. However, *Alveolina* spp. can also be found at a depth of up to 60 m in the fore-reef of a deep lagoon environment [66]. In the Eocene, *Alveolina* spp. associated with milliolids is typical of shallow water vegetative substrate in a sandy/sea grass environment [60,67–69]. This can be confirmed by the presence of chert in this microfacies, which comes from high water wave action. This indicates inner ramp settings (Figure 13e). The presence of *Alveolina* spp. in this uppermost microfacies indicates a Bartonian age and deposition in a deeper portion of the shallow marine platform [32]. This microfacies is equivalent to FZ8 (platform interior–restricted) of Flügel and Munnecke [34].

## 3.3. Depositional Environment

The Yadgaar Section lies on the margins of the Ceno-Tethys Ocean. A comprehensive study of Paleocene–Oligocene rocks of the Yadgaar Section can reveal the depositional environment of rocks during the mature development and the closing stage of the Ceno-Tethys Ocean. These rock formations include Lockhart Limestone of Late-Paleocene, Margalla Hill Limestone, and Chorgali Formations of Early-Middle Eocene, as well as Kuldana Formation of Middle Eocene-Early Oligocene age. A petrological study of the Late Paleocene Lockhart Limestone reveals its depositional environment as a shallow shelf carbonate platform, typically from the platform margin to the toe of the slope and slope areas [9]. This study indicated that shallow marine platform carbonates were deposited in the final phase of transgression–regression events during the Early–Middle Eocene age. These deposits are represented by the Margalla Hill Limestone and the Chorgali Formation in the UIB [8]. Detailed microfacies studies based on the textural and benthonic foraminiferans revealed their depositional environments on the carbonate platform (Figure 14). Early–Middle Eocene carbonates were categorized as EM-I–IX, belonging to the restricted-marine platform, open-marine platform, platform-edge and platform sand shoals, toe of slope and slope, arid near-coast evaporitic platforms, humid near-coast brackish regions, platform-edge and platform sand shoals, open-marine platform, open-marine platform, restricted-marine platform, and restricted-marine platform areas, respectively. These microfacies are found to be equivalent to the FZ8, FZ7, FZ6–5, FZ5, FZ3, FZ9A–FZ9B, FZ6–7, FZ7–FZ8 and FZ8 (Figure 14b) microfacies of Flügel and Munnecke [34]. The closing stage of the Ceno-Tethys Ocean is marked by the deposition of the Middle Eocene–Early Oligocene Kuldana Formation. A detailed study on the Kuldana Formation indicated a transitional environment from deltaic to continental because of the presence of limestone lenses in the base and sandstone/shale alternations in the Yadgaar Section [8].

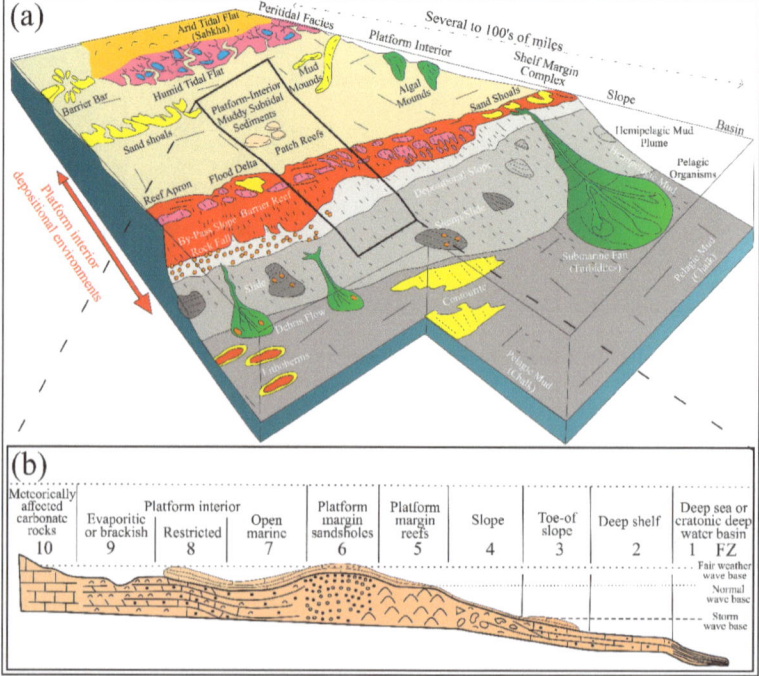

**Figure 14.** (**a**) Figure illustrating a generalized carbonate platform with associated environments. The rectangle corresponds to (**b**). (**b**) Depositional model of Early–Middle Eocene carbonate rock microfacies on specific zones of the carbonate platform [34].

## 3.4. Reservoir Characterization

### 3.4.1. Diagenetic Impact

Early–Middle Eocene carbonate rocks are comprised of well-developed to fused nodular limestone and shales in the Yadgaar Section. To understand diagenetic effects on the carbonate Formations, eight samples of limestone and four samples of shale were collected from the base, middle, and top of the section for SEM analysis (Figure 15a–f). Light grey colored nodular limestone beds are present in the base and middle portions, while fused nodular limestone and shale are present in the top portion of the deposits in the study area deposits.

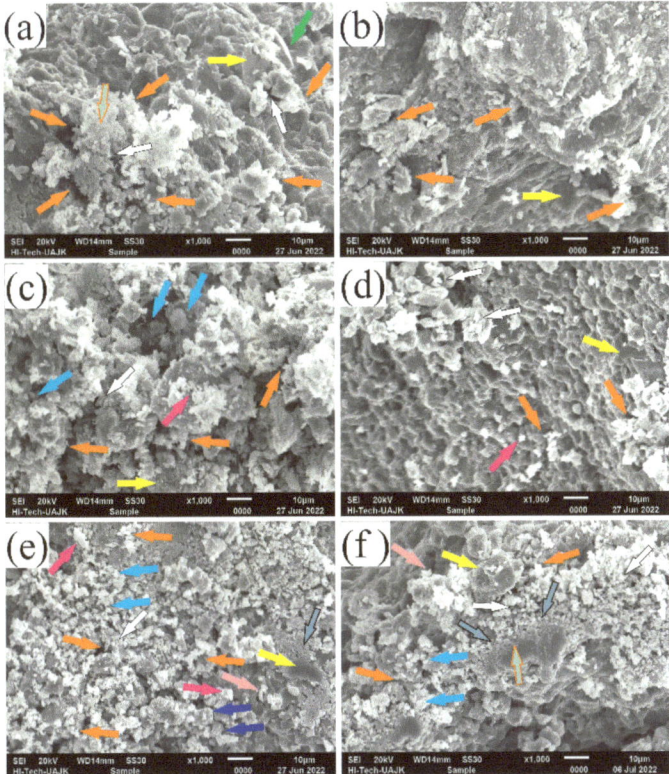

**Figure 15.** SEM image of shale and limestone samples from Early–Middle Eocene strata. (**a**,**b**) Images of the samples obtained from the base of the section; (**c**,**d**) Images of the samples taken from the middle of the Margalla Hill Limestone; (**e**,**f**) Images denote samples obtained from top of Margalla Hill Limestone. Yellow arrows denote calcite crystals, indigo blue arrows denote dolomite crystals, orange–brown arrows denote drusy cement, white arrows denote blocky cement, sky blue arrows denote rim cement, light grey arrows with a red outline denote compacted illite, indigo arrows with a red outline denote montmorillonite clay, and black arrows with red outline denote chert grains as found by Prothero and Schwab [70]. Moreover, green arrows denote a fibrous column of calcite, light pink arrows denote mosaic (drusy) dolomite cement, silver arrows with a black outline denote granular cement, and dark purple arrows denote dog tooth calcite crystals as found by Flügel and Munnecke [34].

SEM evaluations of the limestone (Figure 15a) and shale (Figure 15b) revealed the presence of first-generation drusy cement dominating throughout the Early–Middle Eocene deposits. In the Early–Middle Eocene deposits (Margalla Hill Limestone), first-generation cement is recognized by the presence of drusy cement. This shows the presence of the first stage of post-depositional diagenesis (eogenetic) in the deposits. However, as shown in Figure 2, a minor occurrence of second and third generation cement was observed by the presence of blocky and rim cements, respectively. This demonstrates that mesogenetic and telogenetic diagenesis occurred here. The top section is comprised of middle Eocene strata (Chorgali Formation) (Figure 2), which possess an abundance of second-generation cements by the presence of blocky cement in the base, which starts to decrease towards the middle of the stratigraphic sequence. Meanwhile, first generation drusy cement again starts to increase towards the top of the section. In the middle Eocene strata, third generation rim cements are rarely found in the middle portion of the carbonate deposits.

Based on these results, our study found a greater abundance of cementation and other alterations that occurred during eogenesis compared to during later stages of diagenesis. However, in the middle part of the section (Figure 2), the grade of diagenesis slightly increased, which is revealed by the presence of mesogenetic and telogenetic cements of the-post depositional diagenesis. The minor presence of smectite, calcite crystals, chert, compacted illite, compacted montmorillonite, dog tooth calcite, dolomite, fibrous column of calcite, granular cement, montmorillonite clay, mosaic drusy dolomite, and radiaxial fibrous cement can also be observed in the SEM analysis (Figure 16).

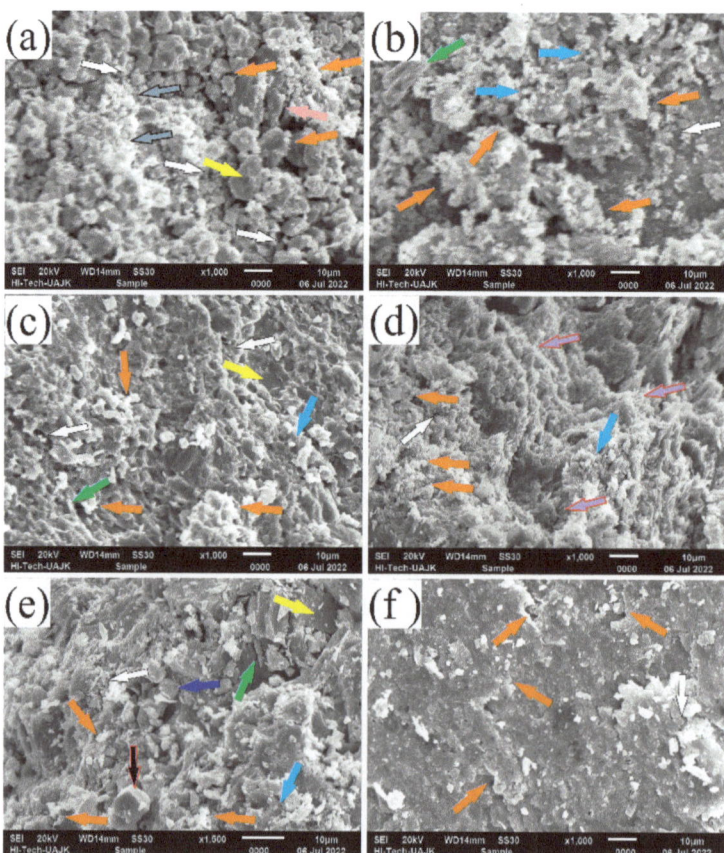

**Figure 16.** SEM image of shales and limestone samples from the Middle Eocene Chorgali Formation. (**a**,**b**) Images of the samples obtained from the base of the Chorgali Formation; (**c**,**d**) Images of the samples taken from the middle of the Chorgali Formation; (**e**,**f**) Images denote samples obtained from top of the Chorgali Formation. Refer to the caption of Figure 15 above for an explanation of the arrow colors.

### 3.4.2. Porosity Types and Reservoir Characteristics

From the base to top in Figure 17, the first microfacies (EMI) possesses a dominant presence of channel (Ch) type porosity along with little occurrence of channel (Ch), vuggy (Vg), and fenestral (Fe) pore types (Figure 17a). Image J software's image threshold enhancement technique calculates the presence of 10% porosity in this microfacies (Figure 17a). The second microfacies (EMII) is composed of nearly equal amounts of vuggy (Vg), fractured (Fr), and moldic (Md) porosity (Figure 17b). The porosity calculated in this microfacies is 8%. The third microfacies (EMIII) is comprised of only fracture (Fr) and shelter (Sh) type porosities (Figure 17c). The calculated porosity in this microfacies is 10%. Vuggy (Vg) and fenestral (Fe) type porosities are also observed in the fourth microfacies EMIV (Figure 17d). The calculated porosity in this microfacies is 5%.

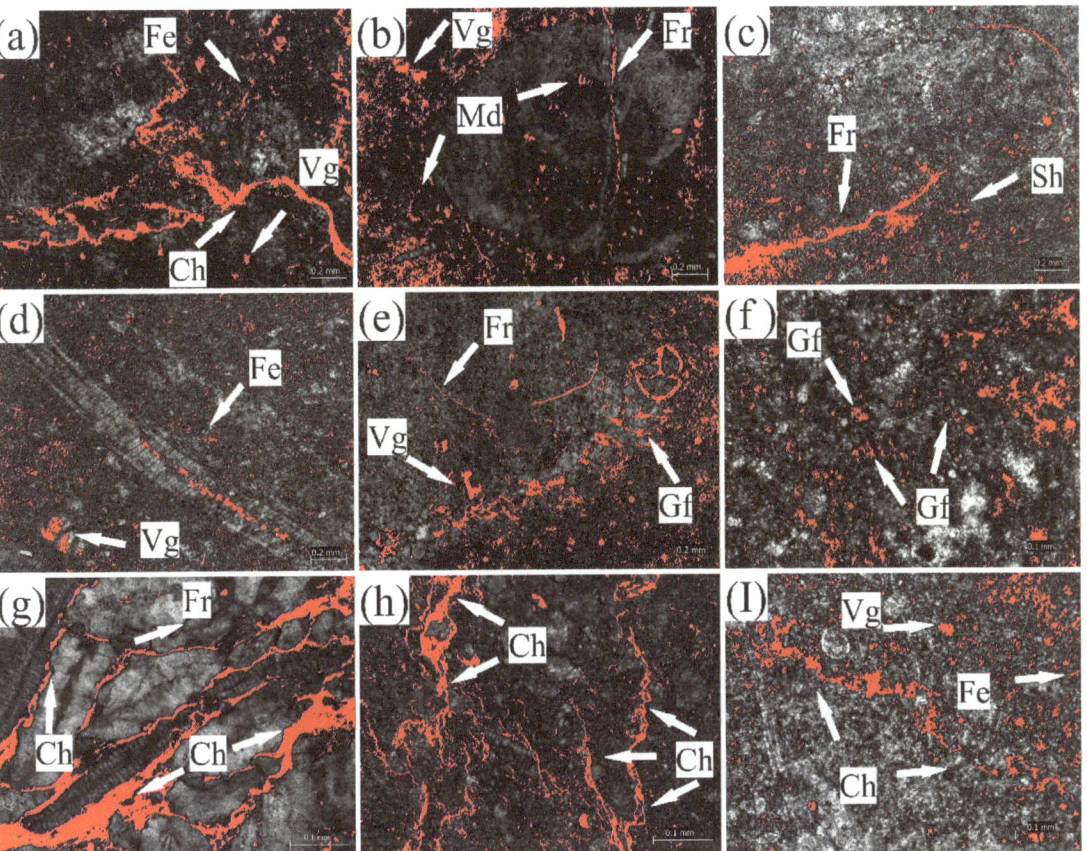

**Figure 17.** The image threshold enhancement technique in Image J software calculates porosity and reveals pore type for all microfacies of the Early–Middle Eocene rocks. (a–I) EMI, EMII, EMIII, EMIV, EMV, EMVI, EMVII, EMVIII, and EMVIX, respectively. Abbreviations of the porosity types shown in the figures: Fe, Fenestral; Ia, Intraparticle; Sh, Shelter; Vg, Vuggy; Ch, Channel; Gf, Growth Framework; Md, Moldic; and Fr, Fracture pore types.

An increase in porosity values has been noted as moving towards EMV (Figure 17e). The average porosity value for this microfacies was determined to be 10% (Figure 9c,d). This microfacies shows that vuggy (Vg) and growth framework (Gf) porosities are most common (Figures 9c and 16e), while fracture (Fr) porosity is less common. EMVI depicted only the growth framework (Gf) type of porosity, with a porosity value of 7% (Figure 17f).

In EMVII, the highest amount of porosity (15%) has been noted (Figure 17g). Frequent amounts of channel (Ch) and fracture (Fr) types of porosities can be observed clearly. The eighth microfacies (EMVIII) also displayed a higher porosity value of 11% (Figure 17h). However, only the channel (Ch) type porosity is observed in this microfacies. At the top of the Early–Middle Eocene carbonate rocks, strata comprising the ninth microfacies (EMIX) display channel (Ch), fenestral (Fe), and vuggy (Vg) types of porosities. This microfacies possesses 9% porosity (Figure 17I). The calculated average porosity from all microfacies of the Early–Middle Eocene carbonate strata in the Yadgaar Section is 7.4%.

Fracture and moldic types of porosities reveal the presence of early diagenetic effects such as the eogenetic alterations. They possess a low porosity but a high permeability range in shelf carbonate rocks, meaning they are of high reservoir quality [71,72]. Vuggy, fracture, and moldic porosities reflected within these sediments indicate they are favorable strata in terms of reservoir potential [72,73]. Fenestral porosity is associated with meteoric diagenesis. It can coexist with microfacies related to packstone, dolomitization, and dolomudstone [72]. The quality of a reservoir mainly depends on permeability, which is enhanced by the processes of dolomitization and de-dolomitization [74]. Channel type porosity is dominantly observed in thin sections of most microfacies. This type of porosity is likely to increase the permeability and reservoir quality. It is developed due to chemical dissolution in high energy conditions associated with diagenetic compaction [71,72,74]. Diagenetic features including fractures, channels, and dissolution are indications of tectonic forces exerted on the rocks after their formation [9,26,38,75]. In general, petrological observations, fracture, channel, growth framework, and moldic type porosities highly facilitate interconnecting pores, creating high permeability. The occurrence of all these types of porosities in the Early–Middle Eocene carbonate rocks of the Yadgaar Section is a positive signature of high reservoir quality.

## 4. Conclusions

On the basis of field, petrographic, and SEM analysis, we have drawn the following conclusions. Early–Middle Eocene carbonate deposits of the Margalla Hill Limestone and the Chorgali Formation are found in the Yadgaar Section. Field observation and petrographic studies revealed nine microfacies (EMI–EMIX) within these deposits. Detailed paleontological and petrological studies indicate that these microfacies are deposited in the restricted marine (platform interior), open marine platform, platform edge, platform margin reef, toe of the slope apron, arid-humid platform interior, platform edge, open marine platform interior, and restricted marine platform interior, respectively, of the Ceno-Tethys Ocean. Benthic foraminifera microfossils revealed that these microfacies were deposited in the Early–Middle Eocene period in the Ceno-Tethys Ocean. Microfossils, including algae and foraminifera, as well as the presence of gypsum crystals, indicate shallow marine methane hydrate settings in complex lagoon conditions. Most of the microfacies (EMI–EMVI) from the base to upper-middle of the section were deposited in the shallow marine, oxygenated, and photic zone, while the top three microfacies (EMVII–EMIX) were deposited in a relatively deeper portion of the carbonate platform with normal salinity. SEM analysis indicated a dominant presence of first generation drusy cement throughout these lower-middle Eocene limestones and shales. This phenomenon indicates an early grade of post-depositional diagenesis (eogenetic) effects on these deposits. However, a minor amount of blocky (second generation) and rim (third generation) types of cement are also observed in the middle of the section, which indicates mesogenetic and telogenetic diagenesis alterations, respectively. It can be observed that these post-depositional diagenetic stages are directly proportional to the cement generations. Petrological observations reveal fracture, channel, growth framework, and moldic types of porosities, which provide favorable conditions for producing high permeability. Regional tectonic stresses also enhance the development of diagenetic features. Moreover, the diagenetic changes, pore types, and porosity values have considerably enhanced the reservoir characteristics of these carbonate rocks. On the basis of fossils such as *Acervulinidae* spp., Eocene sponge, *Milliolid* spp. (*Idalina grelaudae*), *Rotalia* spp., and *Saccammina grzybowskii*, the relative chronological age of these carbonate deposits is suggested as Early–Middle Eocene.

**Author Contributions:** Conceptualization, A.B., M.S.M. and R.Y.; data collection, A.B., M.S.M. and R.Y.; methodology, A.B., M.S.M., R.Y. and H.T.J.; software, A.B., M.S.M., R.Y. and H.T.J., writing—original draft preparation, A.B., M.S.M. and R.Y.; supervision; A.B., M.S.M. and R.Y.; writing—review and editing, M.Z., G.K., H.T.J. and A.B. All authors have read and agreed to the published version of the manuscript.

**Funding:** This study was financially supported by the China-ASEAN Maritime Cooperation Fund Project (grant No. 12120100500017001) and the National Natural Science Foundation of China (grant No. 41972146).

**Institutional Review Board Statement:** Not applicable.

**Informed Consent Statement:** Not applicable.

**Data Availability Statement:** The data used in this work is available on request to the corresponding author(s).

**Conflicts of Interest:** The authors declare no conflict of interest (financial or non-financial).

# References

1. Martín-Martín, M.; Guerrera, F.; Tosquella, J.; Tramontana, M. Middle Eocene carbonate platforms of the westernmost Tethys. *Sediment. Geol.* **2021**, *415*, 105861. [CrossRef]
2. Pomar, L.; Baceta, J.I.; Hallock, P.; Mateu-Vicens, G.; Basso, D. Reef building and carbonate production modes in the west-central Tethys during the Cenozoic. *Mar. Pet. Geol.* **2017**, *83*, 261–304. [CrossRef]
3. Rivero-Cuesta, L.; Westerhold, T.; Alegret, L. The Late Lutetian Thermal Maximum (middle Eocene): First record of deep-sea benthic foraminiferal response. *Palaeogeogr. Palaeoclimatol. Palaeoecol.* **2020**, *545*, 109637. [CrossRef]
4. Janjuhah, H.T. Sedimentology and Origin of Microporosity in Miocene Carbonate Platforms, Central Luconia, offshore Sarawak, Malaysia. Doctoral Dissertation, Universiti Teknologi PETRONAS, Perak, Malaysia, 2018.
5. Tawfik, M.; El-Sorogy, A.; Moussa, M. Metre-scale cyclicity in Middle Eocene platform carbonates in northern Egypt: Implications for facies development and sequence stratigraphy. *J. Afr. Earth Sci.* **2016**, *119*, 238–255. [CrossRef]
6. Fazal, A.G.; Umar, M.; Shah, F.; Miraj, M.A.F.; Janjuhah, H.T.; Kontakiotis, G.; Jan, A.K. Correction: Fazal et al. Geochemical Analysis of Cretaceous Shales from the Hazara Basin, Pakistan: Provenance Signatures and Paleo-Weathering Conditions. *J. Mar. Sci. Eng.* **2022**, *10*, 800. *J. Mar. Sci. Eng.* **2022**, *10*, 1654. [CrossRef]
7. Mateen, A.; Wahid, A.; Janjuhah, H.T.; Mughal, M.S.; Ali, S.H.; Siddiqui, N.A.; Shafique, M.A.; Koumoutsakou, O.; Kontakiotis, G. Petrographic and Geochemical Analysis of Indus Sediments: Implications for Placer Gold Deposits, Peshawar Basin, NW Himalaya, Pakistan. *Minerals* **2022**, *12*, 1059. [CrossRef]
8. Bilal, A.; Mughal, M.S.; Janjuhah, H.T.; Ali, J.; Niaz, A.; Kontakiotis, G.; Antonarakou, A.; Usman, M.; Hussain, S.A.; Yang, R. Petrography and Provenance of the Sub-Himalayan Kuldana Formation: Implications for Tectonic Setting and Palaeoclimatic Conditions. *Minerals* **2022**, *12*, 794. [CrossRef]
9. Bilal, A.; Yang, R.; Fan, A.; Mughal, M.S.; Li, Y.; Basharat, M.; Farooq, M. Petrofacies and diagenesis of Thanetian Lockhart Limestone in the Upper Indus Basin (Pakistan): Implications for the Ceno-Tethys Ocean. *Carbonates Evaporites* **2022**, *37*, 78. [CrossRef]
10. Hottinger, L.; Bassi, D. *Paleogene Larger Rotaliid Foraminifera from the Western and Central Neotethys*; Springer: Berlin/Heidelberg, Germany, 2014.
11. Hallock, P. Symbiont-bearing foraminifera: Harbingers of global change? *Micropaleontology* **2000**, *46*, 95–104.
12. El-Azabi, M. Sedimentological characteristics, palaeoenvironments and cyclostratigraphy of the middle Eocene sequences in Gabal el-Ramliya, Maadi-Sukhna stretch, north eastern Desert. In Proceedings of the Egyptian 8th International Conference on the Geology of Arab World, Cairo, Egypt, 13–16 February 2006; pp. 1–31.
13. Swati, M.A.F.; Haneef, M.; Ahmad, S.; Naveed, Y.; Zeb, W.; Akhtar, N.; Owais, M. Biostratigraphy and depositional environments of the Early Eocene Margalla Hill Limestone, Kohala-Bala area, Haripur, Hazara Fold-Thrust Belt, Pakistan. *J. Himal. Earth Sci.* **2013**, *46*, 65.
14. Shah, S.M.I. Stratigraphy of Pakistan (memoirs of the geological survey of Pakistan). *Geol. Surv. Pak.* **2009**, *22*, 1–8.
15. Muhammad, S.; Khalid, P. Hydrogeophysical investigations for assessing the groundwater potential in part of the Peshawar basin, Pakistan. *Environ. Earth Sci.* **2017**, *76*, 494. [CrossRef]
16. Iqbal, M.F.; Malik, A.H. Investigation of limestone exploitation area and its environmental impacts using GIS/RS techniques: A case study of Margalla Hills National Park, Islamabad. *J. Himal. Earth Sci.* **2010**, *43*, 31.

17. Mughal, M.S.; Zhang, C.; Du, D.; Zhang, L.; Mustafa, S.; Hameed, F.; Khan, M.R.; Zaheer, M.; Blaise, D. Petrography and provenance of the Early Miocene Murree Formation, Himalayan Foreland Basin, Muzaffarabad, Pakistan. *J. Southeast Asian Earth Sci.* **2018**, *162*, 25–40. [CrossRef]
18. Salih, H.D. Larger benthic foraminiferal assemblages from Sinjar Formation, SW Sulaimaniyah City Kurdistan Region, Iraq. *Iraqi Bull. Geol. Min.* **2012**, *8*, 1–17.
19. Mirza, K.; Akhter, N.; Ejaz, A.; Zaidi, S.F.A. Biostratigraphy, microfacies and sequence stratigraphic analysis of the Chorgali Formation, Central Salt Range, northern Pakistan. *Solid Earth Sci.* **2022**, *7*, 104–125. [CrossRef]
20. Ali, A. Sedimentology of the Chor Gali Formation, Central Salt Range Pakistan. Master's Thesis, Institute of Geology, University of the Punjab, Lahore, Pakistan, 2012. Volume 121, Unpublished.
21. Yasin, M.; Umar, M.; Rameez, S.; Samad, R. Biostratigraphy of early eocene margala hill limestone in the muzaffarabad area (Kashmir Basin, Azad Jammu and Kashmir). *Pak. J. Geol. (PJG)* **2017**, *1*, 16–20.
22. Dunham, R.J. *Classification of Carbonate Rocks According to Depositional Textures*; AAPG: Tulsa, OK, USA, 1962.
23. Rahimi, A.; Adabi, M.; Aghanabati, A.; Majidifard, M.; Jamali, A. Dolomitization mechanism based on petrography and geochemistry in the Shotori Formation (Middle Triassic), Central Iran. *Open J. Geol.* **2016**, *6*, 1149–1168. [CrossRef]
24. Gregg, J.M.; Sibley, D.F. Epigenetic Dolomitization and the Origin of Xenotopic Dolomite Texture: REPLY. *J. Sediment. Res.* **1986**, *56*, 735–736.
25. Janjuhah, H.T.; Ahmed Salim, A.M.; Ali, M.Y.; Ghosh, D.P.; Amir Hassan, M.H. Development of carbonate buildups and reservoir architecture of Miocene carbonate platforms, Central Luconia, offshore Sarawak, Malaysia. In Proceedings of the SPE/IATMI Asia Pacific Oil & Gas Conference and Exhibition, online, 12–14 October 2021.
26. Ahmad, I.; Shah, M.M.; Janjuhah, H.T.; Trave, A.; Antonarakou, A.; Kontakiotis, G. Multiphase Diagenetic Processes and Their Impact on Reservoir Character of the Late Triassic (Rhaetian) Kingriali Formation, Upper Indus Basin, Pakistan. *Minerals* **2022**, *12*, 1049. [CrossRef]
27. Sibley, D.F. *Climatic Control of Dolomitization, Seroe Domi Formation (Pliocene), Bonaire, NA*; AAPG: Tulsa, OK, USA, 1980.
28. Coppard, S.E.; Campbell, A.C. Taxonomic significance of test morphology in the echinoid genera Diadema Gray, 1825 and Echinothrix Peters, 1853 (Echinodermata). *ZOOSYSTEMA-PARIS-* **2006**, *28*, 93.
29. Janjuhah, H.T.; Alansari, A.; Santha, P.R. Interrelationship Between Facies Association, Diagenetic Alteration and Reservoir Properties Evolution in the Middle Miocene Carbonate Build Up, Central Luconia, Offshore Sarawak, Malaysia. *Arab. J. Sci. Eng.* **2018**, *44*, 341–356. [CrossRef]
30. Janjuhah, H.T.; Gamez Vintaned, J.A.; Salim, A.M.A.; Faye, I.; Shah, M.M.; Ghosh, D.P. Microfacies and depositional environments of miocene isolated carbonate platforms from Central Luconia, Offshore Sarawak, Malaysia. *Acta Geol. Sin.-Engl. Ed.* **2017**, *91*, 1778–1796. [CrossRef]
31. Kroh, A.; Nebelsick, J.H. Echinoderms and Oligo-Miocene carbonate systems: Potential applications in sedimentology and environmental reconstruction. *Carbonate Syst. Dur. Oligocene–Miocene Clim. Transit.* **2010**, *42*, 201–228.
32. Özcan, E.; Okay, A.; Bürkan, K.; Yücel, A.; Özcan, Z. Middle-Late Eocene marine record of the Biga Peninsula, NW Anatolia, Turkey. *Geol. Acta Int. Earth Sci. J.* **2018**, *16*, 163–187.
33. Feng, Q.; Gong, Y.-M.; Riding, R. Mid-Late Devonian calcified marine algae and cyanobacteria, South China. *J. Paleontol.* **2010**, *84*, 569–587. [CrossRef]
34. Flügel, E.; Munnecke, A. *Microfacies of Carbonate Rocks: Analysis, Interpretation and Application*; Springer: Berlin/Heidelberg, Germany, 2010; Volume 976.
35. Yaseen, A.; Rajpar, A.R.; Munir, M.; Roohi, G. Micropaleontology of Lockhart Limestone (Paleocene), Nilawahan Gorge, Central Salt Range, Pakistan. *J. Himal. Earth Sci.* **2011**, *44*, 9–16.
36. Sameeni, S.J.; Haneef, M.; Shabbir, F.; Ahsan, N.; Ahmad, N. Biostratigraphic studies of Lockhart Limestone, Changlagali area, Nathiagali-Murree road, Hazara, northern Pakistan. *Sci. Int.* **2013**, *25*, 543–550.
37. Ahmad, S.; Kroon, D.; Rigby, S.; Hanif, M.; Imraz, M.; Ahmad, T.; Jan, I.U.; Ali, A.; Zahid, M.; Ali, F. Integrated paleoenvironmental, bio-and sequence-stratigraphic analysis of the late Thanetian Lockhart Limestone in the Nammal Gorge section, western Salt Range, Pakistan. *J. Himal. Earth Sci.* **2014**, *47*, 16–23.
38. Janjuhah, H.T.; Sanjuan, J.; Alquadah, M.; Salah, M.K. Biostratigraphy, Depositional and Diagenetic Processes in Carbonate Rocks form Southern Lebanon: Impact on Porosity and Permeability. *Acta Geol. Sin.-Engl. Ed.* **2021**, *5*, 1668–1683. [CrossRef]
39. Mattes, B.W.; Mountjoy, E.W. *Burial Dolomitization of the Upper Devonian Miette Buildup, Jasper National Park, Alberta*; AAPG: Tulsa, OK, USA, 1980.
40. Janjuhah, H.T.; Salim, A.M.A.; Alansari, A.; Ghosh, D.P. Presence of microporosity in Miocene carbonate platform, Central Luconia, offshore Sarawak, Malaysia. *Arab. J. Geosci.* **2018**, *11*, 204. [CrossRef]
41. Basso, D.; Nalin, R.; Nelson, C.S. Shallow-water Sporolithon rhodoliths from north island (New Zealand). *Palaios* **2009**, *24*, 92–103. [CrossRef]
42. Sarkar, S. Microfacies analysis of larger benthic foraminifera-dominated Middle Eocene carbonates: A palaeoenvironmental case study from Meghalaya, NE India (Eastern Tethys). *Arab. J. Geosci.* **2017**, *10*, 121. [CrossRef]
43. Varrone, D.; d'Atri, A. Acervulinid macroid and rhodolith facies in the Eocene Nummulitic limestone of the Dauphinois Domain (Maritime Alps, Liguria, Italy). *Swiss J. Geosci.* **2007**, *100*, 503–515. [CrossRef]

44. Janjuhah, H.T.; Alansari, A.; Vintaned, J.A.G. Quantification of microporosity and its effect on permeability and acoustic velocity in Miocene carbonates, Central Luconia, offshore Sarawak, Malaysia. *J. Pet. Sci. Eng.* **2019**, *175*, 108–119. [CrossRef]
45. Astibia, H.; Elorza, J.; Pisera, A.; Alvarez-Pérez, G.; Payros, A.; Ortiz, S. Sponges and corals from the Middle Eocene (Bartonian) marly formations of the Pamplona Basin (Navarre, Western Pyrenees): Taphonomy, taxonomy, and paleoenvironments. *Facies* **2014**, *60*, 91–110. [CrossRef]
46. Leszczyński, S.; Kołodziej, B.; Bassi, D.; Malata, E.; Gasiński, M.A. Origin and resedimentation of rhodoliths in the Late Paleocene flysch of the Polish Outer Carpathians. *Facies* **2012**, *58*, 367–387. [CrossRef]
47. Janjuhah, H.T.; Alansari, A. Offshore carbonate facies characterization and reservoir quality of Miocene rocks in the southern margin of South China Sea. *Acta Geol. Sin.-Engl. Ed.* **2020**, *94*, 1547–1561. [CrossRef]
48. Murray, J.W.; Alve, E.; Jones, B.W. Palaeoclimatology, Palaeoecology. A new look at modern agglutinated benthic foraminiferal morphogroups: Their value in palaeoecological interpretation. *Palaeogeogr. Palaeoclim. Palaeoecol.* **2011**, *309*, 229–241. [CrossRef]
49. Stalder, C.; Vertino, A.; Rosso, A.; Rüggeberg, A.; Pirkenseer, C.; Spangenberg, J.E.; Spezzaferri, S.; Camozzi, O.; Rappo, S.; Hajdas, I. Microfossils, a key to unravel cold-water carbonate mound evolution through time: Evidence from the eastern Alboran Sea. *PLoS ONE* **2015**, *10*, e0140223. [CrossRef]
50. Reiss, Z.; Hottinger, L. *The Gulf of Aqaba: Ecological Micropaleontology*; Springer Science & Business Media: Berlin/Heidelberg, Germany, 2012; Volume 50.
51. Renema, W. Larger foraminifera as marine environmental indicators. *Scr. Geol.* **2002**, *124*, 1–260.
52. Baumgartner-Mora, C.; Baumgartner, P.O. Latest Miocene-Pliocene Larger Foraminifera and depositional environments of the carbonate bank of La Désirade Island, Guadeloupe (French Antilles). *Rev. Micropaléontologie* **2011**, *54*, 183–205. [CrossRef]
53. Banerjee, S.; Khanolkar, S.; Saraswati, P.K. Facies and depositional settings of the Middle Eocene-Oligocene carbonates in Kutch. *Geodin. Acta* **2018**, *30*, 119–136. [CrossRef]
54. Lin, Q.; Wang, J.; Algeo, T.J.; Su, P.; Hu, G. Formation mechanism of authigenic gypsum in marine methane hydrate settings: Evidence from the northern South China Sea. *Deep Sea Res. Part I Oceanogr. Res. Pap.* **2016**, *115*, 210–220. [CrossRef]
55. Koo, H.J.; Jang, J.K.; Lee, D.H.; Cho, H.G. Authigenic Gypsum Precipitation in the ARAON Mounds, East Siberian Sea. *Minerals* **2022**, *12*, 983. [CrossRef]
56. Kocherla, M. Authigenic Gypsum in Gas-Hydrate Associated Sediments from the East Coast of India (Bay of Bengal). *Acta Geol. Sin.-Engl. Ed.* **2013**, *87*, 749–760. [CrossRef]
57. Selley, R.C.; Cocks, L.R.M.; Plimer, I.R. *Encyclopedia of Geology*; Elsevier Academic: Amsterdam, The Netherlands, 2005.
58. Pierre, C.; Bayon, G.; Blanc-Valleron, M.-M.; Mascle, J.; Dupré, S. Authigenic carbonates related to active seepage of methane-rich hot brines at the Cheops mud volcano, Menes caldera (Nile deep-sea fan, eastern Mediterranean Sea). *Geo-Marine Lett.* **2014**, *34*, 253–267. [CrossRef]
59. Novikova, S.A.; Shnyukov, Y.F.; Sokol, E.V.; Kozmenko, O.A.; Semenova, D.V.; Kutny, V.A. A methane-derived carbonate build-up at a cold seep on the Crimean slope, north-western Black Sea. *Mar. Geol.* **2015**, *363*, 160–173. [CrossRef]
60. Beavington-Penney, S.J.; Wright, V.P.; Racey, A. The middle Eocene Seeb Formation of Oman: An investigation of acyclicity, stratigraphic completeness, and accumulation rates in shallow marine carbonate settings. *J. Sediment. Res.* **2006**, *76*, 1137–1161. [CrossRef]
61. Mehdi, H.; Vahidinia, M.; Hrabovsky, J. Larger foraminiferal biostratigraphy and microfacies analysis from the Ypresian (Ilerdian-Cuisian) limestones in the Sistan Suture Zone (eastern Iran). *Turk. J. Earth Sci.* **2019**, *28*, 122–145.
62. Anketell, J.; Mriheel, I. Depositional environment and diagenesis of the Eocene Jdeir Formation, Gabes-Tripoli basin, Western Offshore, Libya. *J. Pet. Geol.* **2000**, *23*, 425–447. [CrossRef]
63. Drobne, K.; Cosovic, V.; Moro, A.; Buckovic, D. The role of the Palaeogene Adriatic Carbonate Platform in the spatial distribution of Alveolinids. *Turk. J. Earth Sci.* **2011**, *20*, 721–751. [CrossRef]
64. Langer, M.R.; Hottinger, L. Biogeography of selected "larger" foraminifera. *Micropaleontology* **2000**, *46*, 105–126.
65. Zamagni, J.; Mutti, M.; Košir, A. Evolution of shallow benthic communities during the Late Paleocene–earliest Eocene transition in the Northern Tethys (SW Slovenia). *Facies* **2008**, *54*, 25–43. [CrossRef]
66. Yordanova, E.K.; Hohenegger, J. Taphonomy of larger foraminifera: Relationships between living individuals and empty tests on flat reef slopes (Sesoko Island, Japan). *Facies* **2002**, *46*, 169–203. [CrossRef]
67. Tomás, S.; Frijia, G.; Bömelburg, E.; Zamagni, J.; Perrin, C.; Mutti, M. Evidence for seagrass meadows and their response to paleoenvironmental changes in the early Eocene (Jafnayn Formation, Wadi Bani Khalid, N Oman). *Sediment. Geol.* **2016**, *341*, 189–202. [CrossRef]
68. Tomassetti, L.; Benedetti, A.; Brandano, M. Middle Eocene seagrass facies from Apennine carbonate platforms (Italy). *Sediment. Geol.* **2016**, *335*, 136–149. [CrossRef]
69. Beavington-Penney, S.J. Analysis of the effects of abrasion on the test of Palaeonummulites venosus: Implications for the origin of nummulithoclastic sediments. *Palaios* **2004**, *19*, 143–155. [CrossRef]
70. Prothero, D.R.; Schwab, F. *Sedimentary Geology*; Macmillan: New York, NY, USA, 2004.
71. Zielinski, J.P.T.; Vidal, A.C.; Chinelatto, G.F.; Coser, L.; Fernandes, C.P. Evaluation of pore system properties of coquinas from Morro do Chaves Formation by means of X-ray microtomography. *Braz. J. Geophys.* **2018**, *36*, 541–557. [CrossRef]

72. Amel, H.; Jafarian, A.; Husinec, A.; Koeshidayatullah, A.; Swennen, R.J.M.; Geology, P. Microfacies, depositional environment and diagenetic evolution controls on the reservoir quality of the Permian Upper Dalan Formation, Kish Gas Field, Zagros Basin *Mar. Pet. Geol.* **2015**, *67*, 57–71. [CrossRef]
73. Janjuhah, H.T.; Kontakiotis, G.; Wahid, A.; Khan, D.M.; Zarkogiannis, S.D.; Antonarakou, A. Integrated Porosity Classification and Quantification Scheme for Enhanced Carbonate Reservoir Quality: Implications from the Miocene Malaysian Carbonates. *J. Mar. Sci. Eng.* **2021**, *9*, 1410. [CrossRef]
74. Abu-Hashish, M.F.; Afify, H.M. Effect of petrography and diagenesis on the sandstone reservoir quality: A case study of the Middle Miocene Kareem Formation in the North Geisum oil field, Gulf of Suez, Egypt. *Arab. J. Geosci.* **2022**, *15*, 465. [CrossRef]
75. Ali, S.K.; Janjuhah, H.T.; Shahzad, S.M.; Kontakiotis, G.; Saleem, M.H.; Khan, U.; Zarkogiannis, S.D.; Makri, P.; Antonarakou, A. Depositional Sedimentary Facies, Stratigraphic Control, Paleoecological Constraints, and Paleogeographic Reconstruction of Late Permian Chhidru Formation (Western Salt Range, Pakistan). *J. Mar. Sci. Eng.* **2021**, *9*, 1372. [CrossRef]

*Article*

# Speleothems and Biomineralization Processes in Hot Spring Environment: The Case of Aedipsos (Edipsos), Euboea (Evia) Island, Greece

Christos Kanellopoulos [1,2,*], Vasiliki Lamprinou [3], Artemis Politi [3], Panagiotis Voudouris [1], Ioannis Iliopoulos [2], Maria Kokkaliari [2], Leonidas Moforis [1] and Athena Economou-Amilli [3]

[1] Faculty of Geology and Geoenvironment, National and Kapodistrian University of Athens, Panepistimiopolis, Ano Ilissia, 15784 Athens, Greece
[2] Department of Geology, University of Patras, 26500 Rio, Greece
[3] Faculty of Biology, Department of Ecology and Systematics, National and Kapodistrian University of Athens, Panepistimiopolis, Ano Ilissia, 15784 Athens, Greece
* Correspondence: ckanellopoulos@gmail.com

**Citation:** Kanellopoulos, C.; Lamprinou, V.; Politi, A.; Voudouris, P.; Iliopoulos, I.; Kokkaliari, M.; Moforis, L.; Economou-Amilli, A. Speleothems and Biomineralization Processes in Hot Spring Environment: The Case of Aedipsos (Edipsos), Euboea (Evia) Island, Greece. *J. Mar. Sci. Eng.* 2022, *10*, 1909. https://doi.org/10.3390/jmse10121909

Academic Editor: Antoni Calafat

Received: 30 October 2022
Accepted: 1 December 2022
Published: 5 December 2022

**Publisher's Note:** MDPI stays neutral with regard to jurisdictional claims in published maps and institutional affiliations.

**Copyright:** © 2022 by the authors. Licensee MDPI, Basel, Switzerland. This article is an open access article distributed under the terms and conditions of the Creative Commons Attribution (CC BY) license (https://creativecommons.org/licenses/by/4.0/).

**Abstract:** Caves with hot springs and speleothem deposits are infrequent environments of high scientific interest due to their unique environmental conditions. The selected site is a small open cave with a hot spring and stalactites in the Aedipsos area (NW Euboea Island, Greece), which was studied through an interdisciplinary approach. The mineralogical composition of the speleothems was determined by optical microscopy, XRD, and SEM-EDS microanalysis, and identification of the Cyanobacteria species was made based on morphological characteristics. The main mineral phase in the studied samples is calcite, with several trace elements (i.e., up to 0.48 wt.% $Na_2O$, up to 0.73 wt.% MgO, up to 4.19 wt.% $SO_3$, up to 0.16 wt.% SrO and up to 2.21 wt.% $Yb_2O_3$) in the mineral-chemistry composition. The dominant facies are lamination and shrubs, which are the most common among the facies of the thermogenic travertines of the area. Based on the studied stalactites, twenty-nine different Cyanobacteria species were identified, belonging to the following orders: Synechococcales (28%), Oscillatoriales (27%), Chroococcales (21%) and Nostocales (21%), and Spirulinales (3%). Among them, thermophilic species (*Spirulina subtilissima*) and limestone substrate species (*Chroococcus lithophilus*, *Leptolyngbya perforans*, and *Leptolyngbya ercegovicii*) were identified. The identified Cyanobacteria were found to participate in biomineralization processes. The most characteristic biomineralization activity is made by the endolithic Cyanobacteria destroying calcite crystals in the outer layer. In a few cases, calcified cyanobacterial sheaths were detected. The presence of filamentous Cyanobacteria, along with extracellular polymeric substance (EPS), creates a dense net resulting in the retention of calcium carbonate crystals.

**Keywords:** stalactite; speleothem; biomineralization; facies; mineralogy; hot spring; Cyanobacteria

## 1. Introduction

The speleothems, i.e., stalactites and stalagmites, representing secondary mineral deposits, could have been created by biogenic and abiogenic processes. In the abiogenic case, the mineral precipitation is due to supersaturation of the solution due to pH changes, outgassing, and evaporation. In the case of microorganisms contribution, biomineralization processes might occur ([1] and references within). The microorganisms can biologically mediate mineral formation in several ways, either directly by creating minerals (external or internal) or passively by accelerating the deposition, by encrustation or extracellular polymeric substance (EPS), or by changing the ambient conditions such as pH [2]. The microorganisms contributing to speleothem formation are mainly Archaea, Algae, and Bacteria, including Cyanobacteria. Multicellular organisms, such as fungi, lichens, and mosses, can also contribute. The most common cave-inhabitant microorganisms are chemolithoautotrophic or chemoheterotrophic, i.e., non-photosynthetic [3]. Photosynthetic microorganisms usually inhabit the cave entrance, where sunlight is reaching. However, some Cyanobacteria species, even if in general they are photosynthetic species, can adjust to the darkness and become heterotrophic, such as *Fisherella*, *Calothrix*, *Geitleria calcarea*, and *Scytonema julianum* [4–6]. The adjustment of the photosynthetic Cyanobacteria is important, as they can play a key role in speleogenetic processes since they can contribute to the dissolution or precipitation of the minerals.

Biodiversity in caves and biomineralization processes are subjects that only recently started to be studied and intercorrelated. Caves represent natural laboratories where microbe–mineral interactions under extreme conditions can be studied [7]. In Greece, several studies concerning speleothems and speleothems-climatic changes have been conducted (e.g., [8–11]). Moreover, studies concerning only biodiversity have taken place in the last years (e.g., [12–19]).

This paper aims to study stalactites in a hot spring cave in the Aedipsos area (Euboea Island, Greece) and the involved geomicrobiological processes. The mineralogical composition and mineral chemistry, the environmental conditions, and the Cyanobacteria species diversity will be assessed to evaluate the biomineralization processes of calcium carbonate minerals.

## 2. Geological Setting

The study site is located at NNW edge of Aedipsos (NW Euboea Island, Greece). The geological formations of the area belong to the Pelagonian geotectonic unit of the Hellenides [20–22], and the main geological formations are: a metamorphic crystalline basement (pre-middle to middle Carboniferous age), basic volcanoclastic complex series (Permian–Triassic age), shallow marine carbonate and clastic rocks (middle Triassic age) with volcanic rocks intercalations [23,24], alluvial deposits and thermogenic travertine depositions (Figure 1).

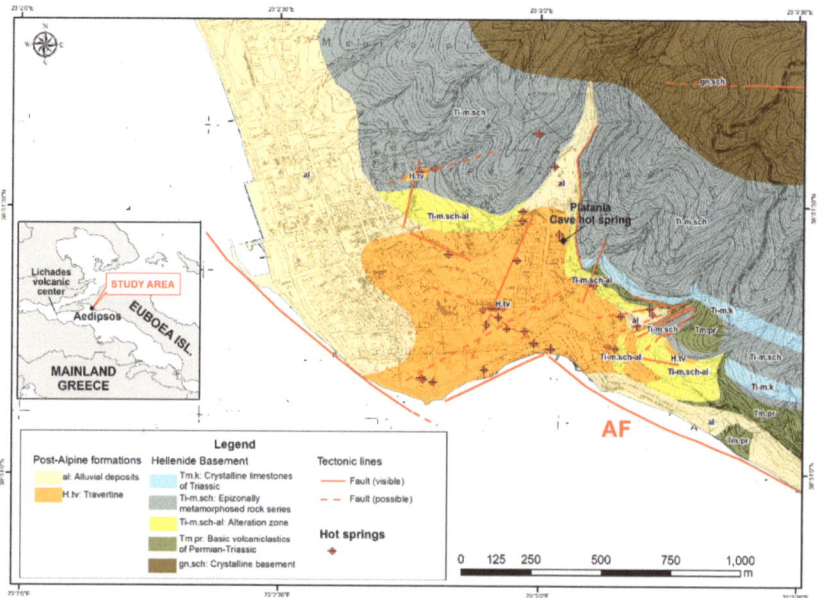

**Figure 1.** Geological map of the Aedipos area, NW of Euboea (AF = Aedipsos Fault; modified after Kanellopoulos et al. [25]). The sampling site is marked with a black dot. The geographical coordinates are in EGSA '87.

Several hot springs occur at the NW Euboea island, mainly in the Aedipsos area, which belongs to seawater-dominated, tectonically controlled, and volcanic-related geothermal systems [25–27]. The Lichades volcanic center, composed by trachyandesite lava, is located several kilometers away; it was dated to 0.5 Ma old (K-Ar method; [28]).

The temperature in the hot springs at Aedipsos reaches up to 84 °C, and the hydrothermal fluids are of sodium-chloride type. Among others, pH is almost neutral, and the springs present chemical similarities [25,27]. They are interpreted as deep-old geothermal fluids migrating from deep basement bedrocks with volcanic origin affinities.

The Aedipsos hot springs commonly deposit thermogenic travertine [27,29–32]. In addition, they present macro- and micro-facies [27,30,31], with bio-mineralization processes resulting in the creation of hybrid travertines [27,29,33–35], i.e., biotic and abiotic contributions.

## 3. Materials and Methods

Samples were extracted from stalactites from a small open cave with a hot spring at the base. During the sampling process, sterile metal tweezers and chisels were used. The unstable water parameters of the hot spring (i.e., temperature, salinity, and pH) were measured in situ once, during sampling, using portable apparatus.

From each sampling site, two sub-samples were collected. The first one was incubated into sterile transparent vials in the field. The second sub-sample was stored in a formaldehyde solution (2.5%). Enriched cultures were obtained in flasks and Petri dishes with BG11 and BG $11_0$ culture media [36]. Cultures were maintained in an incubator (Sanyo, Gallenkamp, Cambridge, UK) under stable conditions and a natural diurnal cycle (north-facing window) at room temperature.

The samples were studied under an optical microscope and a stereo-microscope. For species identification, the classical and recent literature was used ([37–40] and references within) at the Faculty of Biology, National and Kapodistrian University of Athens.

The mineralogical study was conducted on polished sections studied under an optical microscope and powders using X-ray diffraction (Bruker D8 Advanced Diffractometer, using Ni filtered Cu-K$\alpha$ radiation, operating at 40 kV and 40 mA and employing a Bruker Lynx Eye fast detector; Bruker-AXS, Billerica, MA, USA). The XRD results were evaluated using the DIFFRACplus EVA software (Bruker-AXS, Billerica, MA, USA) and the ICDD Powder Diffraction File (2006 version) at the Department of Geology, University of Patras. Selected dehydrated samples in an alcohol series (30–100%), critical point dried, gold-coated, and were studied under SEM (Jeol JSM 5600; JEOL USA, Inc., Peabody, MA, USA) at the Faculty of Geology and Geoenvironment, National and Kapodistrian University of Athens. SEM-EDS analyses were carried out using a Jeol JSM-IT500 SEM instrument (JEOL USA, Inc., Peabody, MA, USA) equipped with an Oxford 100 Ultramax analytical device (Oxford Instruments, Abingdon, UK) at the Hellenic Survey of Geology & Mineral Exploration.

The ArcGIS software was used to modify the geological map presented by Kanellopoulos et al. [25].

## 4. Results

### 4.1. Sampling Sites Description

In Aedipsos, several hot springs occur; very few of them are located inside small caves. The study site is a small open cave with a hot spring at the base (Figure 2). As it was verified from the thermal photos, hot-water circulation occurs at the cave walls (including the roof). The hot spring temperature, just below the stalactites, was 49.2 °C, the pH was 6.05, and the salinity was 20‰. Samples of the stalactites were collected above the hot spring.

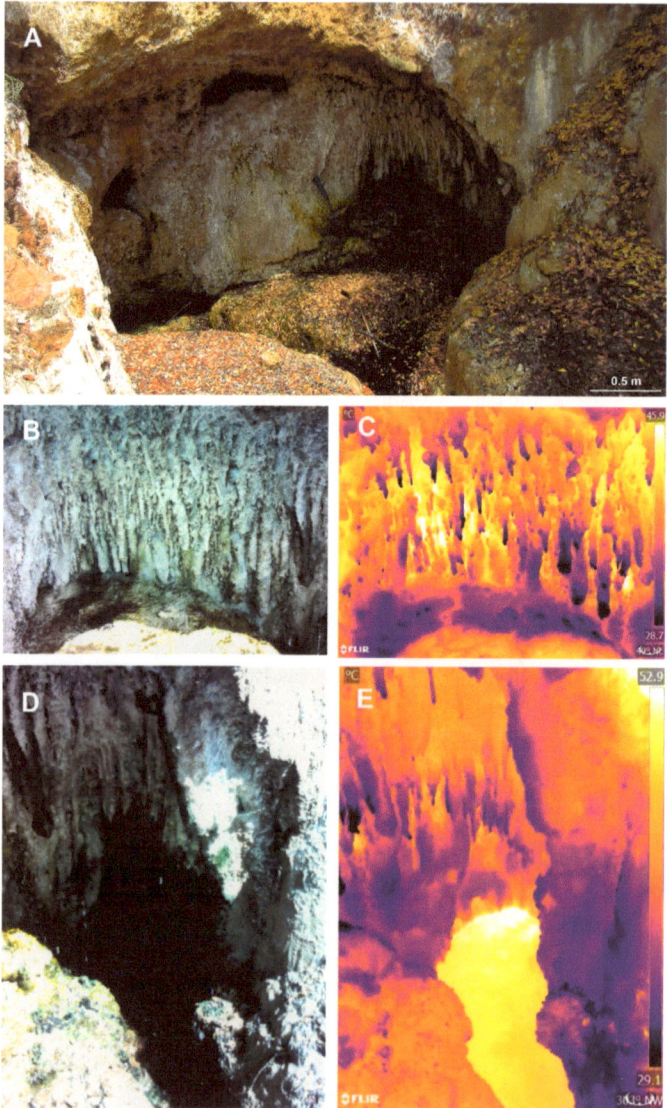

**Figure 2.** (**A**) Overview of the study site. (**B–E**) Paired views of normal images (**B,D**) and corresponding thermal images (**C,E**). A column shows the temperature scale (°C) on the right side of the thermal pictures. (**B,C**) Photo of the stalactites where the samples were collected. (**D,E**) Overview of the hot spring at the bottom of the cave and some stalactites at the top.

*4.2. Mineralogy and Facies*

According to XRD analyses and optical microscopy, the main mineral phase of stalactites is calcite (Figure 3). Based on SEM-EDS microanalyses, the calcite contains up to 0.48 wt.% $Na_2O$, up to 0.73 wt.% $MgO$, up to 4.19 wt.% $SO_3$, up to 0.16 wt.% $SrO$ and up to 2.21 wt.% $Yb_2O_3$ (Table 1).

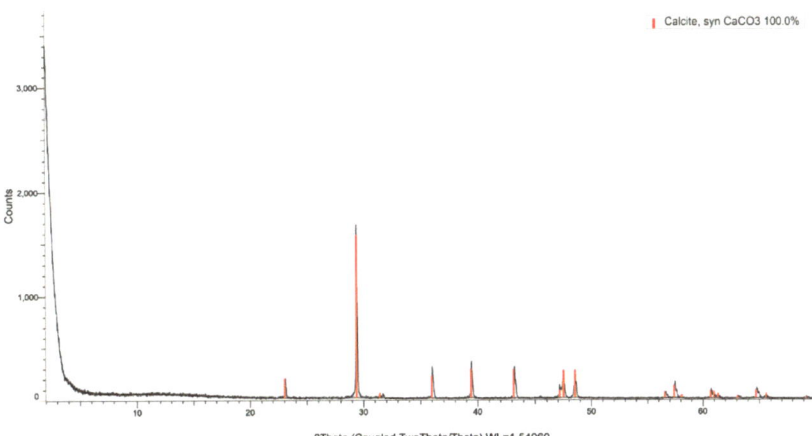

**Figure 3.** Evaluated XRD pattern.

**Table 1.** Representative microanalyses of calcite.

| No. | I | II | III | IV | V | VI | VII | VII | IX | X |
|---|---|---|---|---|---|---|---|---|---|---|
| $Na_2O$ | 0.34 | 0.32 | 0.31 | 0.24 | - | 0.19 | 0.22 | 0.28 | 0.34 | 0.48 |
| MgO | 0.59 | 0.59 | 0.71 | 0.31 | 0.5 | 0.71 | 0.51 | 0.6 | 0.61 | 0.53 |
| $SO_3$ | 3.54 | 2.79 | 2.33 | 3.13 | 0.73 | 1.95 | 2.61 | 2.3 | 4.19 | 2.71 |
| CaO | 51.5 | 52.41 | 53.66 | 52.38 | 53.54 | 52.59 | 52.77 | 53.11 | 52.72 | 51.92 |
| SrO | - | - | - | 0.16 | - | - | 0.16 | - | - | - |
| $Yb_2O_3$ | 1.79 | 2.04 | 2.08 | 1.87 | 2.21 | 1.92 | 2.05 | 1.97 | 2 | 1.9 |
| Total | 57.76 | 58.16 | 59.09 | 58.09 | 56.98 | 57.36 | 58.31 | 58.27 | 59.86 | 57.54 |

The studied samples display mainly lamination (Figures 4A and 5A) and shrub (Figure 5B) facies. The laminas could be from a few micrometers to a few millimeters thick. The laminas usually consist of micritic crystals and alternate with the next laminae, which is similar in mineralogical composition but differs in crystal size and density. Some laminas consist of shrubs (Figure 5B) with thicknesses up to ca. 1 mm. These are stubby, dense crystalline masses of calcite crystals that expand upward by irregular branching.

In a few cases, diatoms are trapped in thin laminas consisting of non-dense micritic crystals (Figure 5C,D). Moreover, in several cases, zones parallel to the lamination (Figure 2C) or nest areas were identified where traces of endolithic Cyanobacteria were present, i.e., holes and grooves, occur.

### 4.3. Cyanobacteria Microflora

In Figure 6, the Cyanobacteria orders are presented based on the latest classification system [41]. As it can be seen, Synechococcales and Oscillatoriales dominate with 28% and 27%, respectively. The orders Chroococcales and Nostocales follow with 21%, and finally, Spirulinales are also present with only 3%.

By studying the fresh and cultured material, a total of twenty-nine (29) different Cyanobacteria species, plus diatoms, were identified (Table 2; Figure 7). Among them, typical thermophilic species were found, such as *Spirulina subtilissima* (Figure 7L). *Chroococcus lithophilus, Leptolyngbya perforans,* and *Leptolyngbya ercegovicii* (Figure 7F) are also present, which are typical limestone substrate Cyanobacteria species.

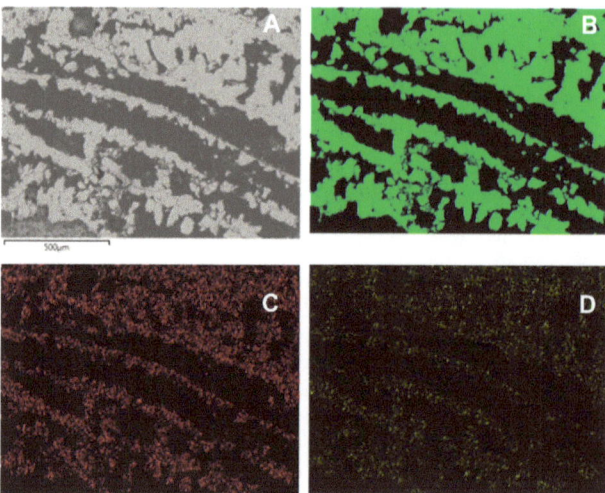

**Figure 4.** Back-scattered electron images (BSEI) of (**A**) laminated stalactite from Aedipsos; (**B–D**) are false color BSEI results of the mapping, displaying the distribution of (**B**) Ca (green), (**C**) S (red), and (**D**) Yb (yellow).

**Table 2.** Identified Cyanobacteria species.

*Anabaena* cf. *iyengarii* Bharadwaja 1935
*Brasilonema* cf. *angustatum* M.A.Vaccarino & J.R.Johansen 2012
*Chroococcus* cf. *mediocris* N.L.Gardner 1927
*Chroococcus lithophilus* Ercegovic 1925
*Chroococcus occidentalis* (N.L.Gardner) Komárek & Komárková-Legnerová 2007
*Chroococcus subnudus* (Hansgirg) G.Cronberg & J.Komárek 1994
*Cyanocohniella calida* J.Kastovský, E.Berrendero, J.Hladil & J.R.Johansen 2014
*Gloeocapsa gelatinosa* Kützing 1843
*Jaaginema thermale* Anagnostidis 2001
*Kamptonema formosum* (Bory ex Gomont) Strunecký, Komárek & J.Smarda 2014
*Leptolyngbya ercegovicii* (Cado) Anagnostidis & Komárek 1988
*Leptolyngbya foveolara* (Gomont) Anagnostidis & Komárek 1988
*Leptolyngbya perforans* (Geitler) Anagnostidis & Komárek 1988
*Leptolyngbya* sp.C
*Nostoc punctiforme* Hariot 1891
*Nostoc* sp.B
*Nostoc* sp.C
Nostocaceae
*Oscillatoria crassa* (C.B.Rao) Anagnostidis 2001
*Oscillatoria* sp.B
*Oscillatoria subbrevis* Schmidle 1901
*Oxynema acuminatum* (Gomont) Chatchawan, Komárek, Strunecky, Smarda & Peerapornpisal 2012
*Phormidium acidophilum* J.J.Copeland 1936
*Phormidium* cf. *abronema* Skuja, 1901
*Phormidium molischii* (Vouk) Anagnostidis & Komárek 1988
*Pseudanabaena galeata* Böcher 1949
*Schizothrix* cf. *lardacea* Gomont 1892
*Schizothrix* sp.A
Diatoms

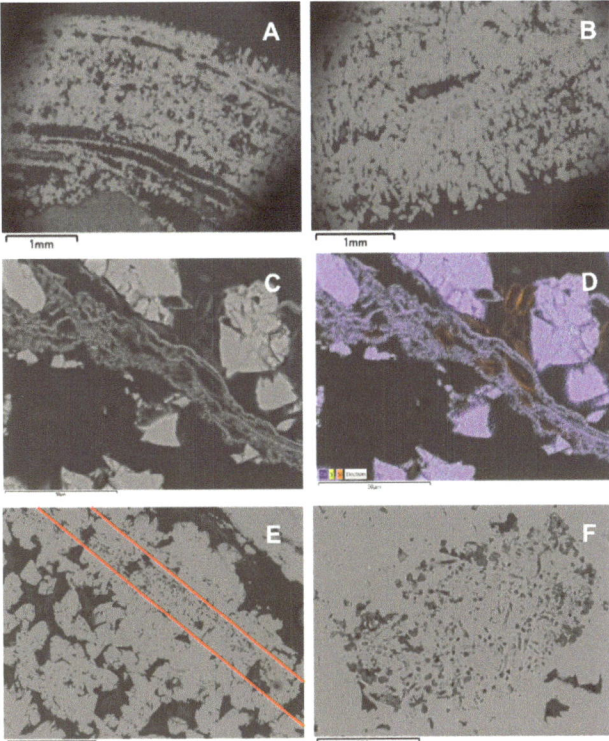

**Figure 5.** Back-scattered electron images (BSEI) presenting (**A**) laminated facies of stalactite, (**B**) laminae with shrubs that expand upward by irregular branching, (**C**) laminae consisting of micritic crystals of calcite and into it diatoms are trapped, (**D**) false color BSEI, derived from the corresponding black and white BSEI (see (**C**)), displaying the distribution of Ca (purple), S (yellow) and Si (orange) where the diatoms are distinct, (**E**,**F**) holes and grooves in calcite crystals, suggesting the presence of endolithic Cyanobacteria.

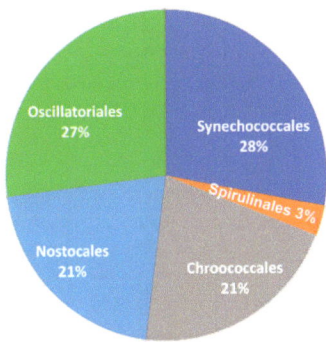

**Figure 6.** Pie diagrams presenting the percentage of each Cyanobacteria order.

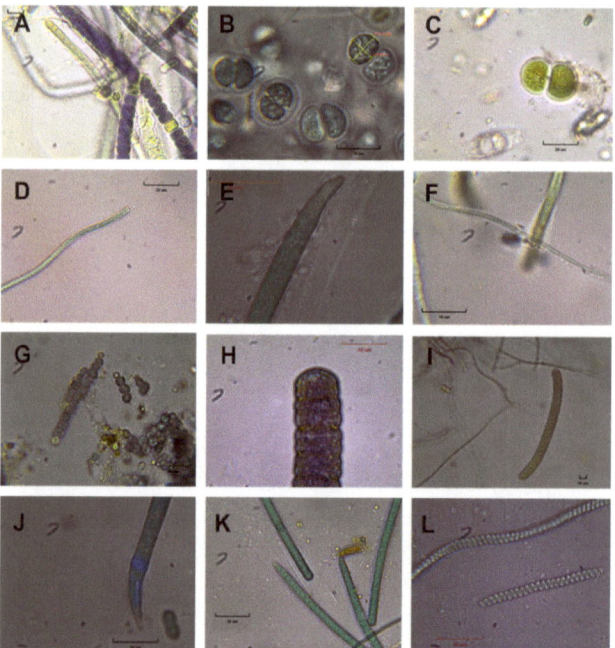

**Figure 7.** Cyanobacterial microflora under an optical microscope. (**A**) 9- *Brasilonema* cf. *angustatum* (Scale bar: 50 μm), (**B**) *Chroococcus* cf. *mediocris* (Scale bar: 50 μm), (**C**) *Chroococcus subnudus* (Scale bar: 20 μm), (**D**) *Jaaginema thermale* (Scale bar: 20 μm), (**E**) *Kamptonema formosum* (Scale bar: 20 μm), (**F**) *Leptolyngbya ercegovicii* (Scale bar: 10 μm), (**G**) *Nostoc punctiforme* (Scale bar: 20 μm), (**H**) *Oscillatoria crassa* (Scale bar: 10 μm), (**I**) *Oscillatoria subbrevis* (Scale bar: 40 μm), (**J**) *Oxynema acuminatum* (Scale bar: 50 μm), (**K**) *Phormidium* cf. *abronema* (Scale bar: 20 μm), (**L**) *Spirulina subtilissima* (Scale bar: 10 μm).

## 5. Discussion

### 5.1. Mineralogical Characterization and Facies

The main mineral phase of the studied samples is calcite (Figure 3), which is the most stable and common $CaCO_3$ polymorph found in speleothems [42]. Calcite is also the most typical main mineral phase in thermogenic travertines. However, the hot spring travertine deposits of Aedipsos have as main mineral phases either calcite, calcite, and aragonite or only aragonite [27,29,30]. In the case of speleothems, it was suggested that aragonite precipitates when the water has Mg/Ca ratios >1 (usually in dolomitic settings [42,43]). However, the Mg/Ca ratio in the study site is less than one [25]; additionally, based on the geological setting of the area, no dolomite occurrences have been testified nearby.

Based on SEM-EDS observations and microanalyses, the calcite, except for $CaCO_3$, contains several trace elements, i.e., Na, Mg, S, Sr, and Yb (Table 1). The incorporation of $Mg^{2+}$ and $Sr^{2+}$ into calcite has been well documented (e.g., [44]). However, it is worth mentioning that high-Mg calcites are usually observed in marine organisms [45]. The Aedipsos hot springs have high Na-Cl content and are characterized as seawater-dominated areas [25]. The incorporation of rare earths elements (REE), including $Yb^{3+}$ in calcite, takes place by adsorption onto calcite surfaces [46]. In the studied samples, the Yb presented equal distribution (Figure 4D) and reached 2.21 wt.% $Yb_2O_3$. Although the presence of Yb-calcite in speleothems is not common, its presence in the studied samples could be explained because they are not typical speleothems but hot-spring related, and the Yb could be attributed to the hydrothermal fluid. The presence of sulfate-containing calcite has been recently proved. A characteristic example comes from LaDuke Yellowstone hot spring, USA [47]. Okumura et al. [47], based on XPS, XRD, and TEM analysis, verified that sulfur was the principal foreign element in synthetic and natural (from LaDuke hot spring, USA) calcite crystals, with a mean atomic ratio of S/Ca around 5%; the chemical form of sulfur was proven to be sulfate ($SO_2^{-4}$). Sulfate is usually incorporated at the carbonate site of the calcite structure (structurally substitute [48,49]). In the studied samples of Aedipsos, the distribution of the S presented equal distribution (Figure 4C) and reached up to 4.19 wt.% $SO_3$.

The studied samples were found to display mainly lamination (Figures 4A and 5A) and shrub (Figure 5B) facies. These two facies are the most common among the thermogenic travertine deposition of Aedipsos [30,33,34]. Moreover, lamination is the most characteristic facies among the speleothems [42].

In a few cases, diatoms are trapped in thin laminas, consisting of not-dense micritic crystals (Figure 5C,D). Similar structures were described in previous studies in the thermogenetic travertine of Aedipsos, and they were attributed to EPS dense net resulting in the retention of calcium carbonate crystals and diatoms [33,34]. Moreover, in several cases, zones parallel to the lamination (Figure 2C) or nest areas were identified, where traces of endolithic Cyanobacteria presence, i.e., holes and grooves, occur.

*5.2. Cyanobacteria Diversity*

The dominant orders of Cyanobacteria are Synechococcales and Oscillatoriales, while Chroococcales, Nostocales, and Spirulinales follow. Kanellopoulos et al. [32] studied the diversity of cyanobacterial microflora of NW Euboea Island hot spring depositions. Based on their results, the summarized cyanobacterial microflora of Aedipsos hot springs is dominated by the orders Oscillatoriales (35.7%) and Synechococcales (31.4%), followed by Chroococcales (15.9%), Spirulinales (10.1%), and Nostocales (6.6%), with Chroococcidiopsidales (0.3%) barely present. Thus, in speleothems, the dominant orders are the same as in most of the hot springs of Aedipsos, i.e., Synechococcales and Oscillatoriales. In addition, the Nostocales and Chroococcales are more abound in the speleothems. The Spirulinales are significantly decreased, and Chroococcidiopsidales are totally absent.

It is very interesting that in the same study [33], the cyanobacterial microflora of the hot spring of the cave and the drainage channel depositions were also studied. In the case of the cave hot spring (T = 49.2 °C, Sal = 20‰, pH = 6.05, only limited access to sunlight), only two orders were identified, i.e., Oscillatoriales (71%) and Synechococcales (29%). While in the samples from the drainage channel, where there is full access to sunlight (T = 43.1–37.2 °C, Sal = 27–24‰, pH = 6.5–6.27), the Oscillatoriales (50–29%) is dominant, followed by Chroococcales (25–12%), Synechococcales (21–19%), Spirulinales (19–11%), Nostocales (11%-not identified), and Chroococcidiopsidales (3%-not identified). Thus, the speleothems present several similarities, but at the same time, also distinct differences concerning the cyanobacterial microflora of the cave hot spring and the drainage channel.

Based on fresh and cultured material, a total number of twenty-nine (29) different Cyanobacteria species, plus diatoms, were identified. By comparing these results with recent extensive studies on the Cyanobacteria diversity of Aedipsos hot spring depositions and the pioneer species [33,34], ca. 32% of the identified Cyanobacteria species presented here, do not appear in the Aedipsos travertine deposits (i.e., *Anabaena* cf. *iyengarii*, *Chroococcus lithophilus*, *Chroococcus subnudus*, *Jaaginema thermale*, *Oscillatoria subbrevis*, *Phormidium molischii*, *Pseudanabaena galeata*, *Schizothrix* cf. *lardacea* and *Schizothrix* sp.A) indicating the peculiarity of the specific environment.

*5.3. Biomineralization Processes*

Cyanobacteria biomineralization processes were identified in the outer layer of the samples (Figure 8). The presence of endolithic cyanobacteria is detrimental for the calcite crystals, i.e., they bore holes and dig channels in the crystals (Figure 8C–E). These structures could be a result of secretion of acidic substances or EPS.

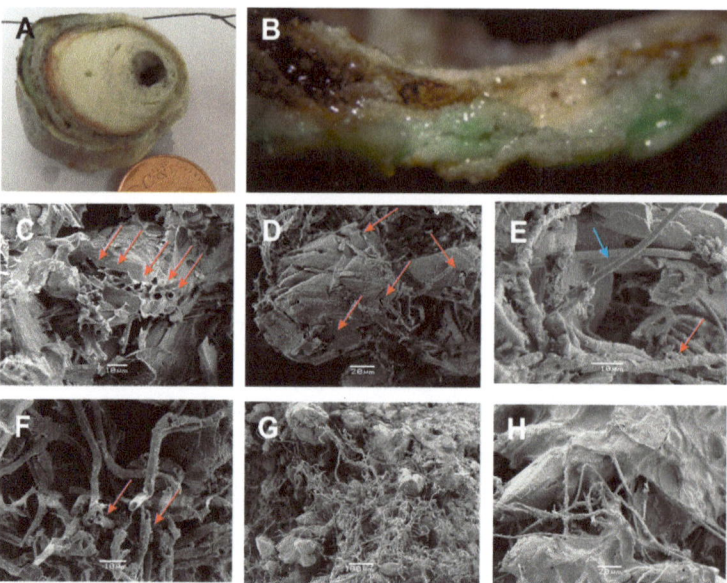

**Figure 8.** Biomineralizing processes by Cyanobacteria under SEM. (**A**) Vertical cut section surface of stalactite. (**B**) Detailed view of the outer periphery of stalactite, presenting fluvial crust with characteristic zoning and filaments of the endolithic *Leptolyngbya perforans* and *L. ercegoviccii* destroying the stalactite; whereas the upper zone is covered by granular epilithic species such as *Chroococcus lithophilus*. (**C,D**) Calcite crystals with distinct holes and dig channels are occupied by filamentous Cyanobacteria. (**E**) Filamentous Cyanobacteria of the endolithic *Leptolyngbya perforans* are coming out of calcite crystal (blue arrow) and calcified sheaths of Cyanobacteria filaments (red arrow). (**F**) Calcified sheaths of Cyanobacteria filaments by micritic calcium carbonate crystals. (**G**) Retention of calcium carbonate crystals by filamentous Cyanobacteria. (**H**) EPS along with filaments.

In some cases, calcified cyanobacterial sheaths were observed in micritic crystals (Figure 8F). The occurrence of sheath structure could be related to oxygenic photosynthesis, i.e., increase in the pH in the cell vicinity leading to carbonate oversaturation and precipitation [50,51], or could be related to the presence of nucleating molecules [52].

In some cases, filamentous Cyanobacteria, along with EPS, create a dense net resulting in the retention of calcium carbonate crystals (Figure 8G,H).

The above-mentioned biomineralization processes are similar to other biomineralization processes identified and are described recently in the thermogenic travertine deposits of Aedipsos [33,34]. Although, the intensity of the destructive biomineralization processes of the endolithic cyanobacteria are characteristic for the speleothems.

## 6. Conclusions

Speleothems are secondary mineral deposits formed under extreme conditions. In the present study, samples were collected from a cave environment where a hot spring is spouting in the Aedipsos area (NW Euboea Island, Greece).

The main mineral phase of the samples is calcite, with several trace elements, i.e., up to 0.48 wt.% $Na_2O$, up to 0.73 wt.% MgO, up to 4.19 wt.% $SO_3$, up to 0.16 wt.% SrO and up to 2.21 wt.% $Yb_2O_3$. The main faces of the studied stalactites are lamination and shrubs, representing the most common among the faces of the thermogenic travertines of the area.

In the outer stalactite layers, thirty (30) different Cyanobacteria species were identified belonging to the orders Synechococcales (28%), Oscillatoriales (27%), Chroococcales (21%) and Nostocales (21%), and also, Spirulinales (3%). Among the identified taxa, thermophilic species (*Spirulina subtilissima*) and limestone substrate species (*Chroococcus lithophilus*, *Lep-*

*tolyngbya perforans* and *Leptolyngbya ercegovicii*) occurred. The ca. 32% of the identified Cyanobacteria species presented here were not found in the Aedipsos travertine deposits.

Based mainly on SEM observations, biomineralization processes were observed in the outer layer of the sample. Similar biomineralization processes were also documented recently in the thermogenic travertine deposits of Aedipsos. The most characteristic biomineralization process of the speleothems is the high-intensity distraction of calcite crystals by endolithic Cyanobacteria. Additionally, in rare cases, calcified cyanobacterial sheaths were found, as well as the presence of filamentous Cyanobacteria and EPS, which create a dense net resulting in the retention of calcium carbonate crystals.

This study highlighted the importance of the geomicrobiological study of speleothems, especially in the extreme environments of hot springs. These sites can be considered as natural labs of unique conditions. Further research ought to be conducted in the area, including additional study sites and DNA metagenomic analysis, in order to fully outline biodiversity in these extreme environments, and the related biomineralization processes.

**Author Contributions:** Conceptualization C.K., V.L. and A.E.-A.; methodology, C.K., V.L. and A.E.-A.; sampling C.K. and V.L.; biological experiments and assessment A.P., V.L. and A.E.-A.; geological experiments and assessment A.P., C.K., P.V., M.K., I.I. and L.M.; SEM analysis and geobiological assessment A.P., C.K. and V.L.; GIS C.K.; visualization C.K., V.L. and A.P.; writing—original draft C.K. and V.L.; writing—review and editing A.E.-A., C.K. and V.L., P.V., I.I., M.K. and L.M.; supervision A.E.-A. All authors have read and agreed to the published version of the manuscript.

**Funding:** This research received no external funding.

**Institutional Review Board Statement:** Not applicable.

**Data Availability Statement:** Not applicable.

**Acknowledgments:** The authors would like to thank the local population and authorities, and especially the Director of the Public Properties Company- Aedipsos branch, Ilias Siakantaris, for their cooperation during the fieldwork. The corresponding author would like to thank George Vougioukalakis from the Greek Geological Survey (IGME, present name Hellenic Survey of Geology and Mineral Exploration, HSGME) for his support and encouragement during this research.

**Conflicts of Interest:** The authors declare no conflict of interest.

# References

1. Reitner, J. Modern cryptic microbialite/metazoan facies from Lizard Island (Great Barrier Reef, Australia): Formation and Concept. *Facies* **1993**, *29*, 3–40. [CrossRef]
2. Lowenstam, H.A.; Weiner, S. *On Biomineralization*; Oxford University Press: New York, NY, USA, 1989; p. 324.
3. Konhauser, K. *Introduction to Geomicrobiology*; Blackwell Publishing: Oxford, UK; Malden, MA, USA, 2007; 425p.
4. Friedman, I. *Geitleria calcarea* n. gen. et n. sp., a new atmophytic lime-encrusting blue-green alga. *Bot. Not.* **1955**, *108*, 439–445.
5. Bourrelly, P.; Depuy, P. Quelques stations françaises de *Geitleria calcarea*, Cyanophycee cavernicole. *Schweiz. Z. Hydrol.* **1973**, *35*, 136–140.
6. Whitton, B.A. The biology of Rivulariaceae. In *The Cyanobacteria—A Comparative Review*; Fay, P., van Baalen, C., Eds.; Elsevier: Amsterdam, The Netherlands, 1987; pp. 513–534.
7. Barton, H.A.; Northup, D.E. Geomicrobiology in cave environments: Past, current and future perspectives. *J. Cave Karst Stud.* **2007**, *69*, 163–178.
8. Psomiadis, D.; Dotsika, E.; Albanakis, K.; Ghaleb, B.; Hillaire-Marcel, C. Speleothem record of climatic changes in the northern Aegean region (Greece) from the Bronze Age to the collapse of the Roman Empire. *Palaeogeogr. Palaeoclimatol. Palaeoecol.* **2018**, *489*, 272–283. [CrossRef]
9. Theodorakopoulou, K.; Kyriakopoulos, K.; Athanassas, C.D.; Galanopoulos, E.; Economou, G.; Maniatis, Y.; Godelitsas, A.; Dotsika, E.; Mavridis, F.; Darlas, A. First Speleothem Evidence of the Hiera Eruption (197 BC), Santorini, Greece. *Environ. Archaeol.* **2021**, *26*, 336–348. [CrossRef]
10. Antonelou, A.; Tsikouras, B.; Papoulis, D.; Hatzipanagiotou, K. Investigation of the formation of speleothems in the Agios Georgios cave, Kilkis (N. Greece). *Bull. Geol. Soc. Greece* **2010**, *43*, 876–885. [CrossRef]
11. Ifanti, E. Petrogenetic Processes and Deposition Conditions of Speleothems at Perama Cave. Master's Thesis, University of Patras, Patras, Greece, 2013. (In Greek)
12. Anagnostidis, K.; Economou-Amili, A.; Pantazidou, A. Studies on the microflora of the cave Perama, Ioannina, Greece. *Bull. Soc. Spéléol. Grece* **1982**, *18*, 458–530.

13. Lamprinou, V.; Danielidis, D.; Economou-Amili, A.; Pantazidou, A. Distribution survey of Cyanobacteria in three Greek caves of Peloponnese. *Int. J. Speleol.* **2012**, *41*, 267–272. [CrossRef]
14. Lamprinou, V.; Danielidis, D.B.; Pantazidou, A.; Oikonomou, A.; Economou-Amilli, A. The show cave of Diros vs. wild caves of Peloponnese, Greece—Distribution patterns of Cyanobacteria. *Int. J. Speleol.* **2014**, *42*, 335–342. [CrossRef]
15. Lamprinou, V.; Hernandez-Marine, M.; Canals, T.; Kormas, K.; Economou-Amilli, A.; Pantazidou, A. Morphology and molecular evaluation of *Iphinoe spelaeobios* gen. nov., sp. nov. and *Loriellopsis cavernicola* gen. nov., sp. nov., two stigonematalean Cyanobacteria from Greek and Spanish caves. *Int. J. Syst. Evol. Microbiol.* **2011**, *61*, 2907–2915. [CrossRef]
16. Lamprinou, V.; Hernández-Mariné, M.; Pachiadaki, M.G.; Kormas, K.A.; Economou-Amilli, A.; Pantazidou, A. New findings on the true-branched monotypic genus *Iphinoe* (Cyanobacteria) from geographically isolated caves (Greece). *Fottea* **2012**, *13*, 15–23. [CrossRef]
17. Lamprinou, V.; Pantazidou, A.; Papadogiannaki, G.; Radea, C.; Economou-Amili, A. Cyanobacteria and associated invertebrates in Leontari cave. *Fottea* **2009**, *9*, 155–164. [CrossRef]
18. Lamprinou, V.; Skaraki, K.; Kotoulas, G.; Anagnostidis, K.; Economou-Amilli, A.; Pantazidou, A. A new species of *Phormidium* (Cyanobacteria, Oscillatoriales) from Greek Caves.—Morphological and Molecular Evaluation. *Fundam. Appl. Limnol.* **2012**, *182*, 109–116. [CrossRef]
19. Lamprinou, V.; Skaraki, K.; Kotoulas, G.; Economou-Amilli, A.; Pantazidou, A. *Toxopsis calypsus* gen. nov., sp. nov. (Cyanobacteria, Nostocales) from cave 'Francthi', Peloponnese, Greece—Morphological and molecular evaluation. *Int. J. Syst. Evol. Microbiol.* **2012**, *62*, 2870–2877. [CrossRef]
20. Mountrakis, D. The Pelagonian zone in Greece: A polyphase-deformed fragment of the Cimmerian continent and its role in the geotectonic evolution of the eastern Mediterranean. *J. Geol.* **1986**, *94*, 335–347. [CrossRef]
21. Vavassis, I. Geology of the Pelagonian zone in northern Evia Island (Greece): Implications for the geodynamic evolution of the Hellenides. Ph.D. Thesis, Univ. de Lausanne, Lausanne, Switzerland, 2001.
22. Jolivet, L.; Faccenna, C.; Huet, B.; Labrousse, L.; Le Pourhiet, L.; Lacombe, O.; Lecomte, E.; Burov, E.; Denèle, Y.; Brun, J.-P.; et al. Aegean tectonics: Strain localization, slab tearing and trench retreat. *Tectonophysics* **2013**, *597–598*, 1–33. [CrossRef]
23. Katsikatsos, G.; Mettos, A.; Vidakis, M.; Bavay, P.; Panagopoulos, A.; Basilaki, A.; Papazeti, E. *Geological study of Aedipos Area—Euboea*; Geothemal Studies (P.E.C.); IGME: Athens, Greece, 1982. (In Greek)
24. Scherreiks, R. Platform margin and oceanic sedimentation in a divergent and convergent plate setting (Jurassic, Pelagonian Zone, NE Evvoia, Greece). *Int. J. Earth Sci.* **2000**, *89*, 90–107. [CrossRef]
25. Kanellopoulos, C.; Xenakis, M.; Vakalopoulos, P.; Kranis, H.; Christopoulou, M.; Vougioukalakis, G. Seawater-dominated, tectonically controlled and volcanic related geothermal systems: The case of the geothermal area in the northwest of the Island of Euboea (Evia), Greece. *Int. J. Earth Sci.* **2020**, *109*, 2081–2112. [CrossRef]
26. Kanellopoulos, C.; Christopoulou, M.; Xenakis, M.; Vakalopoulos, P. Hydrochemical characteristics and geothermometry applications of hot groundwater in Edipsos area, NW Euboea (Evia), Greece. *Bull. Geol. Soc. Greece* **2016**, *50*, 720–729. [CrossRef]
27. Kanellopoulos, C.; Mitropoulos, P.; Valsami-Jones, E.; Voudouris, P. A new terrestrial active mineralizing hydrothermal system associated with ore-bearing travertines in Greece (northern Euboea Island and Sperchios area). *J. Geochem. Explor.* **2017**, *179*, 9–24. [CrossRef]
28. Fytikas, M.; Giuliani, O.; Innocenti, F.; Marinelli, G.; Mazzuoli, R. Geochronological data on recent magmatism of the Aegean Sea. *Tectonophysics* **1976**, *31*, T29–T34. [CrossRef]
29. Kanellopoulos, C. Geochemical research on the distribution of metallic and other elements in the cold and thermal groundwater, soils and plants in Fthiotida Prefecture and N. Euboea. Environmental impact. Ph.D. Thesis, National and Kapodistrian University of Athens, Athens, Greece, 2011. (In Greek with English abstract)
30. Kanellopoulos, C. Distribution, lithotypes and mineralogical study of newly formed thermogenic travertines in Northern Euboea and Eastern Central Greece. *Cent. Eur. J. Geosci.* **2012**, *4*, 545–560. [CrossRef]
31. Kanellopoulos, C. Various morphological types of thermogenic travertines in northern Euboea and Eastern Central Greece. *Bull. Geol. Soc. Greece* **2013**, *47*, 1929–1938. [CrossRef]
32. Kanellopoulos, C. Morphological types, lithotypes, mineralogy and possible bio-mineralization processes in simple and iron-rich travertines from active thermogenic travertine-forming systems in Greece. The cases of Northern Euboea and Eastern Central Greece. In Proceedings of the 19th International Sedimentological Congress, Geneva, Switzerland, 18–22 August 2014; p. 341.
33. Kanellopoulos, C.; Lamprinou, V.; Politi, A.; Voudouris, P.; Economou-Amilli, A. Insights on the biomineralization processes and related diversity of cyanobacterial microflora in thermogenic travertine deposits in Greek hot springs (North-West Euboea Island). *Depos. Rec.* **2022**, *8*, 1055–1078. [CrossRef]
34. Kanellopoulos, C.; Lamprinou, V.; Politi, A.; Voudouris, P.; Economou-Amilli, A. Pioneer species of Cyanobacteria in hot springs and their role to travertine formation: The case of Aedipsos hot springs, Euboea (Evia), Greece. *Depos. Rec.* **2022**, *8*, 1079–1092. [CrossRef]
35. Kanellopoulos, C.; Lamprinou, V.; Politi, A.; Voudouris, P.; Iliopoulos, I.; Kokkaliari, M.; Moforis, L.; Economou-Amilli, A. Microbial Mat Stratification in Travertine Depositions of Greek Hot Springs and Biomineralization Processes. *Minerals* **2022**, *12*, 1408. [CrossRef]
36. Stanier, R.Y.; Kunisawa, R.; Mandel, M.; Cohen-Bazire, G. Purification and properties of unicellular blue-green algae (order Chroococcales). *Bacteriol. Rev.* **1971**, *35*, 171. [CrossRef]

37. Komárek, J.; Anagnostidis, K. Modern approach to the classification system of Cyanophytes 4-Nostocales. *Arch. Hydrobiol. Suppl. Bd. Algol. Stud.* **1989**, *56*, 247–345.
38. Komárek, J.; Anagnostidis, K. Cyanoprokaryota, Part 1: Chroococcales. In *Süßwasserflora von Mitteleuropa, Bd. 19 (1)*; Büdel, B., Gärtner, G., Krienitz, L., Schagerl, M., Eds.; Elsevier GmbH: München, Germany, 1999; pp. 1–548.
39. Komárek, J.; Anagnostidis, K. Cyanoprokaryota, Part 2: Oscillatoriales. In *Süßwasserflora von Mitteleuropa, Bd. 19 (2)*; Büdel, B., Gärtner, G., Krienitz, L., Schagerl, M., Eds.; Elsevier GmbH: München, Germany, 2005; pp. 1–759.
40. Komárek, J.; Kastovsky, J.; Mares, J.; Johansen, J.R. Taxonomic classification of cyanoprokaryotes (cyanobacterial genera), using a polyphasic approach. *Preslia* **2014**, *86*, 295–335.
41. Hauer, T.; Komárek, J. *CyanoDB 2.0—On-Line Database of Cyanobacterial Genera*; World-Wide Electronic Publication, Univ. of South Bohemia & Inst. of Botany AS CR: České Budějovice, Czechia, 2021. Available online: https://www.cyanodb.cz (accessed on 19 December 2021).
42. Fairchild, I.J.; Baker, A. *Speleothem Science: From Process to Past Environments*; John Wiley & Sons: Hoboken, NJ, USA, 2012; p. 432.
43. Feinberg, J.M.; Johnson, K.R. Cave and Speleothem Science: From Local to Planetary Scales. *Elements* **2021**, *17*, 81–86. [CrossRef]
44. Mucci, A.; Mucci, J.W. The incorporation of $Mg^{2+}$ and $Sr^{2+}$ into calcite overgrowths: Influences of growth rate and solution composition. *Geochim. Cosmochim. Acta* **1983**, *47*, 217–233. [CrossRef]
45. Long, X.; Ma, Y.; Qi, L. Biogenic and synthetic high magnesium calcite—A review. *J. Struct. Biol.* **2014**, *185*, 1–14. [CrossRef]
46. Möller, P.; De Lucia, M. Incorporation of Rare Earths and Yttrium in Calcite: A Critical Re-evaluation. *Aquat. Geochem.* **2020**, *26*, 89–117. [CrossRef]
47. Okumura, T.; Kim, H.-J.; Kim, J.-W.; Kogure, T. Sulfate-containing calcite: Crystallographic characterization of natural and synthetic materials. *Eur. J. Mineral.* **2018**, *30*, 929–937. [CrossRef]
48. Kampschulte, A.; Strauss, H. The sulfur isotopic evolution of Phanerozoic seawater based on the analysis of structurally substituted sulfate in carbonates. *Chem. Geol.* **2004**, *204*, 255–286. [CrossRef]
49. Balan, E.; Aufort, J.; Pouillé, S.; Dabos, M.; Blanchard, M.; Lazzeri, M.; Rollion-Bard, C.; Blamart, D. Infrared spectroscopic study of sulfate-bearing calcite from deep-sea bamboo coral. *Eur. J. Mineral.* **2017**, *29*, 397–408. [CrossRef]
50. Riding, R. Cyanobacterial calcification, carbon dioxide concentrating mechanisms, and Proterozoic-Cambrian changes in atmospheric composition. *Geobiology* **2006**, *4*, 299–316. [CrossRef]
51. Jansson, C.; Northen, T. Calcifying cyanobacteria—The potential of biomineralization for carbon capture and storage. *Curr. Opin. Biotechnol.* **2010**, *21*, 365–371. [CrossRef]
52. Merz-Preiss, M.; Riding, R. Cyanobacterial tufa calcification in two freshwater streams: Ambient environment, chemical thresholds and biological processes. *Sediment. Geol.* **1999**, *126*, 103–124. [CrossRef]

Article

# Sedimentary and Diagenetic Controls across the Cretaceous—Paleogene Transition: New Paleoenvironmental Insights of the External Ionian Zone from the Pelagic Carbonates of the Gardiki Section (Epirus, Western Greece)

Leonidas Moforis [1,*], George Kontakiotis [1,*], Hammad Tariq Janjuhah [2], Alexandra Zambetakis-Lekkas [1], Dimitrios Galanakis [3], Panagiotis Paschos [3], Christos Kanellopoulos [3,4], Sotirios Sboras [5], Evangelia Besiou [1], Vasileios Karakitsios [1] and Assimina Antonarakou [1]

[1] Department of Historical Geology and Paleontology, Faculty of Geology and Geoenvironment, National and Kapodistrian University of Athens, Panepistimiopolis, Zografou, 15784 Athens, Greece
[2] Department of Geology, Shaheed Benazir Bhutto University, Sheringal 18050, KPK, Pakistan
[3] H.S.G.M.E.—Hellenic Survey of Geology and Mineral Exploration, 13677 Athens, Greece
[4] Department of Mineralogy & Petrology, Faculty of Geology and Geoenvironment, National and Kapodistrian University of Athens, Panepistimiopolis, Zografou, 15784 Athens, Greece
[5] Institute of Geodynamics, National Observatory of Athens, Lofos Nymphon, Thesio, 11810 Athens, Greece
* Correspondence: leonidasmoforis@gmail.com (L.M.); gkontak@geol.uoa.gr (G.K.)

**Citation:** Moforis, L.; Kontakiotis, G.; Janjuhah, H.T.; Zambetakis-Lekkas, A.; Galanakis, D.; Paschos, P.; Kanellopoulos, C.; Sboras, S.; Besiou, E.; Karakitsios, V.; et al. Sedimentary and Diagenetic Controls across the Cretaceous—Paleogene Transition: New Paleoenvironmental Insights of the External Ionian Zone from the Pelagic Carbonates of the Gardiki Section (Epirus, Western Greece). *J. Mar. Sci. Eng.* **2022**, *10*, 1948. https://doi.org/10.3390/jmse10121948

Academic Editor: Antoni Calafat

Received: 31 October 2022
Accepted: 29 November 2022
Published: 8 December 2022

**Publisher's Note:** MDPI stays neutral with regard to jurisdictional claims in published maps and institutional affiliations.

**Copyright:** © 2022 by the authors. Licensee MDPI, Basel, Switzerland. This article is an open access article distributed under the terms and conditions of the Creative Commons Attribution (CC BY) license (https://creativecommons.org/licenses/by/4.0/).

**Abstract:** Field investigation, biostratigraphic, paleoecological, and sedimentary microfacies analyses, as well as diagenetic processes characterization, were carried out in the Epirus region (Western Ionian Basin) to define the depositional environments and further decipher the diagenetic history of the Late Cretaceous–Early Paleocene carbonate succession in western continental Greece. Planktonic foraminiferal biostratigraphy of the studied carbonates revealed that the investigated part of the Gardiki section covers the Cretaceous–Paleogene (K-Pg) transition, partly reflecting the Senonian limestone and calciturbidites formations of the Ionian zone stratigraphy. Litho-and bio-facies analyses allowed for the recognition of three distinct depositional facies: (a) the latest Maastrichtian pelagic biomicrite mudstone with in situ planktonic foraminifera, radiolarians, and filaments, (b) a pelagic biomicrite packstone with abundant planktonic foraminifera at the K-Pg boundary, and (c) an early Paleocene pelagic biomicrite wackestone with veins, micritized radiolarians, and mixed planktonic fauna in terms of in situ and reworked (aberrant or broken) planktonic foraminifera. The documented sedimentary facies characterize a relatively low to medium energy deep environment, representing the transition from the deep basin to the deep shelf and the toe of the slope crossing the K-Pg boundary. Micropaleontological and paleoecological analyses of the samples demonstrate that primary productivity collapse is a key proximate cause of this extinction event. Additional petrographic analyses showed that the petrophysical behavior and reservoir characteristics of the study deposits are controlled by the depositional environment (marine, meteoric, and burial diagenetic) and further influenced by diagenetic processes such as micritization, compaction, cementation, dissolution, and fracturing.

**Keywords:** microfacies analysis; Senonian limestone formation; siliceous nodules; diagenetic processes; slope-to-basin pelagic carbonates; Ionian calciturbidites; hydrocarbon reservoirs; K-T foraminiferal extinction; stratigraphic correlations; paleoenvironmental reconstruction

## 1. Introduction

Carbonate systems are a major component of the Earth system since they host more than 25% of the marine life [1], play an essential role in the global carbon cycle, and thus the regulation of atmospheric $CO_2$ concentration [2], and further represent major reservoir rocks for water and hydrocarbon resources [3–5]. Furthermore, marine biogenic carbonates are among the most important archives of the climate [6,7], since their biotic constituents reflect changes in ocean chemistry and associated hydroclimate parameters (e.g., sea surface temperature and salinity; [8–10]). In particular, deeply buried carbonate successions in foreland basinal settings overlain by thick siliciclastic sediments are prospective targets for hydrocarbon and geothermal exploration [11]. Therefore, prediction and investigation of carbonate distributions through space and time represents a challenging scientific issue [6,12], which is critical for carbonate reservoir studies and the understanding of past and future climate changes at both a regional and global scales.

During the Late Cretaceous–Early Paleocene, the Mediterranean Tethys (Neo-Tethys) Ocean was characterized by the dominance of carbonate sedimentation, as witnessed by a spread of pelagic marine carbonates in the deeper parts of the basin (including the slope deposits in the platform margins) and shallow limestones and dolomites in the platform belts [13–16]. These carbonate sediments show different biotic associations, sedimentary facies, and stratigraphic architectures depending on the variable environmental conditions across the entire basin. During that time, within the tectonostratigraphic regime of western Greece, the carbonate deposits of the Ionian basin provide an excellent example of the evolution of depositional sequences ranging from deep basin to rimmed carbonate shelf settings. Moreover, the role of biogenic silica in the formation of Late Cretaceous–Early Paleocene pelagic carbonates is critical [17] and gives a great economic and strategic importance to these deposits [18–20]. They have been considered the main reservoir successions and exploration targets for oil and gas in western Greece [19–24]. Due to their increasing importance as reservoir rocks, the investigation of carbonate-derived thin sections gave substantial impetus to facies analysis development, making parallel progress in related topics such as sequence stratigraphy and sedimentology. However, the Late Cretaceous to early Eocene evolution of this setting, including the nature and distribution of these deposits, along with their depositional mechanism and diagenetic processes, are still poorly constrained.

In the present study, we investigate the depositional and diagenetic processes as well as the microfacies types of the Late Cretaceous–Early Paleocene carbonates of the Gardiki section, which is in the Epirus region (western Ionian basin, Ioannina, Greece). This study defines, for the first time, the Cretaceous–Paleogene (K-Pg) boundary depositional evolution in this area, based mainly on litho-stratigraphic, reservoir petrophysical, and diagenetic characteristics in the external Ionian domain, considered as a significant hydrocarbon prolific basin in Western Greece [20–22,25]. This was accomplished by extensive sedimentological and microfacies analyses of the carbonate succession, considering the synthetic paleoenvironmental reconstruction of the area. Finally, this work has further implications for regional geology and a better understanding of the Ionian basin in Western Greece.

## 2. Study Area
### 2.1. Regional Geological Setting

Western Greece is dominated by the external zones of the Hellinides fold-and-thrust belt, divided into three tectonostratigraphic zones, namely the pre-Apulian, Ionian, and Gavrovo-Tripolis zones. At a regional scale, this Alpine belt records the initiation, development, and final destruction of the southeastern margin of the Tethys Ocean and the consequent continent–continent collision between the Apulian and the Pelagonian microcontinents to the east [26,27]. On a local scale, the various sub-basins of the Hellenic Tethys margin have been inverted to produce the main Hellenic thrust sheet folded zones [28–30]. The Ionian zone, bounded westwards by the Ionian thrust and eastwards by the Gavrovo thrust, extends from Albania to the north, forms most of the Epirus region and parts of the Ionian islands, and continues southwards to Central Greece, Crete, and the Dodecanese (Figure 1). According to Aubouin [31] and Igrs-Ifp [32], the Ionian basin was subdivided into the Internal, Middle, and External sub-basins.

**Figure 1.** Geological map of the external Hellinides in Western Greece, illustrating the principal tectonostratigraphic zones: Pre-Apulian, Ionian, Gavrovo-Tripolis, and Pindos. The white square shows the location of the study section in external Ionian domain. Legend interpretations are presented in the inset.

### 2.2. Tectonostratigraphic Evolution of the Ionian Basin

The tectonostratigraphic evolution of the Ionian basin is reflected by the deposition of three distinct stratigraphic sequences indicative of different tectonic regimes. According to Karakitsios [22], these sequences are: the pre-rift, syn-rift, and post-rift (Figure 2).

The oldest formation known in the pre-rift sequence is represented by the Lower to Middle Triassic evaporites, with a thickness greater than 2000 m, composed of anhydrites, gypsum, and halite, usually with interbeds of limestone and dolomite. The sequence is completed by the Late Ladinian-Rhaetian Foustapidima limestone [33] and by the overlying shallow water limestone of the Pantokrator formation of the Lower Jurassic (Hettangian to Sinemurian) age [34,35].

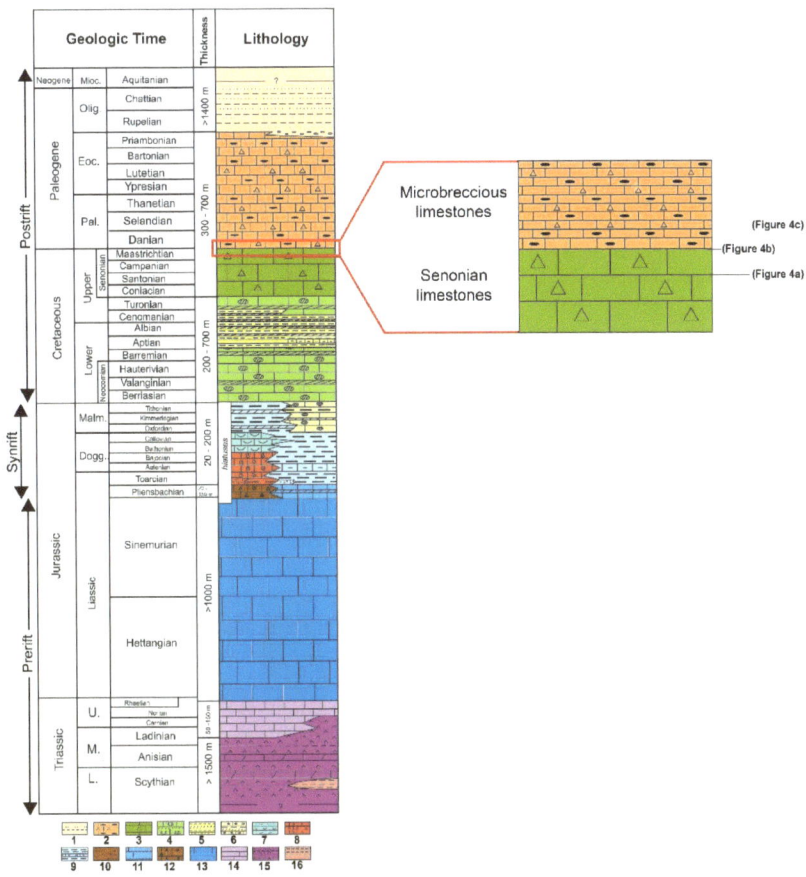

**Figure 2.** Synthetic lithostratigraphic column of the Ionian zone modified after [32], along with the correspondence of study section stratigraphy, as depicted in the enlarged view at the right, in which the position of the different litho-phases characteristic of the Cretaceous–Paleogene transition (corresponding to Figure 4a–c) is also highlighted. The colors in the lithostratigraphic column are consistent with the relevant colors of the International Chronostratigraphic Chart (v2020/01). (1) Shales and sandstones; (2) limestones with rare cherty intercalations, occasionally microbreccious; (3) pelagic limestones with clastic platform elements; (4) pelagic limestones with cherts; (5) cherty beds with shale and marl intercalations; (6) pelagic limestones with cherty nodules and marls; (7) pelagic limestones with bivalves; (8) pelagic, nodular red limestones with ammonites; (9) marly limestones and laminated marls; (10) conglomerates-breccias and marls with ammonites; (11) pelagic limestones with rare cherty intercalations; (12) external platform limestones with brachiopods and small ammonites in upper part; (13) platform limestones; (14) thin-bedded black limestones; (15) evaporites; (16) shales.

The syn-rift sequence begins with the Lower Jurassic (Pliensbachian) pelagic Siniais limestones and their lateral equivalent Louros limestones [36], overlain by Ammonitico Rosso and Limestones with filaments, laterally replaced and overlain by Posidonia beds [37]. The boundary between the Pantokrator and Siniais limestones is gradational. These formations correspond to the general deepening of the Ionian domain with the formation of the Ionian Basin. The structural differentiation separated the initial basin into smaller paleogeographic units with a half-graben geometry and remarkable different basin depths; in most cases, these units do not exceed 5 km across [37,38].

The post-rift sequence begins with the Lower Cretaceous (Berriasian-Turonian) pelagic Vigla limestones, whose deposition was synchronous throughout the Ionian Basin [35,39] and the laterally equivalent Vigla shales. The Vigla limestones cover the syn-rift structures [35], and in some cases, directly overlie the pre-rift units. Consequently, the base of the Vigla limestones formation represents the break-up unconformity of the post-rift sequence in the Ionian Basin. The Senonian limestones overlie the Vigla Formation and comprise two facies: (a) limestones with Globotruncanidae, and (b) microbrecciated intervals with limestones and rudist fragments within a calcareous cement containing pelagic fauna [20,21,23]. During the Paleocene-Eocene, the erosion of the Cretaceous carbonates from the adjacent Gavrovo (to the East) and Apulian platforms (to the West) provided the Ionian Basin with microbreccia or brecciated materials, known as Microbreccious limestones [19,20].

The Late Mesozoic to Eocene carbonate succession passes upwards to the flysch synorogenic sedimentation (siliciclastic turbidites), which began at the Eocene–Oligocene boundary and revealed progressively diminishing thicknesses from the internal to the external areas [34,35]. Until the Early Miocene, the basin was filled with submarine fan deposits in response to the movement of Pindos thrust, the compressional structures, and the deformation of the external Hellenides. As a result, it migrated westwards, uplifted the entire Hellenides orogenic belt, and developed a foreland basin at the edge of the Apulian microcontinent.

### 2.3. Senonian and Microbreccious Limestone Formations

In the present study, we exclusively focus on the K-Pg pelagic deposits of the Ionian basin consisted partly of the Senonian and microbreccious limestones (red box in Figure 2). These formations comprise of pelagic limestones with variable clastic platform elements (e.g., fragments of rudists and fine-grained microbrecciated intervals) and corresponds to a period of basinal sedimentation with significant variations of its lithological characteristics and sedimentary facies from the external (western) to the internal (eastern) parts of the Ionian basin [19,20] during the Late Cretaceous–Early Paleocene. According to Skourtsis-Coroneou et al. [40], their heterogeneity could be attributed to the presence of clastic limestones (e.g., floatstones-rudstones with a micritic matrix, biomicritic intercalations and rare cherty nodules) in the most external Ionian sub-basins, bioclastic wackestone to packstone intercalated with biomicrites in the middle sub-basins, and massive microbreccia containing rudists and coral fragments in the internal ones. Notably, the facies distribution of the Senonian reflects the separation of the Ionian Basin into a central area (the middle and outer parts of the Ionian Zone), characterized by deeper water sedimentation and two surrounding talus slopes issued from the western Gavrovo and Apulian platforms which provided the clastic carbonate material that was transported by turbidity currents into the Ionian basin [37]. Stratigraphically, although in some places these calciturbidites can be intercalated with the uppermost horizons of the underlying Vigla Limestone formation, in general they comfortably overlay the lower Cretaceous Vigla limestones and shales.

## 3. Materials and Methods

A total of 90 samples were collected from the Gardiki section (lat: 39°20′22.09″ N, long: 20°33′5.63″ E) in Epirus (western Greece), corresponding to the passage from the Late Cretaceous to the Paleocene. The fieldwork was conducted in October 2019, including field logging, measuring, and sampling, as well as additional observations made on the lithostratigraphic properties of the carbonates (i.e., color, lithological and textural variations, bed thickness, syn- and post-depositional features). The sampling resolution was variable along the section depending on the outcrop conditions, thickness, and lateral extent of each carbonate unit. For samples B1 up to B39, the sampling resolution was 2.5 m, while for the rest of the samples (B40–B90) it was 0.2 m.

All selected samples were analyzed for microfacies (lithofacies and biofacies) and diagenetic characteristics to determine their precise age, paleoenvironmental depositional conditions, and diagenetic history. Thin sections were prepared both in the Historical Geology-

Paleontology Laboratory (National and Kapodistrian University of Athens; NKUA) and at the Hellenic Survey of Geology and Mineral Exploration (H.S.G.M.E.), biostratigraphically and sedimentologically studied under a polarized LEICA DM LP microscope, and photos were performed with OLYMPUS UC30 Microscope Digital Camera. All thin sections (90 in total) were examined to determining different types of diagenetic processes performed on these deposits, and finally to evaluate the diagenetic history of the study formations.

The microfacies' definition and textural characters analysis of the carbonate rocks were defined according to Dunham [41] classification scheme, which were later modified by Embry and Klovan [42] and Flügel [43], based on the Standard Microfacies Types (SMF) in the Facies Zone (FZ) of the rimmed carbonate platform model. Depositional paleoenvironments were reconstructed based on the observed sedimentological characteristics during fieldwork and interpreted sedimentary facies analysis, and through comparison with additional outcropped data known from the existing literature on time equivalent deposits of the Ionian basin [18–20,44,45], as well as other environmental studies [46–54].

## 4. Results

### 4.1. Description of the Section and Field Observations

Figure 3 illustrates the location of the study part of Gardiki section, regarding the K-Pg carbonates in Epirus area.

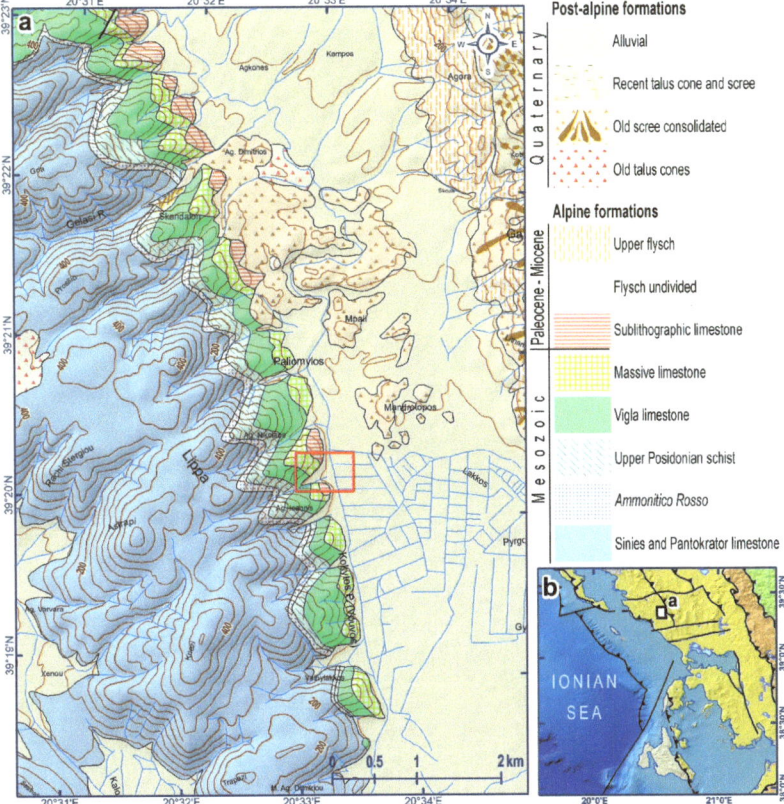

**Figure 3.** (**a**) General view of the study area based on the existing geological map. The red box indicates the study section. (**b**) Referenced map with the study lithostratigraphic units of the Ionian zone, as depicted in Figure 1. Geological data modified after [55].

The study part of Gardiki section consists of Late Cretaceous–Early Paleocene pelagic deposits, such as pelagic limestones and radiolarian chert horizons. The Senonian limestone formation in this area is mainly characterized by light-grey-to-yellowish limestones with abundant planktonic foraminifera, and more rare radiolarians. At the base of the section, thin-to-medium-bedded carbonate deposits are evident in rhythmic alternations with centimeter- to decimeter-thick cherts (Figure 4a). In the middle part of the section, the limestones become even thinner (Figure 4b), with the decrease in their thickness to be more evident when approaching the K-Pg boundary. Moreover, onwards up to the top of the section, the siliceous intervals occur very rarely as scarce thin chert intercalations (up to 5 cm thickness) and/or lenses within the dominant pelagic limestone unit. At the upper part of the section, a more uniform massive bioclastic limestone succession appears (Figure 4c). The lithology of this part of the section is described as solid, thick-bedded limestones that can be easily separated from the underlain medium-to-thin-bedded ones.

In the uppermost part of the section, reddish-to-brownish chert horizons, both nodular and bedded siliceous concretions, were observed. The siliceous nodules differ in size (up to 35 cm in diameter) and color (red to black), and appear in a variety of shapes, mostly sub-spherical, elongated-to-flattened, or mushroom ones (Figure 5a–c).

It is remarkable that all observed siliceous nodules are characterized by two or three distinct colors from the rim to the core, such as black to the periphery, rusty red brown to light red in the main body, and a thin blackish layer at the interior. This development in levels is probably due to differential mineralogic (quartz, moganite, chalcedony) and/or biogenic source (diatoms, radiolarian, or siliceous sponges) composition. The acute red color should be indicative of the advancement of certification, while the thin black layer at the interior and exterior is possibly related to diagenetic processes. Under such circumstances, the calcium carbonate can be replaced by waters rich in silica flowing through the rock and form diagenetic siliceous beds. Cherts formed in this way usually occurs as nodules within the carbonate succession, like the dark-colored nodules (Figure 5a), or thin siliceous beds as ribbon cherts (Figure 5b), mushroom-shaped masses (Figure 5c), lenses, or even the irregular in structure and size siliceous layers reported here. During field investigation, bedding parallel stylolites were also observed, which possess dissolution induced open longitudinal vugs and cavity-filling carbonate cement. Such features probably created during the nodules development and potentially act as a conduit for fluid (even for hydrocarbons) migration along their amplitude. Their presence should further reflect prolific source rocks through the reservoir properties (i.e., secondary porosity and permeability increase) increase of the hosted deposits through the fracturing.

*4.2. Biostratigraphy and Paleoecology*

The biostratigraphic analysis based on planktonic foraminifera showed that the study part of Gardiki section covers the boundary between the Late Cretaceous (Maastrichtian) and Early Paleocene (Danian) time span, exhibiting a well-preserved transition across the boundary. The age determination is in accordance with the geological map concerning the Ionian rock exposures in the Epirus area [55].

As described in detail in Table 1, the base of the section has a latest Maastrichtian age revealed by the presence of Globotruncaniids and Heterohelicids in samples B1–B10. The middle part of the section (samples B11–B32) corresponds to the K-Pg boundary based on the coexistence of *Globotruncana stuarti* de Lapparent [56], *Globotruncana aegyptiaca* Nakkady [57], and Globigerinidae (Figure 7). Upwards up to the top (samples B33–B90), Globigeriniids are consistency present in higher abundance, while some Globotruncaniids co-exist. However, the latter has a sporadic and highly variable presence and reveals a strong proliferation of altered tests (i.e., broken, dissolved) or aberrant forms, which are generally smaller in size within the Danian samples (samples B33–B90). In accordance with previous studies related to Danian reworking [58,59], such stratigraphic and size distribution peculiarities after the K-Pg boundary evidence that the Maastrichtian specimens are probably reworked. We note that additional stable isotopes on these species and subsequent comparison of the isotopic values obtained below and above the K-Pg boundary will ensure if these species are in fact reworked or real Cretaceous survivors. However, such geochemical analyses could be considered as motivational for future work.

**Figure 4.** Outcrop images showing the different parts of Gardiki section corresponding to the Cretaceous–Paleogene transition. (**a**) Alterations of thin-to-medium-bedded limestones and cherts of Late Maastrichtian age, (**b**) Thin-bedded pelagic limestones corresponding to the K-Pg boundary, (**c**) Massive bioclastic limestone succession of early Paleocene age.

**Figure 5.** Reddish to brownish chert horizons. Siliceous nodules appear in a variety of shapes, mostly (**a**) sub-spherical, (**b**) elongated-to-flattened, or (**c**) mushroom types.

**Table 1.** Detailed description of the analyzed samples (thin sections), where depositional facies, lithology, formation, age, and depositional environments are presented.

| Samples | Formation | Lithology | Facies Description | Figures (Lithology/Facies) | Depositional Environment | Energy Conditions | Chrono-Stratigraphy |
|---|---|---|---|---|---|---|---|
| B33–B90 | Limestone with microbreccia | Massive bioclastic limestones with siliceous nodules | Bioclastic packstone enriched mostly on planktonic foraminifera or micritized radiolarians (SMF 3-4,10) | Figures 4c and 6c | Toe of slope (FZ3-4) | Medium energy | Early Paleocene (Danian) |
| B11–B32 | Senonian | Thin-bedded limestones | Biomicrite wackestone–packstone with abundant planktonic foraminifera (SMF 2-3) | Figures 4b and 6b | Deep basin (FZ1)Deepshelf (FZ2) | Low to Medium energy | K-Pg boundary |
| B1–B10 | Senonian | Thin- to medium-bedded limestones with cherts intercalations | Pelagic biomicrite mudstone (SMF 1-3) | Figures 4a and 6a | Deepbasin (FZ1) | Low Energy | Late Cretaceous (Late Maastrichtian) |

*4.3. Sedimentary Microfacies Analysis*

Within the Senonian limestone formation, we observed three distinct microfacies: (a) a pelagic biomicrite mudstone to wackestone with sparce planktonic foraminifera, radiolarians and filaments, which correspond to thin-to-medium-bedded deposits, (b) a pelagic biomicrite wackestone–packstone with abundant in situ planktonic foraminifera (both carenate and non-carenate forms), representing the thin-bedded bioclastic limestones of the K-Pg boundary, and (c) a pelagic biomicrite wackestone with both in situ and altered planktonic foraminifera and calcite veins of the massive carbonate unit (Figure 6).

*4.4. Diagenetic Pathways and Processes*

Diagenesis begins at the sediment–water interface, even if most of the changes occur after burial. Several distinct diagenetic characteristics were recognized in the K-Pg deposits based on sedimentological analysis. A specific diagenetic alteration condition was indicated by the diagenetic characteristics, which have a long-term effect on sedimentary rock. Micritization, compaction, cementation, dissolution, and fractures are all prevalent diagenetic processes. Micritization, cementation, compaction (physical and chemical), dissolution, and dolomitization are among the shallow diagenetic processes (Figure 8) [60–62].

According to the evolution of the environment, these processes display a distinct set of modification conditions. The next section explains the many diagenetic processes that affected the studied carbonates.

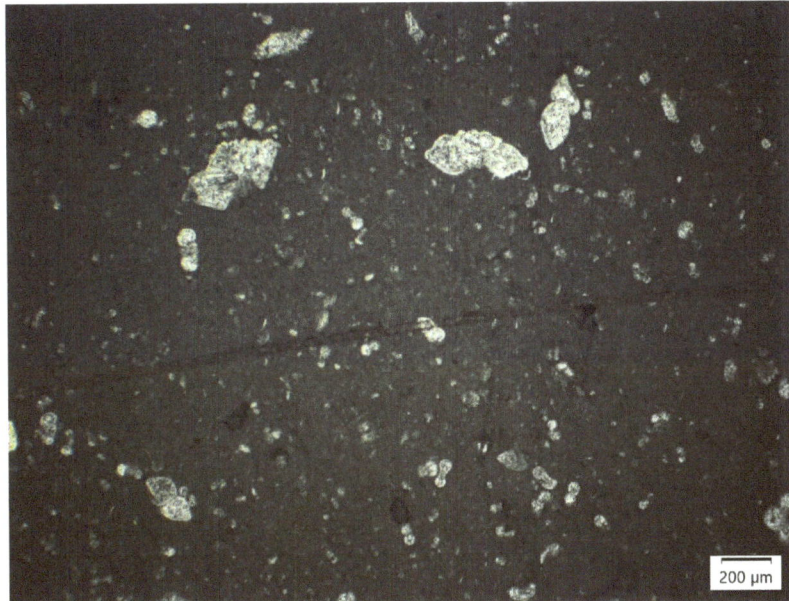

**Figure 6.** Thin section photomicrograph depicts the co-existence of *Globotruncana stuarti* de Lapparent [56], *Globotruncana aegyptiaca* Nakkady [57], and Globigerinidae determining the K-Pg boundary.

### 4.4.1. Micritization

The micrite envelopes provide evidence of simultaneous alteration and deposition [63–66]. In general, the micritized skeletal components survive disintegration and act as a surface for late precipitation (Figure 8a). Micrite envelopes are present in all portions of the investigated carbonate samples. The presence of bioclasts in the examined samples indicates a high level of microbial micritization (Figure 8a,b,d). Figure 7a,d show that a micrite envelope covers the grains in the early stages of micritization, whereas other grains have an uneven internal structure and calcite cement filling (Figure 8a). Allochems may be protected from further deterioration by micritization, which has no direct impact on reservoir quality.

### 4.4.2. Cementation

According to sedimentological features, multiple generations of calcite cement were observed with a modest input of dolomite. The predominant cement type seen was equant calcite spar, which filled interstitial pore spaces, skeletal chambers, and certain pore types (Figure 8b,e,f). In the observation of the epitaxial overgrowth cement (Figure 8e), it is crucial to remark that in certain samples, the overgrowth reaches a length of up to 100 microns. Calcites ranging in size from silty to fine crystalline equants that partly or fully filled fractures (Figure 8c). Additionally, porous, small to medium crystalline, and euhedral dolomite cement fill pore spaces and fractures to a significant extent (Figure 8c,d).

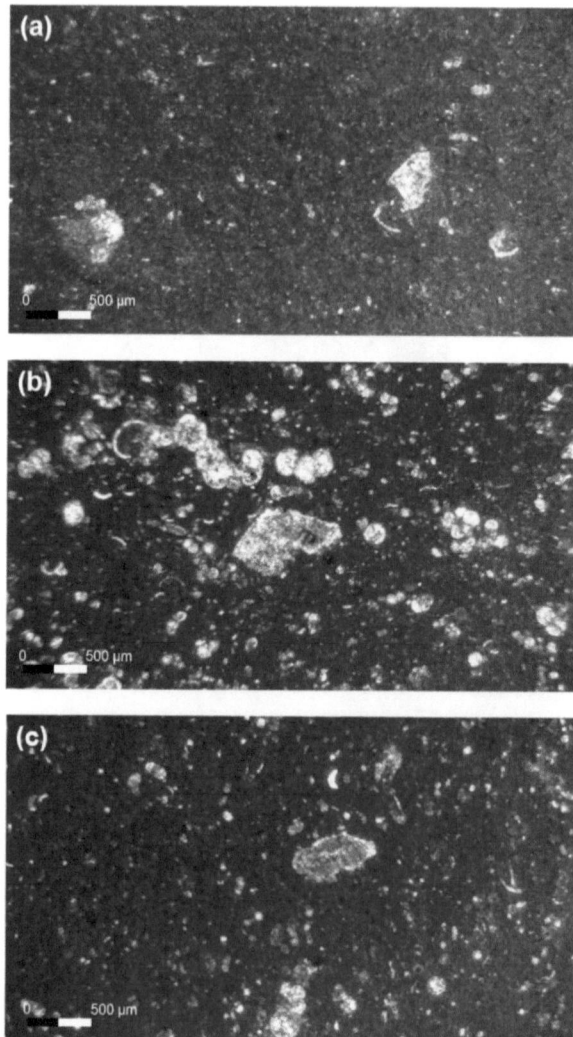

**Figure 7.** Main microfacies types recognized: (**a**) Pelagic biomicrite mudstone with planktonic foraminifera, radiolarians, and filaments, (**b**) Pelagic wackestone–packstone with in situ planktonic foraminifera, (**c**) Pelagic biomicrite wackestone with altered planktonic foraminifera and/or radiolarians.

4.4.3. Dissolution

Dissolution has been defined as the leaching of metastable bioclasts due to the presence of meteoric water by Budd [67], Morse and Arvidson [68], Lambert et al. [69], Tucker and Wright [70], and Janjuhah et al. [71]. The study rocks are characterized by secondary pore spaces, such as moldic and vuggy pores, developed by dissolution. High-magnesium calcite (HMC) replaces aragonite in the process of dissolution because of the minerals' solubility [72,73]. They also show the most fabric-selective dissolving, with a little proportion originating from non-fabric dissolution (Figure 8a,b,e,f). Late-stage porosity development is indicated by the dissolution that is often seen (Figure 8a,b,e,f). Moldy, vuggy, and fractured porosity are all associated with dissolution (Figure 8a,b,d–f).

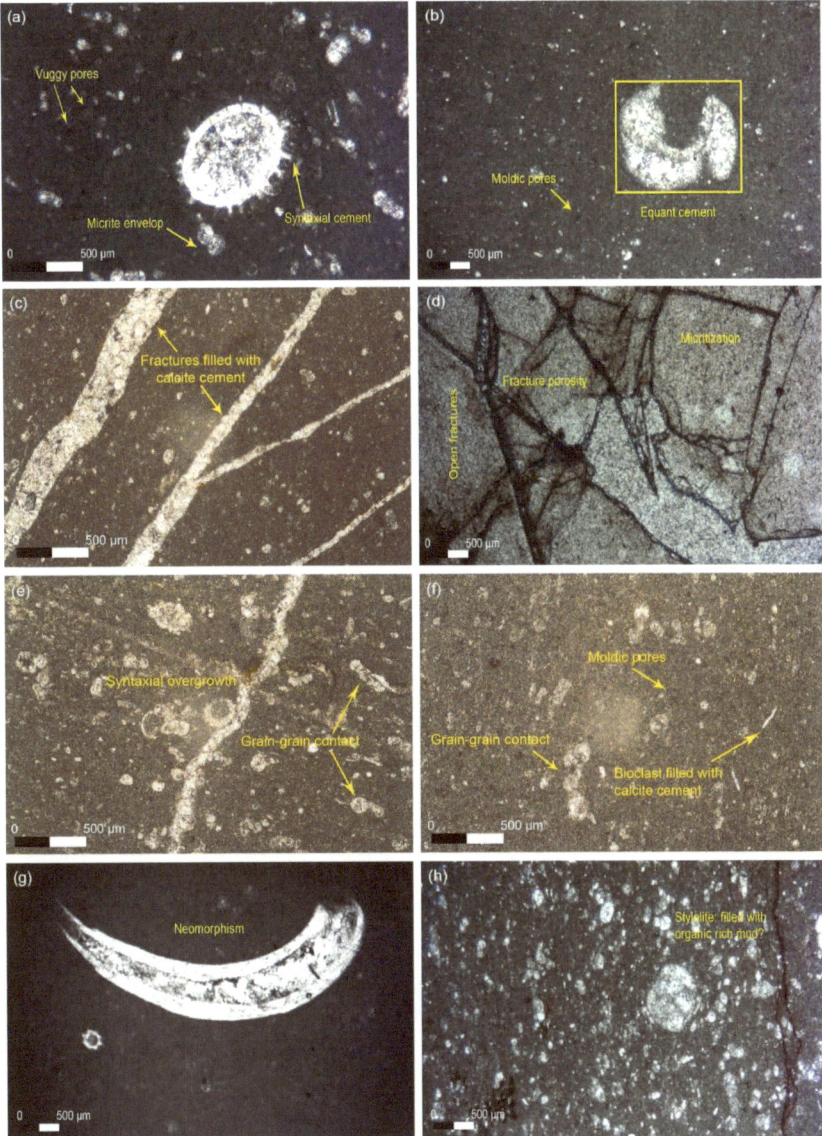

**Figure 8.** Petrographic analysis showing the different diagenetic processes in carbonate samples collected from Gardiki section (Epirus, Western Greece), (**a**) high micritization, vuggy pores, and syntactical overgrowth cement, (**b**) micritization, Junctions Equant spars of marine moldic pores, (**c**) micritization, late stage of diagenesis-fractured filled with calcite cement, (**d**) early stage of micrite envelop, late stage of open fracture, (**e**) epitaxial overgrowth cement, moldic pores, (**f**) moldic pores, grain-grain contact (physical compaction), (**g**) coarse pale brown neomorphic spar replacement of aragonite shell, (**h**) and stylolites filled with organic-rich mud.

4.4.4. Fracturing

Reservoir quality has improved dramatically due to repeated tectonic stress releases throughout time. Using thin sections, the effectiveness, orientations, and cross-cutting relationships of carbonate fractures can be examined. Their widths vary from a few micrometers to a few centimeters, and their orientations shift from vertical to horizontal.

Vertical to sub-vertical fractures dominate the early set, and each fracture is either partly or completely calcite-cement-filled (Figure 8c,d). In contrast, horizontal to sub-horizontal fractures cross-cut the earlier set in the late set. Most of them lack cementation, which improves the reservoirs' quality (Figure 8c,d). Fractures occurred in the burial environment because dissolution seams were generally earlier than the time of formation of the fractures.

4.4.5. Neomorphism

Diagenesis begins immediately upon deposition and continues at all levels until metamorphism takes over. Burial may also be expressed in terms of temperature, pressure, and fluid composition, as well as physical and chemical changes, such as neomorphism. Certain carbonate samples exhibit neomorphism, as seen by the replacement of lime mudstone with finer crystalline calcite spars and the substitution of low magnesium calcite (LMC) for aragonite or high magnesium calcite (HMC) in shell fragments, altering the original texture and structure (Figure 8a,g). Recrystallization of the skeletal particles, either partially or completely, results in aggradation neomorphism (Figure 8g). This process is aided by the presence of calcite with a high magnesium content, which most likely accumulates in the shells of skeletal components until they reach their stable phase under meteorological circumstances [74]. This investigation demonstrated subaerial diagenesis, since numerous skeletal grains have micrite envelopes and exhibit evidence of aggradation neomorphism (Figure 8a).

4.4.6. Compaction

Grain fracture and porosity are reduced because of compaction. Compaction may be caused by mechanical stress, but the chemical reactions are highly dependent on temperature, pressure, and water concentration in the pore space [75]. Compaction may decrease porosity by more than 40% in certain circumstances and is more effective than cementation in reducing porosity [5,76]. The Gardiki carbonates have been affected by both mechanical and chemical compactions. The sediment overburden caused concave–convex grain interactions, which resulted in grain form distortion and breakage (Figure 8e,f).

## 5. Discussion

*5.1. Depositional Conditions and Paleoenvironmental Implications during the K-Pg Transition*

The described sedimentary facies distribution reflects a relatively deep environment with some internal differentiations in terms of the environmental deposition and energy conditions (Figure 9). In the Gardiki section of Epirus area, the Senonian post-rift sequence of Ionian basin contains the pelagic biomicrite mudstone (SMF 1-3), whose deposition characterizes a low-energy, deep, basinal environment (FZ1). Even though radiolarians were rarer compared to the planktonic foraminifera in these deposits, the presence of small-sized filaments within the micritic matrix also denotes a relatively deep, low-energy environment. The passage from mixed carbonate-silicious (alterations of limestones and cherts) to almost carbonate material (thin-bedded limestones with sparsely observed siliceous horizons or nodules) during the latest Maastrichtian is indicative of slightly shallower depositional environments close enough to the carbonate compensation depth, such as those of the deep shelf (FZ2). Biomicrite wackestone–packstone with abundant planktonic foraminifera (Globoruncaniids and Globigeriniids; SMF 2-3) characterizes the deposits across the K-Pg boundary. The Early Paleocene massive limestones include in situ bioclastic packstone enriched mostly on planktonic foraminifera (both in situ and reworked) or radiolarians (SMF 3). The micritized nature of the radiolarians (SMF10), where evident, along with some broken foraminifera (SMF 4), indicates a medium-energy environment, possibly due to their transportation to the basin margin. Such carbonate systems can be usually accumulated in relatively deep parts of the basin, such as the toe of the slope (FZ3-4), and can be further shed from shelf margins.

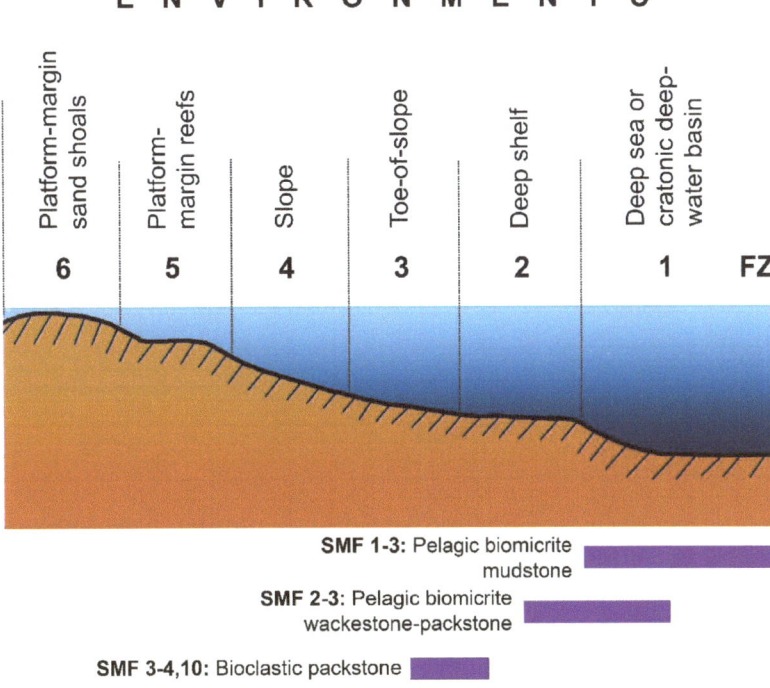

**Figure 9.** Depositional distribution model for Gardiki section based on microfacies analysis.

Overall, the main depositional facies observed were characterized by pelagic biomicrite mudstone with some planktonic foraminifera, radiolarians, and/or filaments at the base (latest Maastrichtian), packstone with abundant planktonic foraminifera at the middle, and wackestone up to the top of the succession (early Paleocene), implying that the depositional environments across the K-Pg transition did not change significantly. Only a gradual environmental shallowing from the deep basin to the slope can be inferred by the sedimentary facies' interpretation. On the contrary, the transition from the lower to the upper stratigraphically deposits in the study area documents a significant diversification reflected by quite different faunal assemblages, in terms of both their content and abundance. A gradual decrease of in situ radiolarians and increase of planktonic foraminiferal abundance was observed through the Late Cretaceous period, while an abrupt extinction of planktonic foraminifera and the presence of micritized radiolarians were evident at the first samples examined just above the K-Pg boundary (interpreted as a reduction of primary productivity). This kind of differentiation in the fauna was further followed by the subsequent reworking of Globotruncaniids (Cretaceous species), as evidenced by broken and aberrant, smaller specimens. Similar abnormal variations in morphology and size of planktonic foraminiferal tests have long been reported globally during the late Maastrichtian crisis [77–79] or immediately above the K-Pg boundary [80–82], and can be compared with similar records documented during other extreme geological events (e.g., Cretaceous ocean anoxic events [83–85]; Paleocene-Eocene Thermal Maximum [86]; Messinian Salinity Crisis [87–89]; Late Pliocene [90]; Sapropel S1 [91]), and seem to reflect mainly stressful environmental conditions in response to changes in ecological parameters (i.e., temperature, salinity, dissolved oxygen, acidification, changeable connectivity; [91–93]) or rapid environmental perturbations (i.e., extreme fluctuations in climate and sea level, intense volcanism, meteoric pollution) [78,82]. As evidenced globally in all extinction events through the Earth history, their duration and the corresponding differences in extinction percentages of foraminiferal taxa inhabiting different levels in the oceanic water column during extinction time intervals are associated with specific environmental parameters, which in turn reveal the causes of these events [94–96]. Particularly for the Cretaceous–Tertiary (K-T) turnover, the most plausible scenario seems to be the alternating climate (global warming during the latest Maastrichtian and Deccan-volcanism induced cooling) and subsequent sea level fluctuations (sea level rise during warming and sea level fall during cooling events), as well as productivity changes [95–97], during which the planktonic foraminiferal species experienced high stress and, in response, reduced their sizes and/or developed malformations on their tests [98,99]. However, besides these key environmental perturbations, additional ecological stress factors could be sufficient to account for the increase of aberrant forms; in some cases, this could even be characterized as lethal for planktonic foraminifera likely contributed to the K-T boundary mass extinction. For instance, the increased concentrations of toxins produced by phytoplankton blooms [100] in surface waters could possibly be related to the proliferations of abnormal planktonic foraminiferal tests above the K-Pg boundary [101,102]. Although such geographically heterogeneous spikes of phytoplankton groups have been reported in the earliest Danian [103], more studies need to established the above cause–effect relationship. Overall, the observed aberrant forms can be characterized as likely "victims" of the K-T event, highlighting the gradual or sudden foraminiferal extinctions and the related evolutionary and/or ecological crises occurred globally during that time [78,92,104,105].

Particularly for the K-T event, the largest event during the last 100 M.y. [106] is controversial regarding the character (sudden versus gradual) and severity (proportion of surviving species). The biotic turnovers of this mass extinction were highly variable among different fossil groups. Planktic foraminifera were the "losers", with all tropical and subtropical species (2/3 of all species) extinct, while a few survivors died out during the first 200–300 ky of the Danian [78,81]. Additional quantitative calculations on the decline of diversity of planktonic foraminifera based on the probabilistic stratigraphy technique showed that only a maximum of 22% of Cretaceous species survived the boundary, and 11% passed into the planktonic foraminiferal zone P1a [107]. On the contrary, although calcareous nannoplankton similarly suffered strong declines, several species thrived into the Tertiary, and all other microfossil groups (i.e., dinocysts, benthic foraminifera) reveal minor biotic effects linked with temporal decreases of their diversity and abundance [59,108–112]. The differential planktonic–benthic extinction rate is further supported by the vertical marine carbon isotope gradients ($\Delta\delta^{13}C$) between planktic and benthic species [113]. Our data exhibit that plankton productivity declined significantly at the K/T boundary, which is consistent with previous studies in a global context, indicating that low productivity levels were more severe in the low latitudes and lasted for a few hundred thousand years before the restoration of normal ecosystems [114–116] through the final recovery of the marine carbon cycle [113,117]. These data may indicate that reduced primary productivity in the surface waters was a main cause of this event, with subsequent reduction in food supply for animals at higher trophic levels [118]. Typically, this low productivity state continued for several hundred thousand years and was associated with widespread stunting of marine organisms (Lilliput effect) and low-biomass ecosystems [95,98,119,120].

At that time in the broad external Ionian zone, allochthonous carbonates were transported to the outer shelf by turbidity currents (calciturbidites) formed by the gravitational collapse of the platform margin [19,20,24]. In the study section, no coarser brecciated horizons were observed within these deposits, as usually happened within the Ionian basin in other locations of the western Greece [19,24,44]. Their absence from the studied part of the Gardiki section could be attributed to the foreland basin evolution, as it was less influenced by their greater distance from the thrusts between the tectonic zones, sediment thickness, and overall depositional conditions. It is most likely that the submarine shelfs characterized the nearby Apulian carbonate platform margins in this setting were not so steep and the instability was not so intense to lead to significant erosion. The smoother slope could provide evidence of slightly eroded platforms internally to the Ionian Basin, thus supporting the half-graben Late Cretaceous concept described in detail by Karakitsios [22]. Overall, this study provides evidence that the supply of clastic material was significantly reduced in this setting during that time, also supporting the findings of Bourli et al. [19] about a differential tectonic activity in the geographic area between Epirus and Kefalonia Island. On the contrary, the deposition of the overlying microbreccious limestones during the Paleocene-Eocene shows that intense debris flow formed by the slumping of the platform margin, possibly due to tectonics, and redeposited the platform edge sediments in the deeper parts of the basin. Consequently, the study deposits reflect the pelagic carbonates that pass into the Eocene calciturbidites.

## 5.2. Diagenetic History—Paragenesis

Following an investigation of sedimentological relations, we can ascertain the sequence in which diagenetic events took place. The diagenetic history of carbonate rocks in Gardiki section is complex. The K-Pg reservoir deposits of the Ionian zone are divided into three types: marine, meteoric, and burial diagenetic environments (Figure 10).

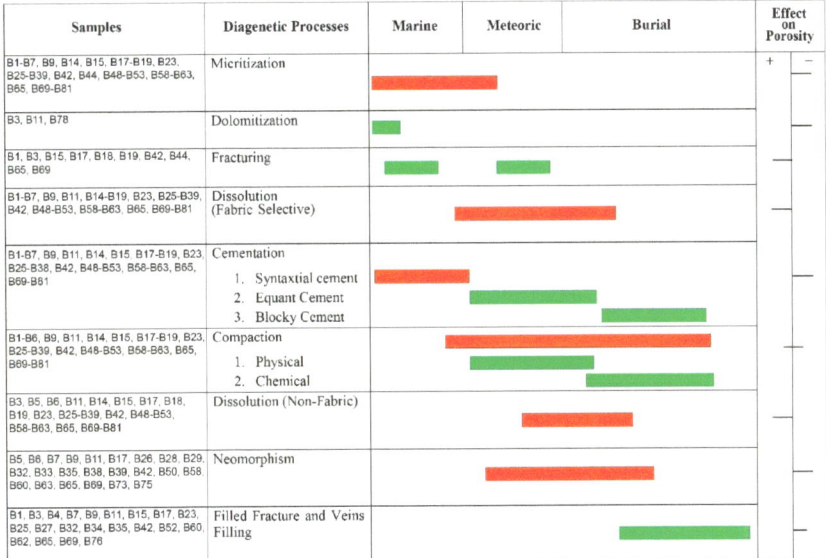

**Figure 10.** The diagenetic history of K-Pg deposits of Gardiki section based on petrographic observations. Red boxes indicate a strong effect on porosity, and green boxes indicate a low effect on porosity.

### 5.2.1. Marine Diagenetic Environment

The diagenetic environment of the Late Cretaceous–Early Paleocene reservoir sediments of the Ionian basin (Epirus, western Greece) was in deep basin to shelf deposits. Micritization is quite prevalent during the early stages of diagenesis (Figure 10). During the first stages of micritization, referred to as micritic envelopes, the skeletal grains' surrounding regions were destroyed (Figure 8a,b,d, and Figure 10). This occurs when microboring organisms destroy allochems at or near the sediment–water interface. Brett and Brookfield [121] defined syn-sedimentary micritization as the presence of micritic bioerosion fringes extending from the grain's surface to the grain's core. A marine low-energy environment seems to be favorable for grain micritization and the formation of micritic envelopes around bioclasts in certain samples (Figure 8a,b,d). Additionally, Adams [122] noted that micritization is the first diagenetic phase that occurs at the sediment–water interface in low-energy shallow marine environments with slow sedimentation rates. According to Karakitsios et al. [123], seawater served as the primary diagnostic fluid in the Ionian basin, resulting in positive/extremely high $\delta^{13}C$ values for calcites in pores. In the deep slope setting, the early sediment underwent penecontemporaneous and seepage–reflex dolomitization, producing microcrystalline dolomites. Physical compaction, calcite cementation, and dolomitization of the Late Cretaceous reservoir rocks resulted in a considerable reduction of primary porosity (Figure 8a,b,d,e).

### 5.2.2. Meteoric Diagenetic Environment

The rifting phase occurred during the Late Jurassic–Early Cretaceous period, and the Vigla limestone formation was deposited in the Ionian basin shortly after the rifting phase [34]. Cretaceous strata in western Greece were considerably changed during the Late Cretaceous epoch [124]. As a consequence, during the Late Cretaceous, a meteoric water diagenetic environment may have formed in connection with the global sea-level fall [125,126] $\delta^{13}C$ value for crystalline limestone is negative, which suggests that recrystallization may also be linked to meteoric freshwater as reported by Tsikos et al. [127].

Meanwhile, dissolution was detected in the rocks examined (Figure 8). Aragonitic organisms, such as bioclasts, had their internal structure partly or entirely disintegrated (Figure 8d). Aragonite's early dissolution might have happened in saltwater or under conditions of shallow burial [128,129]. Two distinct phases of dissolution have been identified. Stage 1 is characterized by the breakdown of aragonitic biota (Figure 8d), implying partial to total bioclast dissolution (HMC grains). Janjuhah et al. [130] noted that meteoric interfaces such as a subaerial exposed surface and the water table are probable locations for significant $CO_2$ inflow, resulting in enhanced HMC grain dissolving and preventing LMC grain dissolution. Stage 2 is a non-fabric selective matrix dissolution, and a critical diagenetic process in the formation of the study limestones (Figure 8a). Non-fabric selective dissolution occurred between a deep meteoric and a deep burial environment. Following that, the void areas were filled with cement (syntextial, equant, and blocky cement) (Figure 8). They often occur during the same diagenetic period but on distinct substrata. The syntaxial overgrowth cement is often used to enclose bioclast fragments in packstone textures (Figure 8e), while the equant calcite cement is used in both fabric- and non-fabric-selective dissolution (Figure 8c). It is possible to observe equant calcite cement filling the fracture component of the pore spaces between grains (Figure 8c). This cement may be the result of precipitation in a shallow diagenetic environment. Additionally, Flügel [131] reported that equant calcite cement is present in meteoric and burial contexts. It should be noted at this stage that further evidence is required to ascertain if this cement is meteoric- or burial-derived.

### 5.2.3. Burial Diagenetic Environment

Other diagenetic processes observed in the examined samples include physical and chemical compactions. Physical compaction is defined by rearranged and broken grains that are compressed vertically. Due to early marine cementation and overburden pressure, certain open fractures of mechanical compaction are readily evident (Figure 8d). Chemical compaction is characterized by the formation of a pressure solution of grains and sediments, which serves as an important source of calcium carbonate for burial cementation. The chemical composition of lime mud was identified, which resulted in the invention of stylolites (Figure 8h). The stylolites detected have a relatively small amplitude. Late-stage neomorphism has occurred, affecting the rock. Neomorphism has completely altered the texture and structure of carbonate rocks in the late stage, post-dating early cementation. This process is promoted by the early decomposition of aragonite and HMC in seawater or under very shallow burial circumstances [62]. The most recent diagenetic event was characterized by fractures that crossed other diagenetic products, including the prior cement and neomorphosed calcite cement (Figure 8e). Later phases, with deep burial (telogenesis), fracturing, and blocky cementation vein filling, occurred [132].

## 6. Conclusions

An integrated sedimentological analysis of the K-Pg carbonate succession representing the topmost part of the Senonian limestone and the lower part of the microbreccious limestone formation in Gardiki section of Epirus (western Greece) provides new insights into the depositional environment and diagenetic history of the Western Ionian Basin, as well as the synthetic paleoenvironmental reconstruction of the area. The biostratigraphic analysis based on planktonic foraminifera showed that the study interval of the Gardiki section covers the boundary between Late Cretaceous (Maastrichtian) and Paleocene (Danian) time span based on the coexistence of *Globotruncana* spp. and Globigerinidae. Lithostratigraphically, the Senonian limestone formation in this area is mainly characterized by light-grey-to-yellowish limestones, with some siliceous intervals to occur as thin reddish-to-brownish chert intercalations and/or lenses within the pelagic limestone unit. Their presence of both nodular and bedded siliceous concretions is of crucial importance on enhancing the reservoir properties (i.e., secondary porosity and permeability increase) of the hosted deposits through the fracturing and stylolites created during the nodule's development and reveals the potentiality to act as open pathways for the fluids (even for hydrocarbons). The main depositional facies (pelagic biomicrite mudstone to packstone) observed into the K-Pg transition are indicative of a low-energy, basinal to toe-of-slope environmental sedimentation. Minor internal differentiation among them implies that the depositional environment across the K-Pg transition did not change significantly in the external Ionian basin. However, significant differences documented in the content and the abundance of the faunal assemblages testify to the stressful environmental conditions in terms of primary productivity decline that prevailed at that time. Furthermore, the petrographic analysis identified several distinct diagenetic features among the study pelagic carbonates, which are categorized into marine, meteoric, and burial diagenetic settings. Micrite envelop, cementation, fracture, compaction, and dissolution are the dominant diagenetic parameters identified in the study samples, attributing both positive and negative effects on porosity. Overall, this study provides a basis for the further evaluation of the hydrocarbon potential in western continental Greece, which contains proven reserves and is of crucial economic and strategic importance.

**Author Contributions:** Conceptualization, L.M. and G.K.; methodology, L.M., G.K. and H.T.J.; software, L.M., G.K., H.T.J. and A.Z.-L.; validation, L.M., G.K., H.T.J. and A.Z.-L.; formal analysis, L.M., G.K., H.T.J., A.Z.-L.; investigation, L.M., G.K., H.T.J., A.Z.-L., D.G., P.P., C.K., S.S., E.B., V.K. and A.A.; resources, L.M., G.K., V.K.; data curation, L.M., G.K., H.T.J., A.Z.-L. and C.K.; writing—original draft preparation, L.M., G.K. and H.T.J.; writing—review and editing, L.M., G.K., H.T.J., A.Z.-L., D.G., P.P., C.K., S.S., E.B., V.K. and A.A.; visualization, L.M. and G.K.; supervision, A.A. and V.K.; project administration, G.K., A.A.; funding acquisition, G.K. All authors have read and agreed to the published version of the manuscript.

**Funding:** This research received no external funding.

**Institutional Review Board Statement:** Not applicable.

**Informed Consent Statement:** Not applicable.

**Data Availability Statement:** The data used in this work is available on request to the corresponding authors.

**Conflicts of Interest:** The authors declare no conflict of interest.

## References

1. Knowlton, N.; Brainard, R.E.; Fisher, R.; Moews, M.; Plaisance, L.; Caley, M.J. Coral Reef Biodiversity. In *Life in the World's Oceans*; John Wiley & Sons: Hoboken, NJ, USA, 2010; pp. 65–78. [CrossRef]
2. Falkowski, P.; Scholes, R.J.; Boyle, E.; Canadell, J.; Canfield, D.; Elser, J.; Gruber, N.; Hibbard, K.; Högberg, P.; Linder, S.; et al. The Global Carbon Cycle: A Test of Our Knowledge of Earth as a System. *Science* **2000**, *290*, 291–296. [CrossRef] [PubMed]
3. Burchette Trevor, P. Carbonate rocks and petroleum reservoirs: A geological perspective from the industry. *Geol. Soc. Lond. Spec. Publ.* **2012**, *370*, 17–37. [CrossRef]

4. Ahmad, I.; Shah, M.M.; Janjuhah, H.T.; Trave, A.; Antonarakou, A.; Kontakiotis, G. Multiphase Diagenetic Processes and Their Impact on Reservoir Character of the Late Triassic (Rhaetian) Kingriali Formation, Upper Indus Basin, Pakistan. *Minerals* **2022**, *12*, 1049. [CrossRef]
5. Janjuhah, H.T.; Kontakiotis, G.; Wahid, A.; Khan, D.M.; Zarkogiannis, S.D.; Antonarakou, A. Integrated Porosity Classification and Quantification Scheme for Enhanced Carbonate Reservoir Quality: Implications from the Miocene Malaysian Carbonates. *J. Mar. Sci. Eng.* **2021**, *9*, 1410. [CrossRef]
6. Laugié, M.; Michel, J.; Pohl, A.; Poli, E.; Borgomano, J. Global distribution of modern shallow-water marine carbonate factories: A spatial model based on environmental parameters. *Sci. Rep.* **2019**, *9*, 16432. [CrossRef]
7. Michel, J.; Laugié, M.; Pohl, A.; Lanteaume, C.; Masse, J.-P.; Donnadieu, Y.; Borgomano, J. Marine carbonate factories: A global model of carbonate platform distribution. *Int. J. Earth Sci.* **2019**, *108*, 1773–1792. [CrossRef]
8. Kontakiotis, G.; Karakitsios, V.; Mortyn, P.G.; Antonarakou, A.; Drinia, H.; Anastasakis, G.; Agiadi, K.; Kafousia, N.; De Rafelis, M. New insights into the early Pliocene hydrographic dynamics and their relationship to the climatic evolution of the Mediterranean Sea. *Palaeogeogr. Palaeoclimatol. Palaeoecol.* **2016**, *459*, 348–364. [CrossRef]
9. Kontakiotis, G.; Besiou, E.; Antonarakou, A.; Zarkogiannis, S.D.; Kostis, A.; Mortyn, P.G.; Moissette, P.; Cornée, J.J.; Schulbert, C.; Drinia, H.; et al. Decoding sea surface and paleoclimate conditions in the eastern Mediterranean over the Tortonian-Messinian Transition. *Palaeogeogr. Palaeoclimatol. Palaeoecol.* **2019**, *534*, 109312. [CrossRef]
10. Kontakiotis, G.; Butiseacă, G.A.; Antonarakou, A.; Agiadi, K.; Zarkogiannis, S.D.; Krsnik, E.; Besiou, E.; Zachariasse, W.J.; Lourens, L.; Thivaiou, D.; et al. Hypersalinity accompanies tectonic restriction in the eastern Mediterranean prior to the Messinian Salinity Crisis. *Palaeogeogr. Palaeoclimatol. Palaeoecol.* **2022**, *592*, 110903. [CrossRef]
11. Goldscheider, N.; Mádl-Szőnyi, J.; Erőss, A.; Schill, E. Review: Thermal water resources in carbonate rock aquifers. *Hydrogeol. J.* **2010**, *18*, 1303–1318. [CrossRef]
12. Kiessling, W.; FlÜGel, E.; Golonka, J.A.N. Patterns of Phanerozoic carbonate platform sedimentation. *Lethaia* **2003**, *36*, 195–225. [CrossRef]
13. Pohl, A.; Laugié, M.; Borgomano, J.; Michel, J.; Lanteaume, C.; Scotese, C.R.; Frau, C.; Poli, E.; Donnadieu, Y. Quantifying the paleogeographic driver of Cretaceous carbonate platform development using paleoecological niche modeling. *Palaeogeogr. Palaeoclimatol. Palaeoecol.* **2019**, *514*, 222–232. [CrossRef]
14. Philip, J.; Masse, J.-P.; Camoin, G. Tethyan Carbonate Platforms. In *The Tethys Ocean*; Nairn, A.E.M., Ricou, L.-E., Vrielynck, B., Dercourt, J., Eds.; Springer: Boston, MA, USA, 1996; pp. 239–265. [CrossRef]
15. Bernoulli, D.; Jenkyns, H.C. Alpine, Mediterranean, and Central Atlantic Mesozoic Facies in Relation to the Early Evolution of the Tethys. In *Modern and Ancient Geosynclinal Sedimentation*; Dott, R.H., Jr., Shaver, R.H., Eds.; SEPM Society for Sedimentary Geology: Tulsa, OK, USA, 1974; Volume 19. [CrossRef]
16. Belghouthi, F. The Upper Cretaceous-Lower Eocene carbonate platforms of the Mateur-Beja area (NW of Tunisia): A pattern of isolated platform. *J. Afr. Earth Sci.* **2022**, *187*, 104453. [CrossRef]
17. Jurkowska, A.; Świerczewska-Gładysz, E.; Bąk, M.; Kowalik, S. The role of biogenic silica in the formation of Upper Cretaceous pelagic carbonates and its palaeoecological implications. *Cretac. Res.* **2019**, *93*, 170–187. [CrossRef]
18. Bourli, N.; Kokkaliari, M.; Iliopoulos, I.; Pe-Piper, G.; Piper, D.J.W.; Maravelis, A.G.; Zelilidis, A. Mineralogy of siliceous concretions, cretaceous of ionian zone, western Greece: Implication for diagenesis and porosity. *Mar. Pet. Geol.* **2019**, *105*, 45–63. [CrossRef]
19. Bourli, N.; Pantopoulos, G.; Maravelis, A.G.; Zoumpoulis, E.; Iliopoulos, G.; Pomoni-Papaioannou, F.; Kostopoulou, S.; Zelilidis, A. Late Cretaceous to early Eocene geological history of the eastern Ionian Basin, southwestern Greece: A sedimentological approach. *Cretac. Res.* **2019**, *98*, 47–71. [CrossRef]
20. Kontakiotis, G.; Moforis, L.; Karakitsios, V.; Antonarakou, A. Sedimentary Facies Analysis, Reservoir Characteristics and Paleogeography Significance of the Early Jurassic to Eocene Carbonates in Epirus (Ionian Zone, Western Greece). *J. Mar. Sci. Eng.* **2020**, *8*, 706. [CrossRef]
21. Karakitsios, V.; Rigakis, N. Evolution and petroleum potential of Western Greece. *J. Pet. Geol.* **2007**, *30*, 197–218. [CrossRef]
22. Karakitsios, V. Western Greece and Ionian petroleum systems. *AAPG Bull.* **2013**, *97*, 1567–1595. [CrossRef]
23. Zelilidis, A.; Maravelis, A.G.; Tserolas, P.; Konstantopoulos, P.A. An overview of the petroleum systems in the Ionian zone, onshore NW Greece and Albania. *J. Pet. Geol.* **2015**, *38*, 331–348. [CrossRef]
24. Zoumpouli, E.; Maravelis, A.G.; Iliopoulos, G.; Botziolis, C.; Zygouri, V.; Zelilidis, A. Re-Evaluation of the Ionian Basin Evolution during the Late Cretaceous to Eocene (Aetoloakarnania Area, Western Greece). *Geosciences* **2022**, *12*, 106. [CrossRef]
25. Tserolas, P.; Maravelis, A.G.; Tsochandaris, N.; Pasadakis, N.; Zelilidis, A. Organic geochemistry of the Upper Miocene-Lower Pliocene sedimentary rocks in the Hellenic Fold and Thrust Belt, NW Corfu island, Ionian sea, NW Greece. *Mar. Pet. Geol.* **2019**, *106*, 17–29. [CrossRef]
26. Papanikolaou, D. Tectonostratigraphic models of the Alpine terranes and subduction history of the Hellenides. *Tectonophysics* **2013**, *595–596*, 1–24. [CrossRef]
27. Papanikolaou, D.I. Organization and Evolution of the Tethyan Alpine System. In *The Geology of Greece*; Papanikolaou, D.I., Ed.; Springer International Publishing: Cham, Switzerland, 2021; pp. 9–24. [CrossRef]

28. Aubouin, J.; Le Pichon, X.; Winterer, E.; Bonneau, M. *Les Hellénides dans l'optique de la tectonique des plaques, 6th Colloquium on the Geology of the Aegean Region, Athens, 1977*; Reprinted from proceedings vol. III; IGME: Athens, Greece, 1979; Volume 3, pp. 1333–1354.
29. Papanikolaou, D. The tectonostratigraphic terranes of the Hellenides. *Ann. Geol. Des. Pays Hell.* **1997**, *37*, 495–514.
30. Papanikolaou, D.I. Greece Within the Alpine Orogenic System. In *The Geology of Greece*; Papanikolaou, D.I., Ed.; Springer International Publishing: Cham, Switzerland, 2021; pp. 1–8. [CrossRef]
31. Aubouin, J. Contribution à l'étude géologique de la Grèce septentrionale: Les confins de l'Epire et de la Thessalie; Place des Hellénides parmi les édifices structuraux de la Méditerranée orientale. *Ann. Geol. Des. Pays Hell.* **1959**, *10*, 1–483.
32. IGRS-IFP. *Étude Géologique de l'Epire (Grèce Nord—Occidentale)*; Technip & Ophrys Editions: Paris, France, 1966.
33. Dragastan, O.; Papanikos, D.; Papanikos, P. Foraminifères, Algues et Micrproblematica du Trias de Messopotamos, Epire (Grèce continentale). *Rev. Micropaléontologie* **1985**, *27*, 244–248.
34. Karakitsios, V. Chronologie et Geometrie de L'ouverture d'un Bassin et de son Inversion Tectonique: Le Bassin Ionien (Epire, Grece). *Mem. Sc. Terre Univ.* **1990**, *6*, 91–94.
35. Karakitsios, V. Ouverture et inversion tectonique du Bassin Ionien (Epire, Grèce). *Ann. Géologiques Des Pays Helléniques* **1992**, *35*, 185–318.
36. Karakitsios, V.; Tsaila-Monopolis, V. Donnees nouvelles sur les niveaux inferieurs Trias superieur de la serie calcaire ionienne en Epire Grece continentale Consequences stratigraphiques. *Rev. Paleobiol.* **1988**, *91*, 139–147.
37. Karakitsios, V. The Influence of Preexisting Structure and Halokinesis on Organic Matter Preservation and Thrust System Evolution in the Ionian Basin, Northwest Greece1. *AAPG Bull.* **1995**, *79*, 960–980. [CrossRef]
38. Bernoulli, D.; Renz, O. Jurassic carbonate facies and new ammonite faunas from western Greece. *Eclog. Eclogae Geol. Helv.* **1970**, *63*, 573–607.
39. Karakitsios, V.; Koletti, L. Critical Revision of the Age of the Basal Vigla Limestones (Ionian Zone, Western Greece), Based on Nannoplankton and Calpionellids, with Paleogeographical Consequences. In Proceedings of the 4th International Nannoplankton Association Conference, Prague, Czech Republic, 1 September 1992; pp. 165–177.
40. Skourtsis-Coroneou, V.; Solakius, N.; Constantinidis, I. Cretaceous stratigraphy of the Ionian Zone, Hellenides, western Greece. *Cretac. Res.* **1995**, *16*, 539–558. [CrossRef]
41. Dunham, R.J. *Classification of Carbonate Rocks According to Depositional Textures*; American Association of Petroleum Geologists: Tulsa, OK, USA, 1962.
42. Embry, A.F.; Klovan, J.E. A late Devonian reef tract on northeastern Banks Island, NWT. *Bull. Can. Pet. Geol.* **1971**, *19*, 730–781. [CrossRef]
43. Flügel, E. Microfacies analysis of carbonate rocks. In *Analysis, Interpretation and Application*; Springer: Berlin/Heidelberg, Germany, 2004.
44. Bourli, N.; Iliopoulos, G.; Papadopoulou, P.; Zelilidis, A. Microfacies and Depositional Conditions of Jurassic to Eocene Carbonates: Implication on Ionian Basin Evolution. *Geosciences* **2021**, *11*, 288. [CrossRef]
45. Bourli, N.; Kokkaliari, M.; Dimopoulos, N.; Iliopoulos, I.; Zoumpouli, E.; Iliopoulos, G.; Zelilidis, A. Comparison between Siliceous Concretions from the Ionian Basin and the Apulian Platform Margins (Pre-Apulian Zone), Western Greece: Implication of Differential Diagenesis on Nodules Evolution. *Minerals* **2021**, *11*, 890. [CrossRef]
46. Ali, S.K.; Janjuhah, H.T.; Shahzad, S.M.; Kontakiotis, G.; Saleem, M.H.; Khan, U.; Zarkogiannis, S.D.; Makri, P.; Antonarakou, A. Depositional Sedimentary Facies, Stratigraphic Control, Paleoecological Constraints, and Paleogeographic Reconstruction of Late Permian Chhidru Formation (Western Salt Range, Pakistan). *J. Mar. Sci. Eng.* **2021**, *9*, 1372. [CrossRef]
47. Fazal, A.G.; Umar, M.; Shah, F.; Miraj, M.A.; Janjuhah, H.T.; Kontakiotis, G.; Jan, A.K. Geochemical Analysis of Cretaceous Shales from the Hazara Basin, Pakistan: Provenance Signatures and Paleo-Weathering Conditions. *J. Mar. Sci. Eng.* **2022**, *10*, 1654. [CrossRef]
48. Rahim, H.-u.; Qamar, S.; Shah, M.M.; Corbella, M.; Martín-Martín, J.D.; Janjuhah, H.T.; Navarro-Ciurana, D.; Lianou, V.; Kontakiotis, G. Processes Associated with Multiphase Dolomitization and Other Related Diagenetic Events in the Jurassic Samana Suk Formation, Himalayan Foreland Basin, NW Pakistan. *Minerals* **2022**, *12*, 1320. [CrossRef]
49. Zaheer, M.; Khan, M.R.; Mughal, M.S.; Janjuhah, H.T.; Makri, P.; Kontakiotis, G. Petrography and Lithofacies of the Siwalik Group in the Core of Hazara-Kashmir Syntaxis: Implications for Middle Stage Himalayan Orogeny and Paleoclimatic Conditions. *Minerals* **2022**, *12*, 1055. [CrossRef]
50. Bilal, A.; Mughal, M.S.; Janjuhah, H.T.; Ali, J.; Niaz, A.; Kontakiotis, G.; Antonarakou, A.; Usman, M.; Hussain, S.A.; Yang, R. Petrography and Provenance of the Sub-Himalayan Kuldana Formation: Implications for Tectonic Setting and Palaeoclimatic Conditions. *Minerals* **2022**, *12*, 794. [CrossRef]
51. Ruidas, D.K.; Pomoni-Papaioannou, F.A.; Banerjee, S.; Gangopadhyay, T.K. Petrographical and geochemical constraints on carbonate diagenesis in an epeiric platform deposit: Late Cretaceous Bagh Group in central India. *Carbonates Evaporites* **2020**, *35*, 94. [CrossRef]
52. Zambetakis-Lekkas, A.; Pomoni-Papaioannou, F.; Alexopoulos, A. Biostratigraphical and sedimentological study of Upper Senonian–Lower Eocene sediments of Tripolitza Platform in central Crete (Greece). *Cretac. Res.* **1998**, *19*, 715–732. [CrossRef]
53. Bilal, A.; Yang, R.; Mughal, M.S.; Janjuhah, H.T.; Zaheer, M.; Kontakiotis, G. Sedimentology and Diagenesis of the Early–Middle Eocene Carbonate Deposits of the Ceno-Tethys Ocean. *J. Mar. Sci. Eng.* **2022**, *10*, 1794.

54. Bilal, A.; Yang, R.; Fan, A.; Mughal, M.S.; Li, Y.; Basharat, M.; Farooq, M. Petrofacies and diagenesis of Thanetian Lockhart Limestone in the Upper Indus Basin (Pakistan): Implications for the Ceno-Tethys Ocean. *Carbonates Evaporites* **2022**, *37*, 78. [CrossRef]
55. IGME. *Geological Map of Greek Series, Paramythia Sheet, Scale 1:50.000*; Institute for Geology and Subsurface Research: Athens, Greece, 1966.
56. de Lapparent, J. *Etude Lithologique des Terrains Crétacés de la Région d'Hendaye*; Imprimerie Nationale: Paris, France, 1918.
57. Nakkady, S. A new foraminiferal fauna from the Esna shales and upper Cretaceous Chalk of Egypt. *J. Paleontol.* **1950**, *24*, 675–692.
58. Huber, B.T.; MacLeod, K.G.; Norris, R.D. Abrupt extinction and subsequent reworking of Cretaceous planktonic Foraminifera across the Cretaceous-Tertiary boundary: Evidence from the subtropical North Atlantic. In *Catastrophic Events and Mass Extinctions: Impacts and Beyond*; Koeberl, C., MacLeod, K.G., Eds.; Geological Society of America: Boulder, CO, USA, 2002; Volume 356.
59. Beiranvand, B.; Zaghbib-Turki, D.; Ghasemi-Nejad, E. Integrated biostratigraphy based on planktonic foraminifera and dinoflagellates across the Cretaceous/Paleogene (K/Pg) transition at the Izeh section (SW Iran). *Comptes Rendus Palevol* **2014**, *13*, 235–258. [CrossRef]
60. Scholle, P.A.; Ulmer-Scholle, D.S. *A Color Guide to the Petrography of Carbonate Rocks: Grains, Textures, Porosity, Diagenesis*; American Association of Petroleum Geologists: Tulsa, OK, USA, 2003. [CrossRef]
61. Janjuhah, H.T.; Alansari, A. Offshore Carbonate Facies Characterization and Reservoir Quality of Miocene Rocks in the Southern Margin of South China Sea. *Acta Geol. Sin.—Engl. Ed.* **2020**, *94*, 1547–1561. [CrossRef]
62. Janjuhah, H.T.; Alansari, A.; Ghosh, D.P.; Bashir, Y. New approach towards the classification of microporosity in Miocene carbonate rocks, Central Luconia, offshore Sarawak, Malaysia. *J. Nat. Gas Geosci.* **2018**, *3*, 119–133. [CrossRef]
63. Tomašových, A.; Gallmetzer, I.; Haselmair, A.; Zuschin, M. Inferring time averaging and hiatus durations in the stratigraphic record of high-frequency depositional sequences. *Sedimentology* **2022**, *69*, 1083–1118. [CrossRef]
64. Sharifi-Yazdi, M.; Rahimpour-Bonab, H.; Tavakoli, V.; Nazemi, M.; Kamali, M.R. Linking diagenetic history to depositional attributes in a high-frequency sequence stratigraphic framework: A case from upper Jurassic Arab formation in the central Persian Gulf. *J. Afr. Earth Sci.* **2019**, *153*, 91–110. [CrossRef]
65. Janjuhah, H.T.; Salim, A.M.A.; Ghosh, D.P.; Wahid, A. Diagenetic Process and Their Effect on Reservoir Quality in Miocene Carbonate Reservoir, Offshore, Sarawak, Malaysia. In Proceedings of the ICIPEG 2016, Singapore, 31 January 2017; pp. 545–558.
66. Janjuhah, H.T.; Salim, A.M.A.; Ghosh, D.P. Sedimentology and reservoir geometry of the Miocene Carbonate deposits in Central Luconia, Offshore, Sarawak, Malaysia. *J. Appl. Sci.* **2017**, *17*, 153–170. [CrossRef]
67. Budd, D.A. The Relative Roles of Compaction and Early Cementation in the Destruction of Permeability in Carbonate Grainstones: A Case Study from the Paleogene of West-Central Florida, U.S.A. *J. Sediment. Res.* **2002**, *72*, 116–128. [CrossRef]
68. Morse, J.W.; Arvidson, R.S. The dissolution kinetics of major sedimentary carbonate minerals. *Earth-Sci. Rev.* **2002**, *58*, 51–84. [CrossRef]
69. Lambert, L.; Durlet, C.; Loreau, J.-P.; Marnier, G. Burial dissolution of micrite in Middle East carbonate reservoirs (Jurassic–Cretaceous): Keys for recognition and timing. *Mar. Pet. Geol.* **2006**, *23*, 79–92. [CrossRef]
70. Tucker, M.E.; Wright, V.P. *Carbonate Sedimentology*; Wiley-Blackwell: New York, NY, USA, 2009; p. 496.
71. Janjuhah, H.T.; GÁMez Vintaned, J.A.; Salim, A.M.A.; Faye, I.; Shah, M.M.; Ghosh, D.P. Microfacies and Depositional Environments of Miocene Isolated Carbonate Platforms from Central Luconia, Offshore Sarawak, Malaysia. *Acta Geol. Sin.—Engl. Ed.* **2017**, *91*, 1778–1796. [CrossRef]
72. Morse, J.W.; Arvidson, R.S.; Lüttge, A. Calcium Carbonate Formation and Dissolution. *Chem. Rev.* **2007**, *107*, 342–381. [CrossRef] [PubMed]
73. Janjuhah, H.T.; Salim, A.M.A.; Shah, M.M.; Ghosh, D.; Alansari, A. Quantitative interpretation of carbonate reservoir rock using wireline logs: A case study from Central Luconia, offshore Sarawak, Malaysia. *Carbonates Evaporites* **2017**, *32*, 591–607. [CrossRef]
74. Amel, H.; Jafarian, A.; Husinec, A.; Koeshidayatullah, A.; Swennen, R. Microfacies, depositional environment and diagenetic evolution controls on the reservoir quality of the Permian Upper Dalan Formation, Kish Gas Field, Zagros Basin. *Mar. Pet. Geol.* **2015**, *67*, 57–71. [CrossRef]
75. Xi, K.; Cao, Y.; Jahren, J.; Zhu, R.; Bjørlykke, K.; Haile, B.G.; Zheng, L.; Hellevang, H. Diagenesis and reservoir quality of the Lower Cretaceous Quantou Formation tight sandstones in the southern Songliao Basin, China. *Sediment. Geol.* **2015**, *330*, 90–107. [CrossRef]
76. Janjuah, H.T.; Sanjuan, J.; Alqudah, M.; Salah, M.K. Biostratigraphy, Depositional and Diagenetic Processes in Carbonate Rocks from Southern Lebanon: Impact on Porosity and Permeability. *Acta Geol. Sin.—Engl. Ed.* **2021**, *95*, 1668–1683. [CrossRef]
77. Abramovich, S.; Keller, G. High stress late Maastrichtian paleoenvironment: Inference from planktonic foraminifera in Tunisia. *Palaeogeogr. Palaeoclimatol. Palaeoecol.* **2002**, *178*, 145–164. [CrossRef]
78. Keller, G.; Adatte, T.; Tantawy, A.A.; Berner, Z.; Stinnesbeck, W.; Stueben, D.; Leanza, H.A. High stress late Maastrichtian—Early Danian palaeoenvironment in the Neuquén Basin, Argentina. *Cretac. Res.* **2007**, *28*, 939–960. [CrossRef]
79. Omaña, L.; Alencáster, G.; Torres Hernández, J.R.; López Doncel, R. Morphological Abnormalities and Dwarfism in Maastrichtian Foraminifera from the Cárdenas Formation, Valles-San Luis Potosí Platform, Mexico: Evidence of paleoenvironmental stress. *Boletín Soc. Geológica Mex.* **2012**, *64*, 305–318. [CrossRef]
80. Luciani, V. Planktonic foraminiferal turnover across the Cretaceous-Tertiary boundary in the Vajont valley (Southern Alps, northern Italy). *Cretac. Res.* **1997**, *18*, 799–821. [CrossRef]

81. Arenillas, I.; Arz, J.A.; Gilabert, V. Blooms of aberrant planktic foraminifera across the K/Pg boundary in the Western Tethys: Causes and evolutionary implications. *Paleobiology* **2018**, *44*, 460–489. [CrossRef]
82. Coccioni, R.; Luciani, V. Guembelitria irregularis bloom at the KT boundary: Morphological abnormalities induced by impact-related extreme environmental stress? In *Biological Processes Associated with Impact Events*; Springer: Berlin/Heidelberg, Germany, 2006; pp. 179–196.
83. Coccioni, R.; Luciani, V.; Marsili, A. Cretaceous oceanic anoxic events and radially elongated chambered planktonic foraminifera: Paleoecological and paleoceanographic implications. *Palaeogeogr. Palaeoclimatol. Palaeoecol.* **2006**, *235*, 66–92. [CrossRef]
84. Verga, D.; Premoli Silva, I. Early Cretaceous planktonic foraminifera from the Tethys: The genus Leupoldina. *Cretac. Res.* **2002**, *23*, 189–212. [CrossRef]
85. Verga, D.; Premoli Silva, I. Early Cretaceous planktonic foraminifera from the Tethys: The small, few-chambered representatives of the genus Globigerinelloides. *Cretac. Res.* **2003**, *24*, 305–334. [CrossRef]
86. Luciani, V.; Giusberti, L.; Agnini, C.; Backman, J.; Fornaciari, E.; Rio, D. The Paleocene–Eocene Thermal Maximum as recorded by Tethyan planktonic foraminifera in the Forada section (northern Italy). *Mar. Micropaleontol.* **2007**, *64*, 189–214. [CrossRef]
87. Corbí, H.; Soria, J.M.; Lancis, C.; Giannetti, A.; Tent-Manclús, J.E.; Dinarès-Turell, J. Sedimentological and paleoenvironmental scenario before, during, and after the Messinian Salinity Crisis: The San Miguel de Salinas composite section (western Mediterranean). *Mar. Geol.* **2016**, *379*, 246–266. [CrossRef]
88. Vasiliev, I.; Karakitsios, V.; Bouloubassi, I.; Agiadi, K.; Kontakiotis, G.; Antonarakou, A.; Triantaphyllou, M.; Gogou, A.; Kafousia, N.; de Rafélis, M.; et al. Large Sea Surface Temperature, Salinity, and Productivity-Preservation Changes Preceding the Onset of the Messinian Salinity Crisis in the Eastern Mediterranean Sea. *Paleoceanogr. Paleoclimatol.* **2019**, *34*, 182–202. [CrossRef]
89. Karakitsios, V.; Roveri, M.; Lugli, S.; Manzi, V.; Gennari, R.; Antonarakou, A.; Triantaphyllou, M.; Agiadi, K.; Kontakiotis, G.; Kafousia, N.; et al. A record of the Messinian salinity crisis in the eastern Ionian tectonically active domain (Greece, eastern Mediterranean). *Basin Res.* **2017**, *29*, 203–233. [CrossRef]
90. Wade, B.S.; Olsson, R.K. Investigation of pre-extinction dwarfing in Cenozoic planktonic foraminifera. *Palaeogeogr. Palaeoclimatol. Palaeoecol.* **2009**, *284*, 39–46. [CrossRef]
91. Antonarakou, A.; Kontakiotis, G.; Zarkogiannis, S.; Mortyn, P.G.; Drinia, H.; Koskeridou, E.; Anastasakis, G. Planktonic foraminiferal abnormalities in coastal and open marine eastern Mediterranean environments: A natural stress monitoring approach in recent and early Holocene marine systems. *J. Mar. Syst.* **2018**, *181*, 63–78. [CrossRef]
92. Pardo, A.; Keller, G. Biotic effects of environmental catastrophes at the end of the Cretaceous and early Tertiary: Guembelitria and Heterohelix blooms. *Cretac. Res.* **2008**, *29*, 1058–1073. [CrossRef]
93. Agiadi, K.; Antonarakou, A.; Kontakiotis, G.; Kafousia, N.; Moissette, P.; Cornée, J.-J.; Manoutsoglou, E.; Karakitsios, V. Connectivity controls on the late Miocene eastern Mediterranean fish fauna. *Int. J. Earth Sci.* **2017**, *106*, 1147–1159. [CrossRef]
94. Kaiho, K. Planktonic and benthic foraminiferal extinction events during the last 100 m.y. *Palaeogeogr. Palaeoclimatol. Palaeoecol.* **1994**, *111*, 45–71. [CrossRef]
95. Twitchett, R.J. The palaeoclimatology, palaeoecology and palaeoenvironmental analysis of mass extinction events. *Palaeogeogr. Palaeoclimatol. Palaeoecol.* **2006**, *232*, 190–213. [CrossRef]
96. Li, L.; Keller, G. Maastrichtian climate, productivity and faunal turnovers in planktic foraminifera in South Atlantic DSDP sites 525A and 21. *Mar. Micropaleontol.* **1998**, *33*, 55–86. [CrossRef]
97. Li, L.; Keller, G. Abrupt deep-sea warming at the end of the Cretaceous. *Geology* **1998**, *26*, 995–998. [CrossRef]
98. Abramovich, S.; Keller, G. Planktonic foraminiferal response to the latest Maastrichtian abrupt warm event: A case study from South Atlantic DSDP Site 525A. *Mar. Micropaleontol.* **2003**, *48*, 225–249. [CrossRef]
99. Twitchett, R.J. Incompleteness of the Permian–Triassic fossil record: A consequence of productivity decline? *Geol. J.* **2001**, *36*, 341–353. [CrossRef]
100. Castle, J.W.; Rodgers, J.H., Jr. Hypothesis for the role of toxin-producing algae in Phanerozoic mass extinctions based on evidence from the geologic record and modern environments. *Environ. Geosci.* **2009**, *16*, 1–23. [CrossRef]
101. Jiang, S.; Bralower, T.J.; Patzkowsky, M.E.; Kump, L.R.; Schueth, J.D. Geographic controls on nannoplankton extinction across the Cretaceous/Palaeogene boundary. *Nat. Geosci.* **2010**, *3*, 280–285. [CrossRef]
102. Schueth, J.D.; Bralower, T.J.; Jiang, S.; Patzkowsky, M.E. The role of regional survivor incumbency in the evolutionary recovery of calcareous nannoplankton from the Cretaceous/Paleogene (K/Pg) mass extinction. *Paleobiology* **2015**, *41*, 661–679. [CrossRef]
103. Brinkhuis, H.; Bujak, J.P.; Smit, J.; Versteegh, G.J.M.; Visscher, H. Dinoflagellate-based sea surface temperature reconstructions across the Cretaceous–Tertiary boundary. *Palaeogeogr. Palaeoclimatol. Palaeoecol.* **1998**, *141*, 67–83. [CrossRef]
104. MacLeod, N.; Keller, G. Comparative biogeographic analysis of planktic foraminiferal survivorship across the Cretaceous/Tertiary (K/T) boundary. *Paleobiology* **2016**, *20*, 143–177. [CrossRef]
105. Pardo, A.; Adatte, T.; Keller, G.; Oberhänsli, H. Paleoenvironmental changes across the Cretaceous–Tertiary boundary at Koshak, Kazakhstan, based on planktic foraminifera and clay mineralogy. *Palaeogeogr. Palaeoclimatol. Palaeoecol.* **1999**, *154*, 247–273. [CrossRef]
106. Raup, D.M.; Sepkoski, J.J. Mass Extinctions in the Marine Fossil Record. *Science* **1982**, *215*, 1501–1503. [CrossRef]
107. Paul, C.R.C. Interpreting bioevents: What exactly did happen to planktonic foraminifers across the Cretaceous–Tertiary boundary? *Palaeogeogr. Palaeoclimatol. Palaeoecol.* **2005**, *224*, 291–310. [CrossRef]

108. Keller, G. The end-cretaceous mass extinction in the marine realm: Year 2000 assessment. *Planet. Space Sci.* **2001**, *49*, 817–830 [CrossRef]
109. MacLeod, N. Impacts and marine invertebrate extinctions. *Geol. Soc. Lond. Spec. Publ.* **1998**, *140*, 217–246. [CrossRef]
110. Alegret, L.; Molina, E.; Thomas, E. Benthic foraminiferal turnover across the Cretaceous/Paleogene boundary at Agost (southeastern Spain): Paleoenvironmental inferences. *Mar. Micropaleontol.* **2003**, *48*, 251–279. [CrossRef]
111. Coccioni, R.; Marsili, A. The response of benthic foraminifera to the K–Pg boundary biotic crisis at Elles (northwestern Tunisia). *Palaeogeogr. Palaeoclimatol. Palaeoecol.* **2007**, *255*, 157–180. [CrossRef]
112. Coccioni, R.; Galeotti, S. What happened to small benthic foraminifera at the Cretaceous/Tertiary boundary? *Bull. Soc. Geol. Fr.* **1998**, *169*, 271–279.
113. D'Hondt, S.; Donaghay, P.; Zachos, J.C.; Luttenberg, D.; Lindinger, M. Organic Carbon Fluxes and Ecological Recovery from the Cretaceous-Tertiary Mass Extinction. *Science* **1998**, *282*, 276–279. [CrossRef]
114. Barrera, E.; Keller, G. Productivity across the Cretaceous/Tertiary boundary in high latitudes. *GSA Bull.* **1994**, *106*, 1254–1266. [CrossRef]
115. Birch, H.S.; Coxall, H.K.; Pearson, P.N.; Kroon, D.; Schmidt, D.N. Partial collapse of the marine carbon pump after the Cretaceous-Paleogene boundary. *Geology* **2016**, *44*, 287–290. [CrossRef]
116. Zachos, J.C.; Arthur, M.A.; Dean, W.E. Geochemical evidence for suppression of pelagic marine productivity at the Cretaceous/Tertiary boundary. *Nature* **1989**, *337*, 61–64. [CrossRef]
117. Coxall, H.K.; D'Hondt, S.; Zachos, J.C. Pelagic evolution and environmental recovery after the Cretaceous-Paleogene mass extinction. *Geology* **2006**, *34*, 297–300. [CrossRef]
118. Schulte, P.; Alegret, L.; Arenillas, I.; Arz, J.A.; Barton, P.J.; Bown, P.R.; Bralower, T.J.; Christeson, G.L.; Claeys, P.; Cockell, C.S.; et al. The Chicxulub Asteroid Impact and Mass Extinction at the Cretaceous-Paleogene Boundary. *Science* **2010**, *327*, 1214–1218. [CrossRef] [PubMed]
119. D'Hondt, S.; Herbert, T.D.; King, J.; Gibson, C. Planktic foraminifera, asteroids, and marine production: Death and recovery at the Cretaceous-Tertiary boundary. In *The Cretaceous-Tertiary Event and Other Catastrophes in Earth History*; Ryder, G., Fastovsky, D.E., Gartner, S., Eds.; Geological Society of America: Boulder, CO, USA, 1996; Volume 307.
120. Smith, A.B.; Jeffery, C.H. Selectivity of extinction among sea urchins at the end of the Cretaceous period. *Nature* **1998**, *392*, 69–71. [CrossRef]
121. Brett, C.E.; Brookfield, M.E. Morphology, faunas and genesis of ordovician hardgrounds from Southern Ontario, Canada. *Palaeogeogr. Palaeoclimatol. Palaeoecol.* **1984**, *46*, 233–290. [CrossRef]
122. Adams, C.G. Neogene larger foraminifera, evolutionary and geological events in the context of datum planes. In *Pacific Neogene Datum Planes*; University of Tokyo Press: Tokyo, Japan, 1984; pp. 47–68.
123. Karakitsios, V.; Tsikos, H.; Van Breugel, Y.; Bakopoulos, I.; Koletti, L. Cretaceous Oceanic Anoxic events in western continental Greece. *Bull. Geol. Soc. Greece* **2004**, *36*, 846–855. [CrossRef]
124. Getsos, K.; Pomoni-Papaioannou, F.; Zelilidis, A. A carbonate ramp evolution in the transition from the Apulia platform to the Ionian Basin during Early to Late Cretaceous (NW Greece). *Bull. Geol. Soc. Greece* **2007**, *40*, 53–63. [CrossRef]
125. Spasojevic, S.; Gurnis, M. Sea level and vertical motion of continents from dynamic earth models since the Late Cretaceous. *AAPG Bull.* **2012**, *96*, 2037–2064. [CrossRef]
126. Miller, K.; Wright, J.; Katz, M.; Browning, J.; Cramer, B.; Wade, B.S.; Mizintseva, S. A view of Antarctic ice-sheet evolution from sea-level and deep-sea isotope changes during the Late Cretaceous–Cenozoic. In *Antarctica: A Keystone in a Changing World*; National Academies Press: Washington, DC, USA, 2008; pp. 55–70.
127. Tsikos, H.; Karakitsios, V.; Breugel, Y.V.; Walsworth-Bell, B.; Bombardiere, L.; Petrizzo, M.R.; Damsté, J.S.S.; Schouten, S.; Erba, E.; Silva, I.P.; et al. Organic-carbon deposition in the Cretaceous of the Ionian Basin, NW Greece: The Paquier Event (OAE 1b) revisited. *Geol. Mag.* **2004**, *141*, 401–416. [CrossRef]
128. Tucker, M.E. *Sedimentary Petrology: An Introduction to the Origin of Sedimentary Rocks*; John Wiley & Sons: Hoboken, NJ, USA, 2009.
129. Wilson, M.E.J.; Moss, S.J. Cenozoic palaeogeographic evolution of Sulawesi and Borneo. *Palaeogeogr. Palaeoclimatol. Palaeoecol.* **1999**, *145*, 303–337. [CrossRef]
130. Janjuhah, H.T.; Alansari, A.; Santha, P.R. Interrelationship Between Facies Association, Diagenetic Alteration and Reservoir Properties Evolution in the Middle Miocene Carbonate Build Up, Central Luconia, Offshore Sarawak, Malaysia. *Arab. J. Sci. Eng.* **2019**, *44*, 341–356. [CrossRef]
131. Flügel, E. *Microfacies Analysis of Limestones*; Springer Science & Business Media: Berlin/Heidelberg, Germany, 2012.
132. Hoang, A.N. Diagenetic evolution recorded from the fractured carbonates of Trang Kenh formation, in the north-eastern area of Vietnam. *Petrovietnam J.* **2017**, *6*, 20–28.

*Article*

# A Potential Beach Monitoring Based on Integrated Methods

Isabella Lapietra [1], Stefania Lisco [1,2,*], Luigi Capozzoli [3], Francesco De Giosa [4], Giuseppe Mastronuzzi [1,2], Daniela Mele [1], Salvatore Milli [5,6], Gerardo Romano [1], François Sabatier [7], Giovanni Scardino [1,2] and Massimo Moretti [1,2]

1. Dipartimento di Scienze della Terra e Geoambientali, Campus Universitario, Università degli Studi di Bari "Aldo Moro", Via Edoardo Orabona 4, 70125 Bari, Italy
2. Interdepartmental Research Centre for Coastal Dynamics, Campus Universitario, Università degli Studi di Bari "Aldo Moro", Via Edoardo Orabona 4, 70125 Bari, Italy
3. CNR-IMAA, National Research Council, Institute of Methodologies for Environmental Analysis, 85050 Tito, Italy
4. Environmental Surveys Srl, Via Dario Lupo 65, 74121 Taranto, Italy
5. Dipartimento di Scienze della Terra, Sapienza Università di Roma, Piazzale Aldo Moro 5, 00185 Rome, Italy
6. CNR-IGAG, National Research Council, Institute of Environmental Geology and Geoengineering, Montelibretti, 00010 Rome, Italy
7. Aix-Marseille Université CNRS, IRD, Coll France, CEREGE, Technopôle de l'Arbois-Méditerranée BP80, 13545 Aix-en-Provence, France
* Correspondence: stefania.lisco@uniba.it

**Abstract:** This study focuses on the analysis of sandy beaches by integrating sedimentological, geomorphological, and geophysical investigations. The beach represents an extremely variable environment where different natural processes act simultaneously with human activities, leading to the gathering of different methodologies of the Earth Sciences to study its evolution in space and time. The aim of this research is to propose a potential procedure for monitoring the morpho-sedimentary processes of sandy beaches by analyzing the textural and compositional characteristics of the sands and quantifying the volumes involved in the coastal dynamics. The study area includes two Apulian sandy beaches (Torre Guaceto and Le Dune beach) that are representative of the coastal dynamics of a large sector of the central/northern Mediterranean Sea involving the southern Adriatic Sea and the northern Ionian Sea. Sedimentological and ecological investigations allowed to describe the textural and compositional characteristics of the beach sands by interpreting their sand provenance and the physical/biological interactions within the beach. The topographic surveys carried out with a Terrestrial Laser Scanner and an Optical Total Station, aimed to quantify the variations of sediment volume over time, whereas the Delft3d software was applied to analyze the effects of the dominant wave motion on the sedimentary dynamics. Lastly, the geophysical techniques which included Sub Bottom Profiler procedures, Ground Penetrating Radar investigation, and resistivity models enabled us to calculate the sand sediment thickness above the bedrock.

**Keywords:** pocket beach; beach monitoring; beach dynamics; sediment thickness

## 1. Introduction

Coastal zones can be defined as complex natural ecosystems where hydrodynamic, sedimentary, morphological, and biological conditions and human disturbance interact at different spaces and time scales [1–3]. Coasts are often seen as fundamental resources to be "exploited", especially for touristic and economic purposes. However, social interests, economic investments, and the protection of natural ecosystems must meet the requirements of Integrated Coastal Zone Management (ICZM), which considers the fragility of coastal ecosystems and landscapes.

As reported from the latest United Nations reports, about 37% of the world's population lives within 100 km of the coast. This area is settled by a wide range of activities

**Citation:** Lapietra, I.; Lisco, S.; Capozzoli, L.; De Giosa, F.; Mastronuzzi, G.; Mele, D.; Milli, S.; Romano, G.; Sabatier, F.; Scardino, G.; et al. A Potential Beach Monitoring Based on Integrated Methods. *J. Mar. Sci. Eng.* **2022**, *10*, 1949. https://doi.org/10.3390/jmse10121949

**Academic Editors:** George Kontakiotis, Angelos G. Maravelis and Avraam Zelilidis

Received: 11 November 2022
Accepted: 1 December 2022
Published: 8 December 2022

**Publisher's Note:** MDPI stays neutral with regard to jurisdictional claims in published maps and institutional affiliations.

**Copyright:** © 2022 by the authors. Licensee MDPI, Basel, Switzerland. This article is an open access article distributed under the terms and conditions of the Creative Commons Attribution (CC BY) license (https://creativecommons.org/licenses/by/4.0/).

and infrastructures that have significantly changed the natural response of littoral sectors to extreme events and sea level rise. Indeed, the scientific community agrees on the increasing exposure of coastal areas to the hazard of erosion due to climate change [4,5] intensifying coast vulnerabilities [6–10], and the attention of decision-makers to find sustainable solutions for their protection, disaster risk reduction, and building community resilience [11–14].

The most threatened regions are Europe with 86% and Asia with 69% of their coastal ecosystems at risk [15]. In Europe, most of the impact zones (15,100 km) are actively retreating, some of them despite coastal protection works (2900 km), and another 4700 km have become artificially stabilized [16]. Only the magnitude and nature of erosion change from place to place.

Several European research projects (Eurosion, Conscience, Micore, Peseta, Coastgap) focus on sustainable management along Mediterranean coastal areas. These littoral sectors are mainly wave-dominated and the effect of tides on sedimentation is negligible (microtidal regime). Mediterranean sandy beaches are characterized by terrigenous sediments coming from delta deposits and/or from the erosion of cliffs or headlands occurring in the same coastal site. However, a significant percentage of sandy beaches are constituted by carbonate bioclast components that derive from shells or fragments of the organisms that populate the proximal marine environment of the Mediterranean area [17–19]. Bioclasts become of primary importance in a geological context characterized by the absence of rivers and deltas.

Apulian sandy beaches (Southern Italy) represent a significant Mediterranean example of beach sand composition variability. The composition is connected to the typical coastal geomorphological and sedimentological features in karst areas, often lacking important fluvial courses capable of transporting large quantities of sediments into the sea [20]. At the regional level, many studies analyze the features of Apulian sandy beaches (Figure 1) also showing the processes concerning their susceptibility to erosion. From a granulometric and mineralogical point of view, beach sediments show significantly different features between the Ionian and Adriatic side [21]. Sediments are essentially bioclastic on the Ionian side as the littoral sector of Porto Cesareo (Figure 1 [20,22]) and terrigenous on the Adriatic Seaside [23–26] as it is detected along Torre Mileto, Rosa Marina, Pilone, Torre Canne, and Alimini coastal sectors (Figure 1 [17,21,27]). These beaches are mainly composed of fine- to coarse-grained sands characterized by carbonates (either lithoclasts or bioclasts), quartz, and other minerals such as pyroxene, amphibole, and feldspar [18]. In terms of evolutionary tendency, stable conditions are recorded along the Ionian coast [28–30], whereas retreating tendencies are detected along the Adriatic littoral stretch [17,18,25,31–37].

Evaluating the evolutionary tendency of a beach (accretion/erosion) requires various monitoring techniques, applied to the emerged and submerged sectors, in order to quantify the volume involved in coastal dynamics. In the specific literature, primary studies aimed at the description of sand movements along beaches [38], the recognition of sand textural parameters as a tool to evaluate the health state of the beach [39], the concept of equilibrium profile [40], and the role of dissipative and reflective characteristics [41] for a morphodynamic beach classification [42,43]. Later, the study of sandy beaches experienced significant progress. Table 1 represents an overview of the different studies carried out on the beach environment. The list of authors provides an example of the large number of techniques for beach investigation.

Recently, the implementation of different methodologies contributes to gain relevant information for sandy beach nourishment interventions [83–89] and leads to necessarily integrate the single techniques within an interdisciplinary approach. This type of analysis could provide more manageable and focused actions for safeguarding, protecting, and restoring sandy littoral sectors [17,90–92].

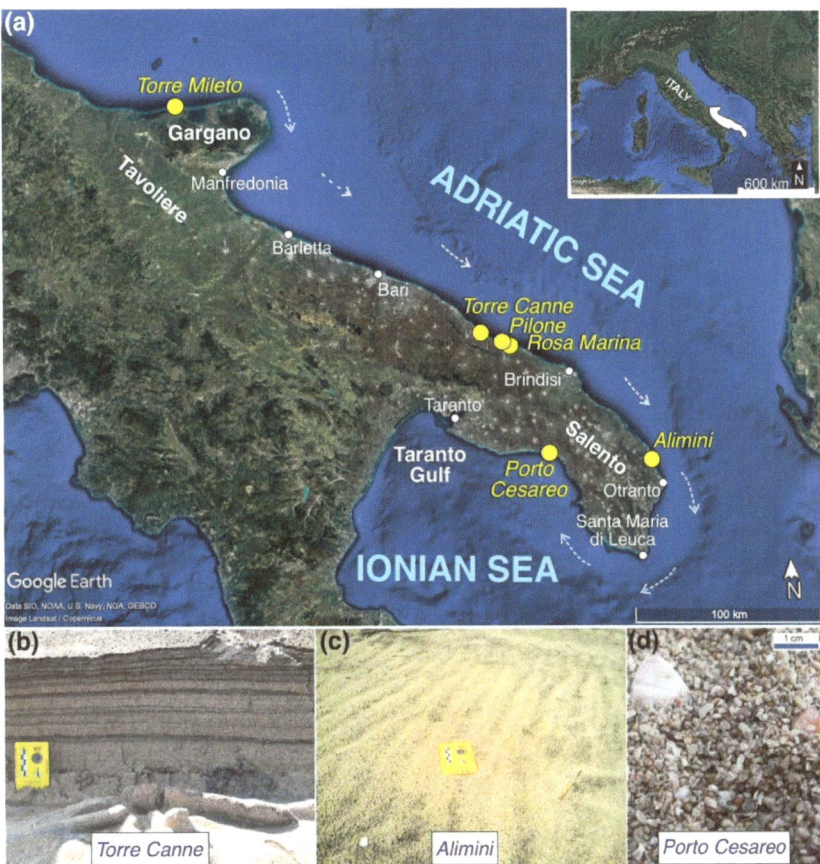

**Figure 1.** (**a**) Geographical location of a few sandy beaches (yellow circle) characterizing the Apulian littoral stretch. The white arrows indicate the general alongshore current influencing the sediment transport on the Adriatic and the Ionian Sea; (**b**–**d**) Macroscopic features variability of the Apulian beach sands in terms of color and grain size (modified from [20]).

For this reason, this work focuses on the mechanism of erosion, transport, and sedimentation of two Apulian sandy littoral sectors through a multidisciplinary approach. We suggest a reliable guideline for monitoring sandy beaches aiming to:

1. a description of the textural and compositional features of beach sands;
2. a short-term erosional/prograding tendency evaluation;
3. an overall sketch scheme of beach dynamics;
4. a measurement of the beach sediment thickness above the bedrock and therefore of the sand amount virtually involved in the beach sedimentary dynamics.

The research proposes a methodological approach that could be useful for scientific communities or private/public coastal management. After a brief description of the study areas, the main part of the research focuses on testing the procedures for beach investigation. In the final paragraphs, a suggestion regarding data management is also provided by using the result of this research as an example of data interpretation.

The possibility to obtain measurements through modern technologies in an interdisciplinary framework represents a development in the reliable understanding of the interactions between complex physical and biological processes occurring in beach environments.

**Table 1.** Techniques for beach monitoring applied in the different sectors of the Earth Sciences. Each color represents a scientific sector with the relative authors. GPR: Ground Penetrating Radar, SSS: Side Scan Sonar; SBP: Sub Bottom Profiler. Geomorphology [44–53]; Sedimentology [54–61]; Sedimentology + Biology [17,27,62–70]; Geophysics [71–82].

| Scientific Sector | Author | Methodology |
|---|---|---|
| Geomorphology | Ciavola et al., 2003<br>Gracia et al., 2005<br>Costas et al., 2006<br>Pranzini, 2008<br>Anfuso et al., 2011<br>Nordstrom et al., 2015<br>Karkani et al., 2017<br>Thom and Hall, 1991<br>Almeida et al., 2010<br>Riazi and Türcker, 2017 | Shoreline variation through aerial photos<br><br>Beach profile variations through topographic surveys |
| Sedimentology | Gao and Collins, 1993<br>Guillen and Palanques, 1996<br>Dawe, 2001<br>Poizot et al., 2008<br>Falk and Ward, 1957<br>Visher, 1965<br>Friedman, 1967<br>Edwards, 2001 | Grain size distribution and sediment trend analysis |
| Sedimentology + Biology | De Falco et al., 2002; 2003<br>Satta et al., 2013<br>Moretti et al., 2016<br>Lisco et al., 2017<br>De Falco et al., 2008<br>Short et al., 2007<br>Brandano et al., 2016<br>Gaglianone et al., 2014;2017<br>Simeone et al., 2018 | Ecological approach for coastal monitoring<br><br>*Posidonia oceanica* monitoring and bioclast evaluation |
| Geophysics | Leatherman, 1987<br>Bristow et al., 2000<br>Neal and Roberts, 2000<br>Neal, 2004<br>Hugenoltz et al., 2007<br>Guillemoteau, 2012<br>Shukla et al., 2013<br>Bristow and Jol, 2020<br>Morang et al., 1997<br>Lubis et al., 2017<br>De Giosa et al., 2019<br>Kim et al., 2020 | GPR procedure for emerged beach and dune environment exploration<br><br>SSS and SBP for sea bottom and sediment thickness investigation |

## 2. Study Area

The study area includes two Apulian sandy littoral stretches (Figure 2a): Torre Guaceto (40,71 lat. N; 17,77 long. E) and Le Dune beach (40,27 lat. N; 17,87 long. E) occurring in the Adriatic and Ionian sectors of the Salento Peninsula, respectively (Brindisi and Lecce provinces). Both beaches are in two of the most important Marine Protected Areas (MPA) which include 15 different habitats of the typical Mediterranean submerged populations and the presence of *Posidonia oceanica* meadows. They can be considered endmembers of the coastal dynamics of a large sector of the central/northern Mediterranean Sea involving the southern Adriatic Sea and northern Ionian Sea with negligible anthropogenic impact.

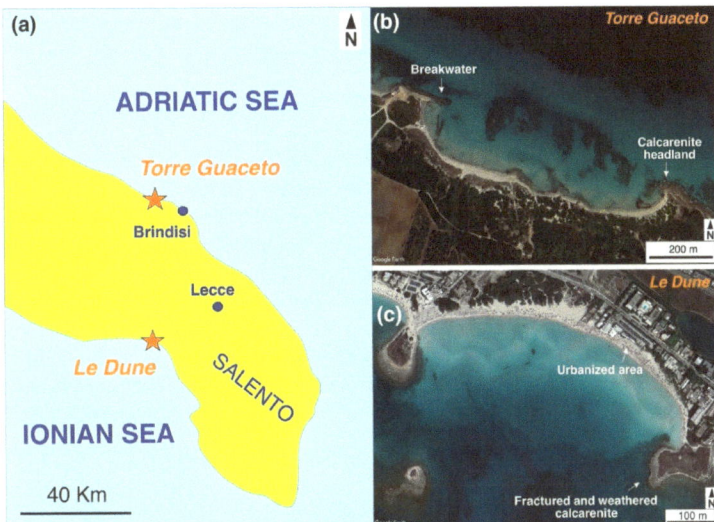

**Figure 2.** Coastal features of Torre Guaceto and Le Dune beach. (**a**) Geographical location of the study areas (orange star); (**b**) satellite image of Torre Guaceto beach showing a landward well developed dune belt; (**c**) satellite image of Le Dune beach showing a large, urbanized area on the southernmost sector and a developed dune environment on the northwestern sector.

Torre Guaceto (Figure 2b) is a pocket beach that stretches for about 1 km. The embayment is delimited by two rocky headlands: the south-eastern Lower Pleistocene shallow-water skeletal calcarenites (Punta Penna Grossa outcrop) and an anthropogenic structure of the 20th century in the north-westward [93]. The area is characterized by high coastal dunes and a rocky and sandy sea bottom. The beach is affected by a slight human impact, and it can be representative of sandy beaches lying on the northern coastal sector of Brindisi in terms of wind and wave weather, sand composition, and granulometric features. Beach sediments range from coarse to fine sands mainly composed of carbonate grains (bioclasts or lithoclasts) and siliciclastic minerals. Wave directions are mainly between 330° (16.79%) and 120° (15.36%). It is also observed that the highest frequency of appearance (22.15%) and the highest significant wave heights (Hs between 3 m and 4 m) derive from the 0° direction [94].

On the contrary, Le Dune beach (Figure 2c) is located along the Ionian Seaside and stretches for about 800 m. The small embayment has a slight Z-shaped form with an increased sediment accumulation on its western limit. Beach sediments range from very coarse to medium-fine sands and they are mainly made up of bioclast fragments. The pocket beach is bounded landward by aeolian dunes and by promontories where Cretaceous limestone crops out. At deeper depths, the shelf extends between the isobaths at −5 m and 100–110 m, and it is predominantly covered by bioclastic sands. Within the inner portion of the shelf, *Posidonia oceanica* meadows form large patches which are replaced at depth by coralligenous platform deposits [95]. Le Dune can be considered representative of the Ionian sandy beaches that have experienced a remarkable urbanization of the coastal areas in the last decades. Generally, wave directions are between 150° (15.91%) and 300° (8.47%), and seas from the south (24.48%) and southeast (15.91%) have the highest frequencies [96]. A rip current phenomenon is often observed during storm events [22].

## 3. Materials and Method

Since this work is based upon a multidisciplinary approach, some of the main techniques used in the field of sedimentology, petrography, biogeology–paleontology, geomorphology, and applied geophysics were used for beach investigation. However, according to the objectives to be achieved, some procedures were constructed within the current research in order to obtain a detailed outline of beach dynamics and sediment thickness quantification.

## 3.1. Beach Sand Analysis

### 3.1.1. Sampling

Sampling procedures were carried out every six months along the backshore, foreshore, and shoreface sectors (example in Figure 3). Ground control points were used as collection points to compare the texture and the composition of the sand and to record the sampling depth variations over time. Samplings were carried out from the base of the foredune until 6 m depth of the shoreface (the local storm-wave base in the Adriatic and the Ionian Sea). Along the shoreline, the collecting points were spaced at around 100–200 m from each other, whereas shoreface samples were collected at each meter depth through diving techniques along two or three cross-shore transects. Around 300 gr of sand was collected between 0 and 2 cm depth down from the water/sediment interface by following the standard sampling procedure for marine sediments proposed by [97].

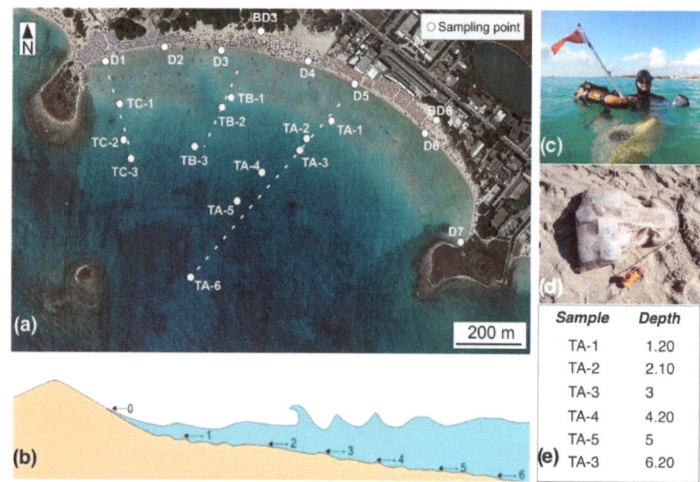

**Figure 3.** (**a**) Map showing the sampling collecting points. D = shoreline sample label, BD = foredune base sample, TA-1 = Sample collected along transect A at 1 m depth; (**b**) Ideal submerged beach profile (not in scale) with the indication of the sampling depths; (**c**) collection of shoreface samples; (**d**) small net for marine samples; (**e**) whiteboard for sample marine marks.

### 3.1.2. Grain-Size Analysis

The grain-size analyses were carried out by using the standard procedures provided by the American Society for Testing and Materials and the British Standard. For the sieving, a set of ASTM sieves with meshes of $\frac{1}{2}$ phi from 2 mm to the minimum granulometric fraction (<0.125 mm) was used. The grain fractions with diameters less than 0.062 mm were excluded from the analysis because they were represented by less than 1–2%. In the laboratory, samples were dried in the oven at a temperature of 80° for 24 h and each individual sample was quartered and set in a sieve column. The sand sediments from 2.0 mm to 0.125 mm were sieved with the vibrating screen for a duration of 20 min. Subsequently, each held fraction was weighed and the results were processed with a specific application for Microsoft Excel (Gradistat© v8, Figure 4), which yield distribution cumulative curves, histograms, and statistically evaluate the main textural parameters: mean size ($M_z$), sorting ($\sigma$), skewness ($S_k$), and Kurtosis ($K_G$) (Figure 4). Lastly, a seasonal comparison of the granulometric parameters was carried out in order to evaluate their temporal changes.

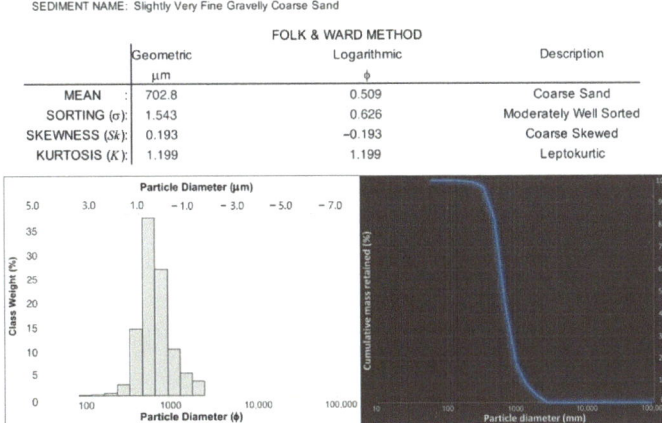

**Figure 4.** Example of a grain-size analysis using Gradistat© (v8).

### 3.1.3. Compositional Analysis and Bioclasts Quantification/Recognition

The most frequent size class of the statistical distribution was investigated through a binocular optical microscope (Figure 5a). The percentages of the main constituents of sands were evaluated to obtain quantitative and qualitative information in terms of petrographic composition. Since the investigation was carried out on Apulian beach sands, which generally include carbonates (lithoclasts or bioclasts) and siliciclastic minerals, the sands were analyzed by considering the diagram proposed by [98,99]. This method provides the recognition of three main classes: CE carbonate extrabasinal (lithoclasts), CI carbonate intrabasinal (bioclasts), and NCE non-carbonate extrabasinal grains. In particular, the sediment fraction was spread on a rigid base divided into five fields (Figure 5b). Each field was analyzed in detail under the microscope (Figure 5c) and for each field, the particles belonging to the NCE, CE, and CI class were counted. Once the counting operation for each field was completed, the results obtained were summed and the percentage of each component was calculated and inserted in the final classification diagram (Figure 5d).

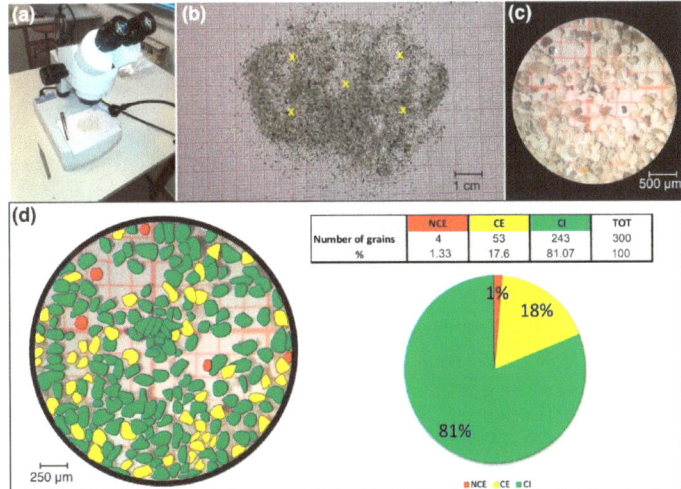

**Figure 5.** Example of hybrid sand composition analysis; (**a**) sand analysis at the microscope; (**b**) sample spread on a rigid base with the field locations; (**c**) field for counting method; (**d**) example of counting for sand classification, green: bioclasts, yellow: carbonate lithoclasts; red: quartz.

Furthermore, the whole shells and fragmented particles were isolated and analyzed by mean of a binocular optical microscope in more detail (example in Figure 6). In this context, a set of tweezers was used to manually separate the fragmented and unrecognizable fraction from the rest of the sample. The percentage was defined among three classes: 0–30%, 30–60%, 60–90%. The whole shells were analyzed and separated on the base of their Phylum to provide a first classification. The class and genus of shells were also evaluated in the case of foraminifers.

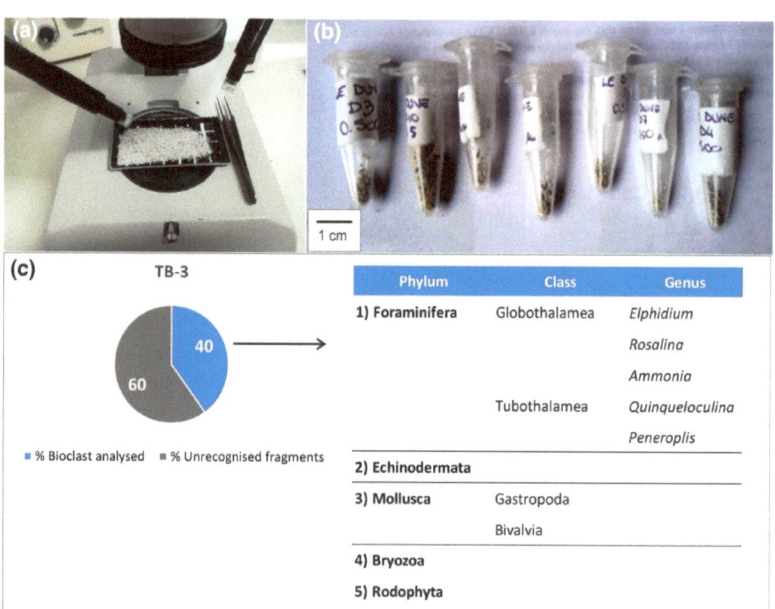

**Figure 6.** Example of bioclast evaluation procedure; (**a**) sample bioclast separation with a set of tweezers; (**b**) isolated bioclasts in test tubes; (**c**) bioclast classification.

The main bioclast classification was carried out on the first sampling season. Successively, the bioclast evaluation was conducted by analyzing any differences in the Phylum already identified.

It is essential to specify that the recognition of the nature of sands was relatively simpler than expected. It recorded a maximum of 3% of grains with uncertain nature mainly related to the distinction between lithoclasts and bioclasts without a recognizable internal structure. The sandy particles with uncertain origin were not considered, while fossils were included in the class of lithoclasts as they were considered part of older deposits.

### 3.2. Wave Simulations and Hydrodynamic Model

Software Delft3D was used to analyze the correlation between sedimentological parameters and wave processes. As shown in Figure 7, GRID, FLOW, and WAVE modules were applied to calculate the hydrodynamic flow and to simulate the wave motion. In particular, coupled simulations were performed by means of Delft3D-WAVE with Delft3D-FLOW to incorporate wave-current interaction. Delft3DFLOW was used to assess the hydrodynamic flow, and Delft3D-WAVE was used to simulate the wave propagation, based on the SWAN (Simulating WAves Nearshore) model. The model was based on two main datasets: bathymetry and climatic information (waves and wind data). The bathymetric data were provided by the EMODnet website (https://portal.emodnet-bathymetry.eu/, accessed on 1 November 2022). The wave data were acquired from SIT Puglia website (http://cartografia.sit.puglia.it/download/PRC/Relazione%20Generale_allegato711.pdf, accessed on 30 October 2022) and from a wave buoy located in the Ionian Sea (Datawell Directional

Waverider—Mk-III property of Autorità di Bacino Distrettuale dell'Appennino Meridionale, located at 40,40 N, 17,18 E). The bathymetric data were provided by the EMODnet website (https://portal.emodnet-bathymetry.eu/, accessed on 25 October 2022).

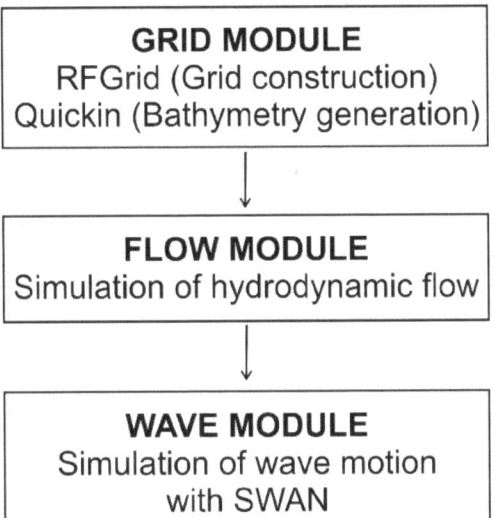

**Figure 7.** Flow chart of the processing steps followed in the application of Delft3D software.

### 3.3. Topographic Surveys

#### 3.3.1. The Terrestrial Laser Scanner (TLS)

The topography of the emerged beach sector was investigated using a Terrestrial Laser Scanner (TLS) that allows constructing 3D elevation models and quantifying the morphological changes over time. In particular, the instrument acquires the spatial coordinates of many points by measuring the distance between the TLS and the object of study in order to build point clouds and successive three-dimensional elevation models. Two repeated survey campaigns were carried out during the summer of 2018 and the summer of 2019 by using a Riegl VZ-400 laser scanner, which is characterized by a theoretical range of 400 m. The instrumental accuracy is ±0.003 m for 50 m (example in Figure 8a).

Due to the length of both beaches, several scans were carried out from four stations with a distance of about 250 m from each other. The point clouds were processed through different stages: scans registration; multi-scan adjustment and georeferencing; 3D point cloud cleaning; triangulation (mesh) and Digital Terrestrial Model DTM creation with a size cell of 50 cm. The Riscan Pro and CloudCompare software were used for point cloud processing (Figure 8b), filtering, and rasterization, while the elevation correction and the comparison of the seasonal DTMs were carried out with ArcMap © 10.1 (Figure 8c).

As the sampling procedure, the TLS acquisition was carried out every six months to detect the main morphological changes in the emerged sectors of the beach.

#### 3.3.2. The Optical Total Station (OTS)

With regards to the shoreface investigation, two profile surveys were performed using the Optical Total Station (OTS) Leica TS15 to collect seasonal bathymetric information. The OTS is composed of an electromagnetic distance measure instrument and electronic theodolite that is also integrated with a computer storage system (Figure 9). The instrument allows the measuring of horizontal and vertical angles as well as the sloping distance between the object and the instrument.

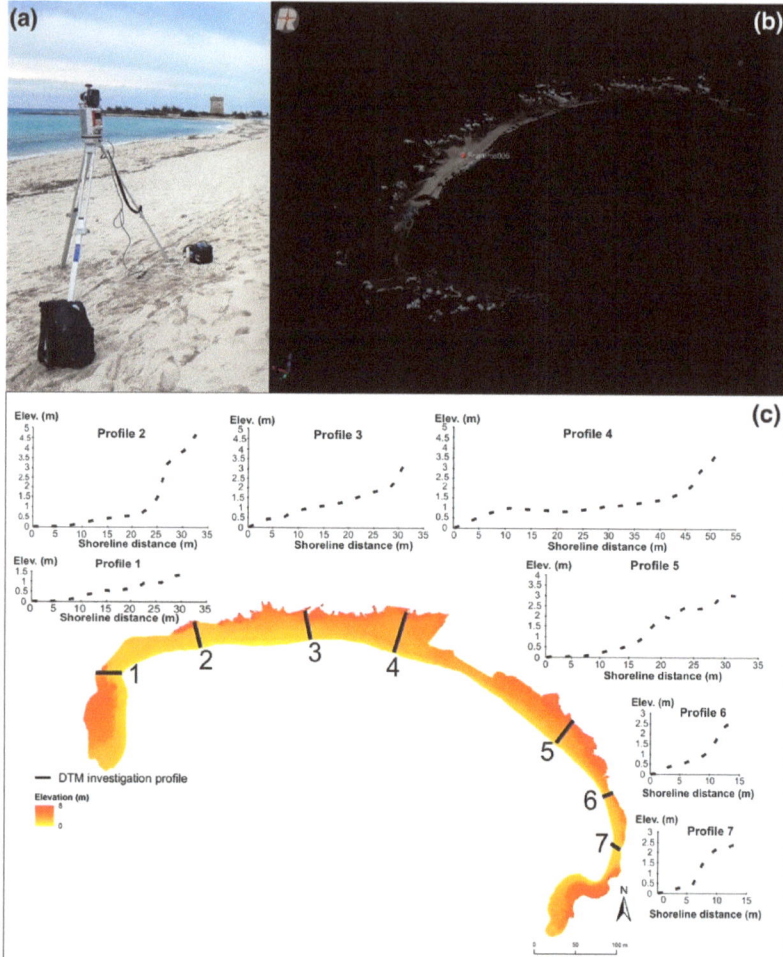

**Figure 8.** Emerged beach topography investigation. (**a**) TLS Riegl VZ-400 instrument during a survey campaign; (**b**) Example of a point cloud during a scan registration with Riscan Pro software: the red point represents the instrument location during the measurement acquisition; (**c**) Digital elevation model (DEM) of the emerged sector in summer 2018 processed with ArcMap© 10.1.

During the seasonal field surveys, the measurements were performed along two cross-shore transects in georeferenced points following the direction of the sampling profiles. About 20 bathymetric points were measured for each transect from the shoreline to a 4 m depth on both beaches.

*3.4. Geophysical Investigation*

3.4.1. Ground Penetrating Radar (GPR)

The GPR investigation was applied for evaluating the sediment thicknesses in the emerged beach sector. The choice of this technique depends on the dynamic nature of this sub-environment that constantly changes with the amount of available material for transport and sedimentation. Generally, these variations are recorded within the sand deposits and particularly in the dune environment, whose stratigraphic horizons can be explored with the GPR technique. The instrument operation is based on the input of high frequency electromagnetic pulses deriving from an antenna, which is placed in contact with the surface to be investigated. The depth of the investigation depends on the frequency used and therefore on the type of antenna. As the signal frequency increases, there is a growth in data resolution but a decreasing in the investigation depth.

**Figure 9.** Example of OTS field surveys during winter 2018/19. (**a**) OTS instrument with the reflective system (prism) for bathymetric point measurement; (**b**) map showing the bathymetric transect location; (**c**) bathymetric profiles extrapolated from the measurement carried out in winter 2018/19.

A survey campaign through a 200 Mhz antenna was carried out at Torre Guaceto and Le Dune beach (Figure 10a). The choice of this frequency depended on some tests carried out previously with 400 and 40 Mhz in order to detect the most significant radar stratigraphy section (Figure 10c). The data acquisition was performed along a cross-shore transect from the swash zone to the dune environment (Figure 10b).

With regards to the data processing, the Reflexw software was used to analyze and elaborate the 2D radar stratigraphic sections.

3.4.2. Sub Bottom Profiler (SBP)

As the GPR investigation, the sub bottom profiler was carried out to quantify the sand thicknesses within the submerged sector of the beach. This technique is widely applied in the field of marine geology. The SBP exploits the elastic properties of the ground to reconstruct the stratigraphic succession of deposits occurring below the seabed. Each surface that marks a lithological transition or any acoustic impedance such as the water/sediment passage represents an elastic discontinuity capable of reflecting part of the seismic energy. The reflected signal is received by a transducer and sent to the visualization program by creating a seismic section. The signal penetration and reflection depend on the frequency and the physical properties of the sediment.

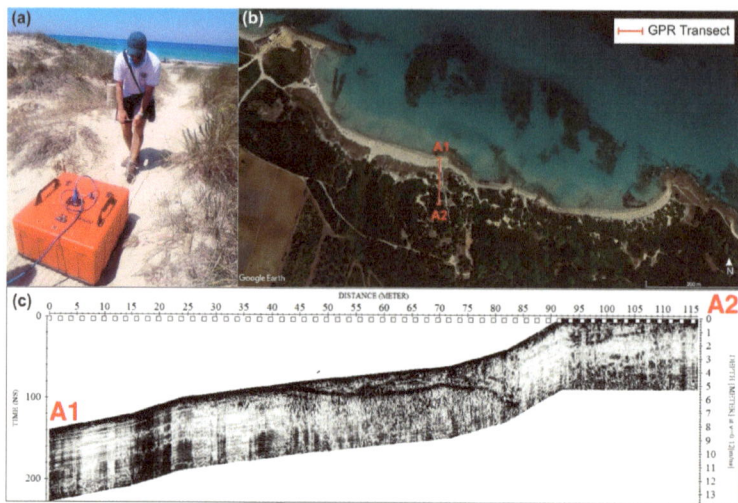

**Figure 10.** GPR procedure example. (**a**) 200 MHz antenna during a GPR field survey; (**b**) map showing the location of the measurement transect; (**c**) example of a radar stratigraphy section deriving from the measurement path A1–A2 by the use of 400 MHz antenna.

In this study case, one marine field survey was carried out for each study area by applying low signal frequencies (85–115 kHz). The investigation was organized by following the same cross-shore transect of the georadar investigation from 6–2 m depth (example in Figure 11).

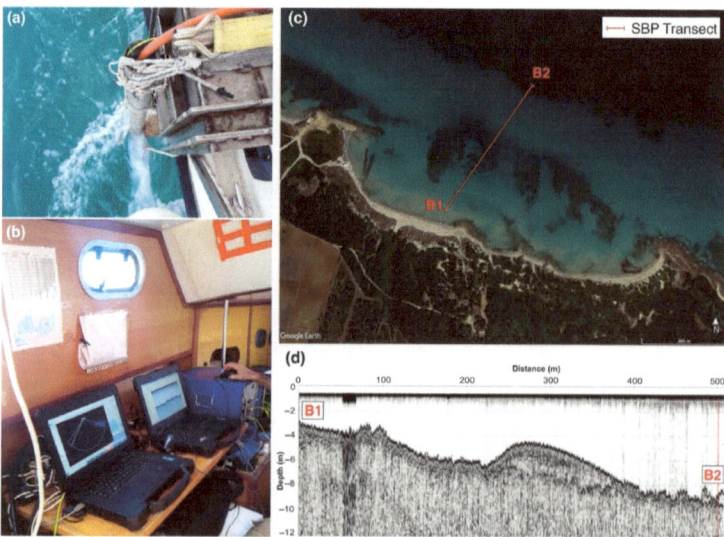

**Figure 11.** SBP field survey and data processing. (**a**) SBP transducer during the data acquisition; (**b**) hardware system and laptops for data monitoring; (**c**) map showing the navigation path; (**d**) 2-dimensional stratigraphic cross sections data processing.

Moreover, a boat capable of reaching shallow waters was utilized for the investigation and the instrumentation was characterized by an Innomar SBP SES 2000 Compact system. The SBP transducer was installed on one side of the boat (Figure 11a), while the hardware system and the data management laptop were positioned inside the boat. Moreover, the whole system was connected to a GPS RTK, and interfaced with the instrument management software to acquire the exact data location. Regarding data acquisition, SES WIN software was utilized to obtain seismic sections in real-time. The program produces a *.SES file which displays the width of the acoustic signal reflected at different depths; it also contains data relating to the longitude, latitude, depth of the seabed, and acquisition parameters. All these data were transferred to Delph Seismic software to analyze the 2-dimensional stratigraphic cross sections, define the bedrock location, and obtain strata thicknesses.

3.4.3. Geoelectric Investigation (ERT)

A geoelectric investigation was applied to analyze the entire beach profile and to gain data about the foreshore subenvironment. The application of a geoelectric investigation provides information relating to the subsoil electrical resistivity distribution.

As shown in Figure 12, the ERT was carried out across the same GPR cross-shore transect. The investigation profile was extended until 3 m depth in the shoreface to reach the SBP profile (Figure 12c). The measurements were carried out by adopting an IRIS Instruments Syscal Pro Georesistive Meter, two multi-electrode cables at 24-channels with 3 m interelectrode spacing for a total length of 141 m. Proceeding from NNE to SSW, the first cable (steel electrodes 1–24) was placed on the subsoil of the emerged beach sector, while the second cable (graphite electrodes 25–48) was placed on the seabed (Figure 12a). The data were acquired both in Wenner-Schlumberger and dipole-dipole configurations. The former provides high vertical resolution while the latter high lateral resolution–although it is characterized by a lower signal-to-noise ratio than other devices [100]. For this reason, reciprocal measures were also carried out to facilitate the data quality control necessary to obtain reliable and well-resolved images. After the quality control (carried out by setting a 10% threshold for repeatability errors, and standard deviation obtained from the stacks), the experimental data were inverted using the RES2DINV code (Geotomo Software, [101]). The resulting model (Figure 12d) was obtained by inverting the dataset relating to the measurements in dipole–dipole configuration, including the GPS-RTK data that were acquired along the same transept for topography correction.

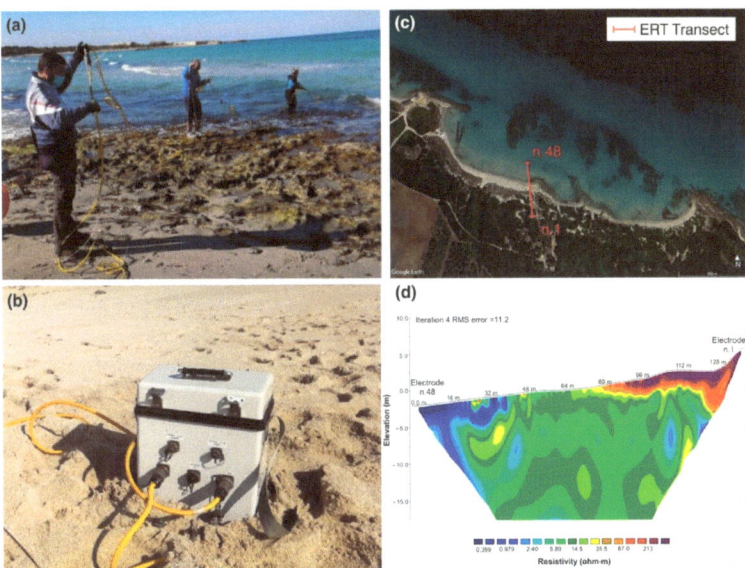

**Figure 12.** Example of ERT field survey. (**a**) Graphite electrodes positioning; (**b**) instrument for sending current into the ground; (**c**) map showing the ERT transect location; (**d**) resistivity model.

Lastly, two excavations were also carried out using shovels along the GPR/ERT transects. This technique was fundamental to calibrate and verify with a direct measurement the exact location of the bedrock extrapolated by the radar stratigraphy sections and the resistivity models.

## 4. Results

### 4.1. Data Comparison

The first step in beach dynamics interpretation is comparing some of the results deriving from different methodologies applied in this research. Figure 13 represents an example of a combination of Delft3D models, DTM comparison, sand classification, and mean size distribution.

This type of comparison enables to highlight the processes concerning the sediment transport and the evolutionary tendency of the beach explained by the connection among the simulation of the wave motion, erosion and accretion phases, sand composition, and granulometric features. For instance, at Torre Guaceto beach, the analysis of the emerged beach underlines the presence of a southernmost sector more exposed to the wave motion than the northern sector during storm events. The northern beach sector is characterized by an erosional tendency, especially at the dune base as reported from the DTM investigation (Figure 13a). Moreover, a significant composition variability is detected along the foreshore samples. The sand of Torre Guaceto beach is classified as "Hybrid intrabasinal sand", but the northern-central sector of the emerged beach is characterized by a higher content of bioclast than the southern sector, which instead is richer in siliciclastic content (Figure 13b). In addition, according to the mean size data, an increasing trend of the grain diameter is recorded from the samples collected close to the outermost promontories towards the central part of the beach (Figure 13c), suggesting a bioclast longshore transport in this direction. Indeed, looking at the main wave direction, the more porous and lighter bioclastic sediments are removed from the southern sector and transported to the central part of the beach. Indirectly, the transport of the bioclast component increases the percentages of the siliciclastic minerals in the southern sector, explaining the sudden longshore compositional variation of the emerged sector of Torre Guaceto beach. Comparing this result with the DTM, we found that the sector with the higher bioclast component (northern central sector) represents the more stable part of the beach in terms of evolutionary tendency, whereas the southernmost sector is eroding.

### 4.2. Beach Dynamics

Another example of data integration can be performed by using the results shown in Figure 14. The main sand characteristics (texture and petrography) coupled with the study of the meteorological events and the interpretation of the sediment transport allows to reconstruct the sedimentary balance and propose an overall sketch scheme of beach dynamics (Figure 14d). In the semi-closed coastal system of Le Dune beach, considering the southerly seas as the main wave direction, the scheme shows how the sediment inputs mainly derive from local sources and the sediment outputs involve small amounts of sediments. The gain sediment includes material coming from the erosion of the dunes, rocky shoreface, and lateral headlands as well as the bioclastic sediment transported from offshore. Therefore, the sedimentary dynamics are controlled by a predominant accretion of sand diffusion by aeolian processes especially in the westernmost part of the coast; a cross-shore sediment transport; a nearshore rip circulation characterized by a longshore sediment transport converging in the middle of the embayment; a weak lateral sediment interchange with adjacent littoral sectors and an offshore sand dispersion during storm events. The current morphological configuration of the sea bottom significantly influences the cross-shore sediment transport. The rocky headlands located in the shallow water and the rocky sea bottom make the beach an almost completely closed bay, which is also characterized by the absence of river basins and the presence of strong rip currents that carry large amounts of sediments towards the offshore during storm events.

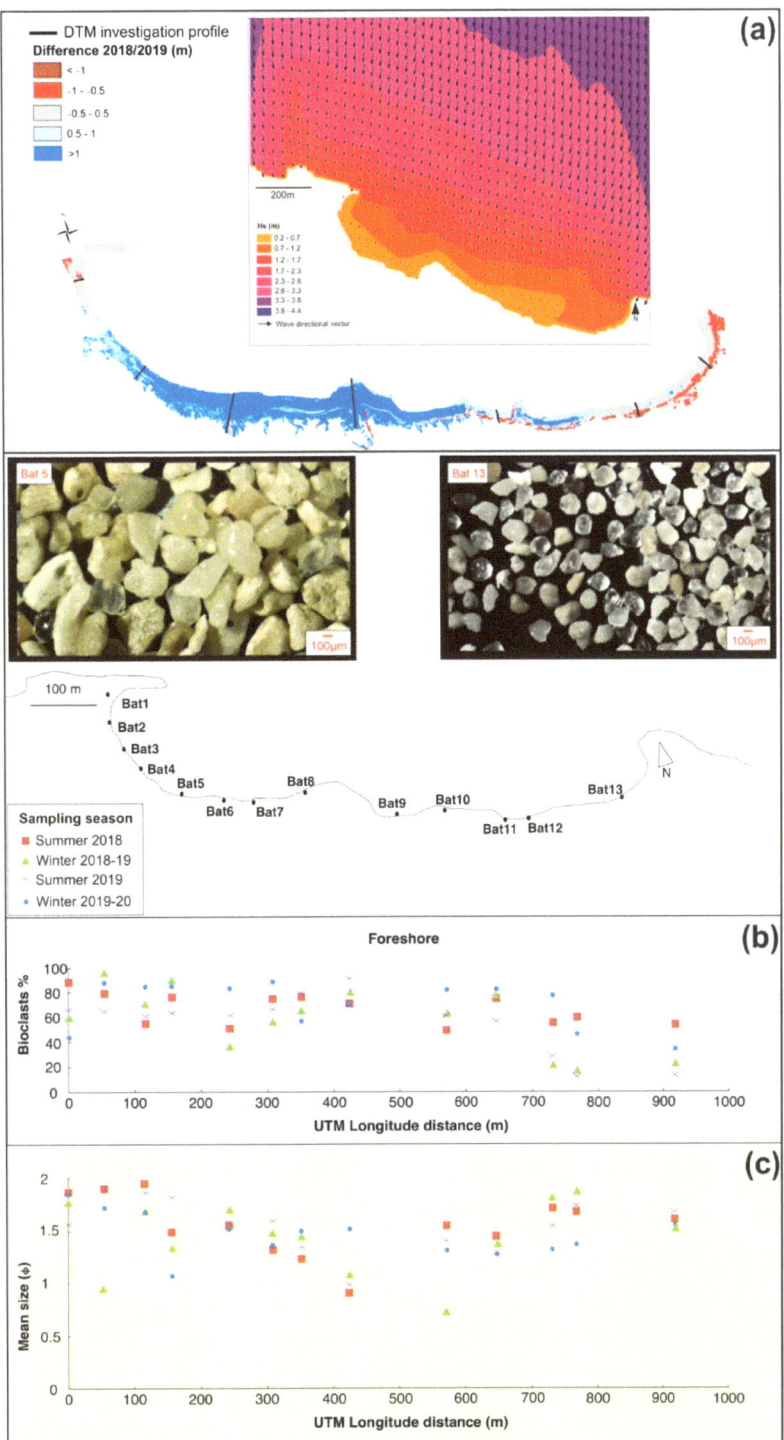

**Figure 13.** Example of emerged beach sector analysis. (**a**) Comparison between DTM investigation and Delft3D Model; (**b**) bioclast analysis and (**c**) mean size investigation for longshore sediment distribution.

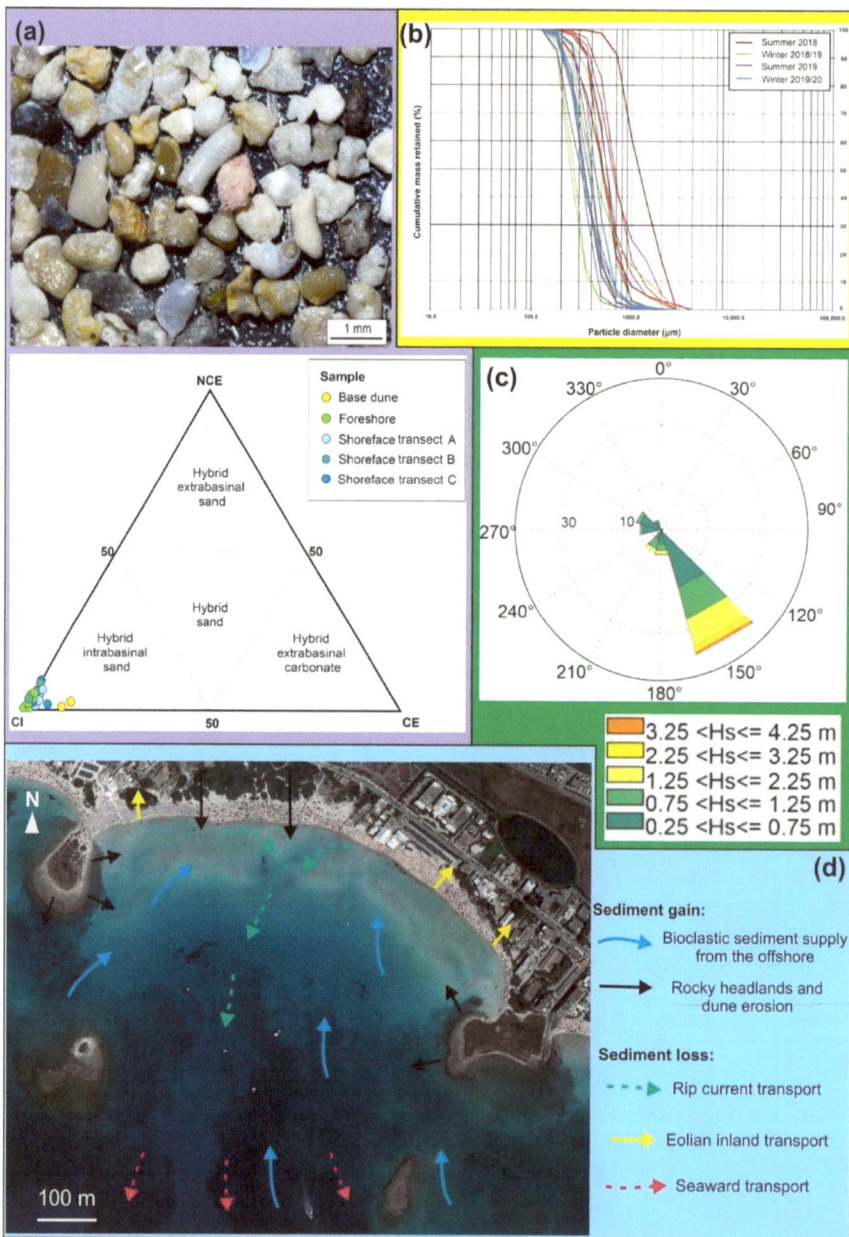

**Figure 14.** Gathering the main results for beach dynamics investigation. (**a**) Petrographical features; (**b**) sand texture of the bioclastic beach sand; (**c**) wave height distribution; (**d**) beach dynamics sketch scheme (modified from [102]).

## 4.3. Sediment Thickness and Beach Profile

One of the main applications deriving from the methodologies tested in this research is the combination of geophysics techniques. This operation is fundamental to reconstruct the entire beach profile from the dune environment to the offshore and to extrapolate the bedrock location. In particular, the merging of the radar section, resistivity model, and sub-bottom profile allows to quantify the sediment thickness within the entire beach environment and to record the location of the sub-environments. For instance, in the case of Torre Guaceto, the beach profile (Figure 15) extends for about 700 m. It represents the connection of the GPR, resistivity model, and SBP navigation transect. The bedrock location ranges between +1 m and −2 m within the emerged beach sector and from −2 m to −10 m in the submerged sector.

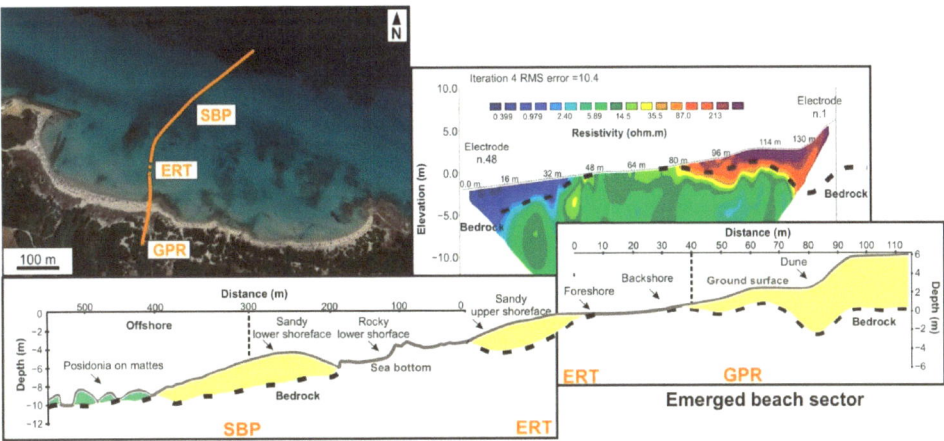

**Figure 15.** Example of beach profile with bedrock location deriving from the merging of GPR + ERT + SBP results. The orange dot line in the map depicts the investigation gap. Within the resistivity model and the GPR profile, the distance axis has been exaggerated for a higher result resolution. As ERT and GPR measurements were performed along the same transect of the emerged beach sector, the figures were overlapped to compare the same bedrock location extrapolated from both models. The yellow area between the ground surface/sea bottom and the bedrock represents the beach sand sediments.

The most relevant reflector resolution (bedrock) occurs below the sediments characterizing the dune environment, the sandy upper shoreface, and beneath the sand bar located between the lower shoreface and the offshore. Moreover, the bedrock visibly emerges in the rocky lower shoreface.

Considering the geology of the study area, the bedrock is certainly associated with the Pleistocene biocalcarenite (packstone and grainstone texture).

## 5. Discussion

### 5.1. Main Findings

The main outcomes highlight and confirm the literature data about the sand composition variability and the morpho-sedimentary dynamics occurring along the Adriatic and Ionian littoral. The findings represent the features of the Mediterranean coastal types.

The results underline the need to improve methodologies for pocket beach investigation which are difficult to preserve due to their low geological records. They are affected by seasonal changes strictly correlated with the main sediment source. The sediment transport mainly occurs between the dune environment and the offshore characterized by the presence of *Posidonia oceanica* meadows. In this context, carbonate factory protection and dune restoration measures are crucial to preserve this geological environment where the anthropic pressure is increasingly growing, especially for tourist activities.

The analysis of Torre Guaceto beach provides a significant example of improvement studies in ancient research. Environments at the same depth could have different compositions due to the main sediment selection caused by the current beach dynamics system.

Indeed, the qualitative and quantitative changes of the bioclast content along the cross-shore result in problematic stratigraphic correlations between adjacent logs in ancient successions located in similar settings. Moreover, this case study points out the presence of distinct dynamics due to the different exposure of the beach sector to wave motion providing information about eroding and prograding phases within the same environment.

The case study of Le Dune beach underlines the existence of many pocket beaches in the world that are only preserved thanks to the bioclast provenance source and the dune environment. In this case, the growth of urbanization, especially along the southern coastal sector, could significantly impact on the amount of sediment available. Although this beach is part of a marine protected area, the lack of collaboration between local decision-makers and the scientific community about the knowledge of pocket beach dynamics has led to the destruction of the dune environment with a consequent establishment of a building area. In this respect, higher restrictive measures should be considered for the protection of these types of environments.

Torre Guaceto and Le Dune represent two different pocket beaches in terms of length, sedimentology, composition, dynamics, and anthropic impact. For this reason, some considerations raised during the investigation. Geophysics provided more detailed results of the submerged sector at Torre Guaceto than at Le Dune beach. The rocky seabed in the latter affected the navigation along the shoreface transect requiring a major support from electrical tomography. The closed beach system, the morphology of the seabed, and the urbanization of the dune environment influenced the progress of the investigations. It would have been interesting to measure the sediment thicknesses along different cross-shore transects, but the southern sector was affected by a significant urbanization.

These results are strictly correlated with a two-year beach monitoring that is only able to evaluate sudden and small-scale beach variations. However, by seasonally applying the proposed methodologies over a longer period of time, it would be possible to recognize the real evolution of our beaches. The study of the evolutionary tendency could provide a significant impact on public decision-makers and current legislation on unnecessary protection and defense works.

*5.2. Final Remarks*

The procedure application highlights the need of comparing data from different sectors of the Earth Sciences to enrich the amount of missing information about complex and dynamic systems such as the beach environment. The multidisciplinary approach enables to collect a wide range of data that can be connected and interpreted with each other. The data combination provides a detailed outline of the beach dynamics in qualitative and quantitative terms.

From a sedimentological point of view, the research allows expanding the knowledge about large-scale sedimentary processes by quantifying the volumes involved in the erosion, transport, and deposition dynamics and the potential preservation of specific sub-environments during the sea-level rise or still-stand periods. Furthermore, this study shows how lateral variability in sand composition (not only in grain size) is a widespread feature in present-day beaches. The lateral composition is mainly influenced by very local processes and the sudden variations can be easily misinterpreted in the geological record.

Moreover, the study of sand composition and in particular the analysis of the bioclast component represents a rising technique in coastal erosion investigation. Indeed, measuring the bioclast percentage in the beach sand allows to hypothesize strategies for provenance marine environment safeguarding and monitoring by avoiding unnecessary or too-restrictive protection measures. Lastly, the beach sediment analysis provides significant progress for compatibility studies and nourishment interventions in terms of textural and compositional features. This research tries to respond to the increasingly pressing need to overcome the conflicts between naturalistic requirements and the use of beaches for social and economic purposes, which represent one of the main land use planning issues of coastal regions.

## 6. Conclusions

This research described the processes of erosion, transport, and sedimentation of two Apulian sandy beaches through a multidisciplinary approach. Table 2 shows the methodological proposal for sandy beach analysis deriving from the multidisciplinary approach tested in this research. This procedure includes topographic, geophysical, sedimentological, and compositional analyses for monitoring and safeguarding these transitional environments that can be useful for research studies or industrial applications.

**Table 2.** Methodological proposal for monitoring sandy beaches. Each color represents the methodological steps for beach monitoring.

| Type of Investigation | Potential Procedure | | | |
|---|---|---|---|---|
| | Method | Beach Sector | Results | Interpretation |
| Topographic surveys | TLS + GPS | Emerged beach | 3D elevation model | Morphological variation |
| | | | DTM profiles | Shoreline changes |
| | OTS + GPS | Shoreface | Bathymetric profiles | Erosion/Accretion |
| | | | Bathymetry map | Bathymetric variation |
| Hydrodynamic model | Delft3D | Shoreface + offshore | Wave motion | Sediment transport |
| Geophysics investigation | GPR + GPS | Emerged beach | Radarstratigraphy section | Beach profile reconstruction |
| | ERT + GPS | Entire beach | Resistivity model | GPR and SBP connection |
| | SBP + GPS | Shoreface + offshore | Lithostratigraphic sections | Bedrock depth and sediment thickness |
| Beach sand analysis | Sampling + GPS | Entire beach | Sand texture | Beach dynamics |
| | Grain size analysis | | Texture lateral variation | Erosion/Accretion |
| | Sand composition | | Sand classification | Sand provenance |
| | Bioclast evaluation | | Bioclast classification | Bioclast provenance |

As shown in Table 2, the procedure for investigating sandy beaches was based upon four main types of analyses: topographic surveys through TLS and OTS; a hydrodynamic model using Delft3D software; a series of geophysics techniques by the use of GPR, ERT and SBP; and a beach sand analysis through sedimentological and compositional applications. Each technique covered a specific beach subenvironment and provided distinct results. Therefore, the procedure could be invalid without the combination of the proposed techniques as the beach dynamics interpretation could be incomplete.

Regarding the geophysical investigation, it was essential to confirm the indirect measures with direct observations such as excavations. The resistivity studies were also correlated to the GPR and SBP surveys for a more detailed outline of the lithology and stratigraphy of the subsoil. Lastly, for the GPR application, the use of 400 and 200 Mhz antennas produced more satisfactory resolutions of the bedrock location than the 40 Mhz antenna.

It is important to emphasize that all the equipment, instrument frequencies, and resolutions were set on the aim of this research and the geological context.

Investigating sandy beaches is not undemanding due to the dynamism of this type of environment. For this reason, as shown in Table 2, the use of GPS supported most of the methodologies in order to respect the exact location of the seasonal investigation. Data comparison should be provided by taking into account the same season of the field measure in order to avoid interpretation issues.

Another suggestion that could improve this type of investigation is the use of the "Sediment transport module" in the application of Delft3D software. Indeed, the Delft3D-SED coupled with the others enables to analyze the effects of dredging on the environment, the sedimentation and resuspension of sediment, and the sand transport. Moreover, a primary phase of data collection on the marine climate of the study area is highly recommended to interpret the beach dynamics.

**Author Contributions:** Conceptualization, I.L., S.L., M.M.; data curation, I.L., S.L., F.D.G., D.M., L.C., G.R., F.S., G.M.; G.S.; writing—original draft preparation, I.L.; writing—review and editing, M.M., S.M.; supervision, M.M.; funding acquisition, M.M. All authors have read and agreed to the published version of the manuscript.

**Funding:** This research was carried out within the Ph.D. project of the "Programma Operativo Nazionale Ricerca e Innovazione 2014–2020 (CCI 2014IT16M2OP005)", Fondo Sociale Europeo, Azione I.1 "Dottorati Innovativi con caratterizzazione industriale" Università degli Studi di Bari "Aldo Moro".

**Institutional Review Board Statement:** Not applicable.

**Informed Consent Statement:** Not applicable.

**Data Availability Statement:** Not applicable.

**Acknowledgments:** This paper is the result of a collaboration among academic, research and industrial activities. We thank all collaborators for their logistic and technic support in every phase of this work. We are thankful to the reviewers for their revisions that allow us to improve the quality of the paper.

**Conflicts of Interest:** The authors declare no conflict of interest. The funders had no role in the design of the study; in the collection, analyses, or interpretation of data; in the writing of the manuscript; or in the decision to publish the results.

## References

1. Thornton, E.; Dalrymple, R.A.; Drake, T.G.; Elgar, S.; Gallagher, E.L.; Guza, R.T.; Hay, A.E.; Holman, R.A.; Kaihatu, J.M.; Lippmann, T.C.; et al. *State of Nearshore Processes Research: II*; Naval Postgraduate School: Monterey, CA, USA, 2000; Naval Postgraduate School Technical Report NPS-OC-00-001.
2. Woodroffe, C.D. *Coasts: Form, Process and Evolution*; Cambridge University Press: Cambridge, UK, 2002.
3. Crossland, C.J.; Baird, D.; Ducrotoy, J.-P.; Lindeboom, H.; Buddemeier, R.W.; Dennison, W.C.; Maxwell, B.A.; Smith, S.V.; Swaney, D.P. The Coastal Zone—A Domain of Global Interactions. In *Coastal Fluxes in the Anthropocene*; Springer: Berlin/Heidelberg, Germany, 2005; pp. 1–37. [CrossRef]
4. Masselink, G.; Russell, P.E. Impacts of climate change on coastal erosion. *MCCIP Sci. Rev.* **2013**, *2013*, 71–86. Available online: http://mccip.cefastest.co.uk/media/1256/2013arc_sciencereview_09_ce_final.pdf (accessed on 12 November 2022).
5. The BACC II Author Team. *Second Assessment of Climate Change for the Baltic Sea Basin*; Regional Climate Studies; Springer: Berlin/Heidelberg, Germany, 2015. [CrossRef]
6. Bonaldo, D.; Antonioli, F.; Archetti, R.; Bezzi, A.; Correggiari, A.; Davolio, S.; De Falco, G.; Fantini, M.; Fontolan, G.; Furlani, S.; et al. Integrating multidisciplinary instruments for assessing coastal vulnerability to erosion and sea level rise: Lessons and challenges from the Adriatic Sea, Italy. *J. Coast. Conserv.* **2019**, *23*, 19–37. [CrossRef]
7. De Serio, F.; Armenio, E.; Mossa, M.; Petrillo, A.F. How to Define Priorities in Coastal Vulnerability Assessment. *Geosciences* **2018**, *8*, 415. [CrossRef]
8. Roig-Munar, F.X.; Martín-Prieto, J.Á.; Rodríguez-Perea, A.; Batista, Ó.O. Environmental Analysis and Classification of Coastal Sandy Systems of the Dominican Republic. In *Beach Management Tools—Concepts, Methodologies and Case Studies*; Botero, C., Cervantes, O., Finkl, C., Eds.; Coastal Research Library; Springer: Cham, Switzerland, 2018; p. 24. [CrossRef]
9. Zhu, Z.-T.; Cai, F.; Chen, S.-L.; Gu, D.-Q.; Feng, A.-P.; Cao, C.; Qi, H.-S.; Lei, G. Coastal Vulnerability to Erosion Using a Multi-Criteria Index: A Case Study of the Xiamen Coast. *Sustainability* **2018**, *11*, 93. [CrossRef]
10. De Andrade, T.S.; de Oliveira Sousab, P.H.G.; Sieglea, E. Vulnerability to beach erosion based on a coastal processes approach. *Appl. Geogr.* **2019**, *102*, 12–19. [CrossRef]
11. National Research Council (NRC). *Measuring and Understanding Coastal Processes*; National Academies Press: Washington, DC, USA, 1989. Available online: http://www.nap.edu/catalog/1445.html (accessed on 10 November 2022).
12. Gombos, M.; Ramsay, D.; Webb, A.; Marra, J.; Atkinson, S.; Gorong, B. Coastal Change in the Pacific Islands: A Guide to Support Community Understanding of Coastal Erosion and Flooding Issues. 2014, 1. Available online: http://www.reefresilience.org/wp-content/uploads/Gombos-et-al.-2014-Coastal-Change-in-the-Pacific-Islands.pdf (accessed on 11 November 2022).
13. Harris, L.; Nel, R.; Holness, S.; Schoeman, D. Quantifying cumulative threats to sandy beach ecosystems: A tool to guide ecosystem-based management beyond coastal reserves. *Ocean. Coast. Manag.* **2015**, *10*, 12–24. [CrossRef]
14. UNISDR. Coastal Erosion Hazard and Risk Assessment. In *Words into Action Guidelines: National Disaster Risk Assessment Hazard Specific Risk Assessment*; UNISDR: Geneva, Switzerland, 2017.
15. UN Environment Programme. Coastal Zone Management. 2021. Available online: https://www.unep.org/explore-topics/oceans-seas/what-we-do/working-regional-seas/coastal-zone-management (accessed on 11 November 2022).
16. European Commission. *A Guide to Coastal Erosion Management Practices in Europe*; Rijkswaterstaat/RIKZ: The Haque, The Netherlands, 2004.

17. Moretti, M.; Tropeano, M.; van Loon, A.J.T.; Acquafredda, P.; Baldacconi, R.; Festa, V.; Lisco, S.; Mastronuzzi, G.; Moretti, V.; Scotti, R. Texture and composition of the Rosa Marina beach sands (Adriatic coast, southern Italy): A sedimentological/ecological approach. *Geologos* **2016**, *22*, 87–103. [CrossRef]
18. Van Loon, A.J.; Moretti, M.; Tropeano, M.; Acquafredda, P.; Baldacconi, R.; Festa, V.; Lisco, S.; Mastronuzzi, G.; Moretti, V.; Scotti, R. Tracing the Source of the Bio/Siliciclastic Beach Sands at Rosa Marina (Apulian Coast, SE Italy). In *Sediment Provenance*; Mazumder, R., Ed.; Elsevier: Amsterdam, The Netherlands, 2017; pp. 25–47. [CrossRef]
19. Pranzini, E. Dinamica e difesa dei litorali. Gestione integrata della fascia costiera. *Studi Costieri* **2019**, *28*, 3–12.
20. Falese, F.G.; Chiocci, F.; Moretti, M.; Tropeano, M.; Mele, D.; Dellino, P.; Lisco, S.; Mastronuzzi, G.; Piscitelli, A.; Sabato, L. *Rapporto di Fase 5*; Autorità di Bacino della Puglia: Bari, Italy, 2016.
21. Mastronuzzi, G.; Sansò, P. La costa senza passato è senza futuro. Il contributo della geomorfologia nella gestione sostenibile delle coste. *Geol. E Territ.* **2013**, *1*, 3–15.
22. Milli, S.; Girasoli, D.E.; Tentori, D.; Tortora, P. Sedimentology and coastal dynamics of carbonate pocket beaches: The Ionian-Sea Apulia coast between Torre Colimena and Porto Cesareo (Southern Italy). *J. Mediterr. Earth Sci.* **2017**, *9*, 29–66.
23. Mastronuzzi, G.; Palmentola, G.; Ricchetti, C. Aspetti della evoluzione olocenica della costa pugliese. *Mem. Della Soc. Geol. Ital.* **1989**, *42*, 287–300.
24. Caldara, M.; Centenaro, E.; Mastronuzzi, G.; Sansò, P.; Sergio, A. Features and present evolution of Apulian Coast (Southern Italy). *J. Costal Res.* **1998**, *26*, 55–64.
25. Mastronuzzi, G.; Palmentola, G.; Sansò, P. Evoluzione morfologica della fascia costiera di Torre Canne (Puglia adriatica). *Studi Costieri* **2001**, *4*, 19–31.
26. Donato, P.; De Rosa, R.; Tenuta, M.; Iovine, R.S.; Totaro, F.; D'Antonio, M. Sr-Nd Isotopic Composition of Pyroxenes as a Provenance Indicator of a Double-Volcanic Source in Sands of the Ofanto River (Southern Italy). *Minerals* **2022**, *12*, 232. [CrossRef]
27. Lisco, S.; Moretti, M.; Moretti, V.; Cardone, F.; Corriero, G.; Longo, G. Sedimentological features of *Sabellaria spinulosa* biocontructions. *Mar. Pet. Geol.* **2017**, *87*, 203–212. [CrossRef]
28. Alfonso, C.; Auriemma, R.; Scarano, T.; Mastronuzzi, G.; Calcagnile, L.; Quarta, G.; Di Bartolo, M. The ancient coastal landscape of the Marine Protected Area of Porto Cesareo (Lecce-ITALY): Recent researches. *Int. J. Soc. Underw. Technol.* **2012**, *30*, 207–215. [CrossRef]
29. Dal Cin, R.; Simeoni, U. Processi erosivi e trasporto dei sedimenti fra S. Maria di Leuca e Taranto (Mare Jonio). Possibili strategie di intervento. *Boll. Soc. Geol. It.* **1987**, *106*, 767–783.
30. Gianfreda, F.; Sansò, P. AMEBA (A parametric Method for Erosion Beach Assessment): Applicazione alle spiagge del Salento leccese. *Period. Soc. Ital. Di Geol. Ambient.* **2007**, *3*, 2–8.
31. Dal Cin, R.; Simeoni, U.; Zamariolo, A.; Mastronuzzi, G.; Sansò, P. Foglio 215 Otranto. In *Atlante delle Spiagge Italiane*; CNR—MURST, S.E.L.C.A.: Firenze, Italy, 1995.
32. Pennetta, L. Ricerche sull'evoluzione recente del delta dell'Ofanto. *Boll. Mus. St. Nat. Lunigiana* **1988**, *6–7*, 41–45.
33. Simeoni, U. I litorali tra Manfredonia e Barletta (basso Adriatico): Dissesti, sedimenti, problematiche ambientali. *Boll. Soc. Geol. It* **1992**, *111*, 367–398.
34. Caldara, M. Aspetti di Geologia ambientale e di morfologia costiera in alcuni tratti del litorale nord-barese. In Atti del Convegno "Cave e coste nel territorio del nord-barese". *Geologi Suppl.* **1996**, *2*, 39–61.
35. Simeoni, U.; Bondesan, M. The role and responsibility of man in the evolution of Italian Adriatic coast. *Bull. Inst. Oceanogr. Monaco* **1997**, *18*, 11–132.
36. Mastronuzzi, G.; Palmentola, G.; Sansò, P. Lineamenti e dinamica della costa pugliese. *Studi Costieri* **2002**, *5*, 9–22.
37. Annese, R.; De Marco, A.; Gianfreda, F.; Mastronuzzi, G.; Sanso, P. Caratterizzazione morfo-sedimentologica dei fondali della baia fra Torre San Leonardo e Torre Canne (costa adriatica, Puglia). *Studi Costieri* **2003**, *7*, 3–19.
38. Ingle, J. The movement of beach sand. An Analysis Using Fluorescent Grains. In *Development in Sedimentology 5*; Elsevier: Amsterdam, The Netherlands, 1966.
39. Dal Cin, R. Distinzione tra spiagge in erosione ed in avanzamento mediane metodo granulometrico. Istituto di Geologia dell'Università di Ferrara con il contributo del C.N.R. *Riv. Ital. Di Geotec.* **1969**, *4*, 227–233.
40. Dalrymple, R.A.; Thompson, W.W. Study of equilibrium beach profiles. *Coast. Eng. Proceeding* **1976**, *15*, 74. [CrossRef]
41. Wright, L.D.; Chappell, J.; Thorn, B.G.; Bradshaw, M.P.; Cowell, P. Morpho-dynamics of reflective and dissipative beach and inshore systems: Southeastern Australia. *Mar. Geol.* **1979**, *32*, 105–140. [CrossRef]
42. Wright, L.D.; Short, A. Morphodynamic variability of surf zones and beaches: A synthesis. *Mar. Geol.* **1984**, *56*, 93–118. [CrossRef]
43. Masselink, G.; Short, A.D. The Effect of Tide Range on Beach Morphodynamics and Morphology: A Conceptual Beach Model. *J. Coast. Res.* **1993**, *9*, 785–800.
44. Ciavola, P.; Gatti, M.; Armaroli, C.; Balouin, Y. Valutazione della variazione della linea di riva nell'area di Lido di Dante (RA) tramite GIS e monitoraggio DGPS cinematico. In Proceedings of the Atti di: Accademia Nazionale dei Lincei, XXI Giornata dell'Ambiente, Aree Costiere, Roma, Italy, 5 June 2003; pp. 113–121.
45. Gracia, F.J.; Anfuso, G.; Benavente, J.; Del Rio, L.; Dominguez, L.; Martinez, J.A. Monitoring coastal erosion at different temporal scales on sandy beaches: Application to the Spanish Gulf of Cadiz coast. *J. Coast. Res.* **2005**, *49*, 22–27.
46. Costas, S.; Alejo, I.; Rial, F.; Lorenzo, H.; Nombela, M.A. Cyclical evolution of a modern transgressive sand barrier in Northwestern Spain elucidated GPR and aerial photos. *J. Sediment. Res.* **2006**, *76*, 1077–1092. [CrossRef]

47. Pranzini, E. Remote sensing in beach erosion monitoring: From the origin, to optimal and further. In *Beach Erosion Monitoring—Results from BEACHMED-e/OpTIMAL Project*; Nuova Grafica Fiorentina: Firenze, Italy, 2008; pp. 49–50.
48. Anfuso, G.; Pranzini, E.; Vitale, G. An integrated approach to coastal erosion problems in Northern Tuscany (Italy): Littoral Morphological Evolution and Cell Distribution. *Geomorphology* **2011**, *129*, 204–214. [CrossRef]
49. Nordstrom, K.F.; Armaroli, C.; Jackson, N.L.; Ciavola, P. Opportunities and constraints for managed retreat on exposed sandy shores: Examples from Emilia-Romagna, Italy. *Ocean. Coast. Manag.* **2015**, *104*, 11–21. [CrossRef]
50. Karkani, A.; Evelpidou, N.; Vacchi, M.; Morhange, C.; Tsukamoto, S.; Frechen, M.; Maroukian, H. Tracking Shoreline Evolution in Central Cyclades (Greece) Using Beachrocks. *Mar. Geol.* **2017**, *388*, 25–37. [CrossRef]
51. Thom, B.G.; Hall, W. Behaviour of beach profiles during accretion and erosion dominated periods. *Earth Surf. Process. Landf.* **1991**, *16*, 113–127. [CrossRef]
52. Almeida, L.P.; Ferreira, Ó.; Pacheco, A. Thresholds for morphological changes on an exposed sandy beach as a function of wave height. *Earth Surf. Process Landf.* **2010**, *36*, 523–532. [CrossRef]
53. Riazi, A.; Türker, U. Equilibrium beach profiles: Erosion and accretion balanced approach. *Water Environ. J.* **2017**, *31*, 317–323. [CrossRef]
54. Gao, S.; Collins, M.B. Analysis of grain size trends, for defining sediment transport pathways in marine environments. *J. Coast. Res.* **1994**, *10*, 70–78.
55. Guillén, J.; Palanques, A. Short- and medium-term grain size changes in deltaic beaches (Ebro Delta, NW Mediterranean). *Sediment. Geol.* **1996**, *101*, 55–67. [CrossRef]
56. Dawe, I.N. Sediment Patterns on a Mixed Sand and Gravel Beach, Kaikoura, New Zealand. *J. Coast. Res.* **2001**, *34*, 267–277.
57. Poizot, E.; Méar, Y.; Biscara, L. Sediment Trend Analysis through the variation of granulometric parameters: A review of theories and applications. *Earth-Sci. Rev.* **2008**, *86*, 15–41. [CrossRef]
58. Folk, R.; Ward, W. Brazos river bar: A study in the significance of grain size parameters. *J. Sediment. Petrol.* **1957**, *27*, 3–26. [CrossRef]
59. Visher, G.S. Fluvial processes as interpreted from ancient and recent fluvial deposits. In *Primary Sedimentary Structures and Their Hydrodynamic Interpretation*; Middleton, G.V., Ed.; Society of Economic Paleontologists and Mineralogists: Tulsa, OK, USA, 1965; Volume 12, pp. 116–132.
60. Friedman, G. Dynamic processes and statistical parameters compared for size frequency distribution of beach and river sands. *J. Sediment. Petrol.* **1967**, *37*, 327–354.
61. Edwards, A.C. Grain size and sorting in modern beach sands. *J. Coast. Res.* **2001**, *17*, 38–52.
62. De Falco, G.; Baroli, M.; Simeone, S.; Piergallini, G. La rimozione della posidonia dalle spiagge: Conseguenze sulla stabilità dei litorali. In *Risultati del Progetto—ARENA—ImpAtto della Rimozione dei BanchEtti di PosidoNia sulla Stabilità degli Arenili*; Fondazione IMC: Oristano, Italy, 2002.
63. De Falco, G.; Molinaroli, E.; Baroli, M.; Bellacicco, S. Grain size and compositional trends of sediments from Posidonia oceanica meadows to beach shore, Sardinia, western Mediterranean. *Estuar. Coast. Shelf Sci.* **2003**, *58*, 299–309. [CrossRef]
64. Satta, A.; Ceccherelli, G.; Cappucci, S.; Carboni, S.; Cossu, A.; Costa, M.; De Luca, M.; Dessy, C.; Farris, E.; Gazale, V.; et al. Linee guida per la gestione integrata delle spiagge. In *I Quaderni della Conservatoria delle Coste 1, Regione Autonoma della Sardegna*; Tipografie Grafiche del Parteolla: Dolianova, Italy, 2013.
65. De Falco, G.; Simeone, S.; Baroli, M. Management of beach-cast Posidonia oceanica seagrass on the island of Sardinia (Italy), Western Mediterranean). *J. Coast. Res.* **2008**, *24*, 69–75. [CrossRef]
66. Short, F.; Carruthers, T.; Dennison, W.; Waycott, M. Global seagrass distribution and diversity: A bioregional model. *J. Exp. Mar. Biol. Ecol.* **2007**, *350*, 3–20. [CrossRef]
67. Brandano, M.; Cuffaro, M.; Gaglianone, G.; Petricca, P.; Stagno, V.; Mateu-Vicens, G. Evaluating the role of seagrass in Cenozoic $CO_2$ variations. *Front. Environ. Sci.* **2016**, *4*, 72. [CrossRef]
68. Gaglianone, G.; Frezza, V.; Mateu-Vicens, G.; Brandano, M. Posidonia oceanica seagrass meadows facies from western Mediterranean Sea. *Rend. Online Soc. Geol. It* **2014**, *31* (Suppl. 1), 187.
69. Gaglianone, G.; Brandano, M.; Guillem, M.-V. The sedimentary facies of Posidonia oceanica seagrass meadows from the central Mediterranean Sea. *Facies* **2017**, *63*, 28. [CrossRef]
70. Simeone, S.; Molinaroli, E.; Conforti, A.; De Falco, G. Impact of ocean acidification on the carbonate sediment budget of a temperate mixed beach. *Clim. Change* **2018**, *150*, 227–242. [CrossRef]
71. Leatherman, S.P. Coastal geomorphological applications of ground-penetrating radar. *J. Coast. Res.* **1987**, *3*, 397–399, ISSN 0749-0208.
72. Bristow, C.S.; Bailey, S.D.; Lancaster, N. The sedimentary structure of linear sand dunes. *Nature* **2000**, *406*, 56–59. [CrossRef] [PubMed]
73. Neal, A.; Roberts, C.L. Applications of ground-penetrating radar (GPR) to sedimentological, geomorphological and geoarchaeological studies in coastal environments. *Geol. Soc. Lond. Spéc. Publ.* **2000**, *175*, 139–171. [CrossRef]
74. Neal, A. Ground-penetrating radar and its use in sedimentology: Principles, problems and progress. *Earth-Sci. Rev.* **2004**, *66*, 261–330. [CrossRef]
75. Hugenoltz, C.H.; Moorman, B.J.; Wolfe, S.A. Ground penetrating radar (GPR) imaging of the internal structure of an active parabolic sand dune. In *Stratigraphic Analyses Using GPR: Geological Society of America Special Paper*; Baker, G.S., Jol, H.M., Eds.; Geological Society of America: Boulder, CO, USA, 2007; Volume 432, pp. 35–45. [CrossRef]

76. Guillemoteau, J.; Bano, M.; Dujardin, J.-R. Influence of grain size, shape and compaction on georadar waves: Example of an Aeolian dune. *Geophys. J. Int.* **2012**, *190*, 1455–1463. [CrossRef]
77. Shukla, S.B.; Chowksey, V.M.; Prizomwala, S.P.; Ukey, V.M.; Bhatt, N.P.; Muraya, D.M. Internal Sedimentary Architecture and Coastal Dynamics as Revealed by Ground Penetrating Radar, Kachchh coast, Western India. *Acta Geophys.* **2013**, *61*, 1196–1210. [CrossRef]
78. Bristow, C.S.; Jol, H.M. An introduction to ground penetrating radar (GPR) in sediments. *Geol. Soc. Lond. Spéc. Publ.* **2007**, *211*, 1–7. [CrossRef]
79. Morang, A.; Larson, R.; Gorman, L. Monitoring the coastal environment; Part III: Geophysical and Research Methods. *J. Coast. Res.* **1997**, *13*, 1064–1085.
80. Lubis, M.Z.; Anggraini, K.; Kausarian, H.; Pujiyati, S. Review: Marine Seismic and Side-Scan Sonar Investigations for Seabed Identification with Sonar System. *J. Geosci. Eng. Environ. Technol.* **2017**, *2*, 166–170. [CrossRef]
81. De Giosa, F.; Scardino, G.; Vacchi, M.; Piscitelli, A.; Milella, M.; Ciccolella, A.; Mastronuzzi, G. Geomorphological Signature of Late Pleistocene Sea Level Oscillations in Torre Guaceto Marine Protected Area (Adriatic Sea, SE Italy). *Water* **2019**, *11*, 2409. [CrossRef]
82. Kim, H.D.; Aoki, S.; Kim, K.H.; Kim, J.; Shin, B.; Lee, K. Bathymetric Survey for Seabed Topography using Multibeam Echo Sounder in Wando, Korea. *J. Coast. Res.* **2020**, *95*, 527–531. [CrossRef]
83. Van der Salm, J.; Unal, O. Towards a common Mediterranean framework for beach nourishment projects. *J. Coast. Conserv.* **2003**, *9*, 35–42. [CrossRef]
84. Nicoletti, L.; Paganelli, D.; Gabellini, M. Aspetti ambientali del dragaggio di sabbie relitte a fini di ripascimento: Proposta di un protocollo di monitoraggio (Environmental aspects of relict sand dredging for beach nourishment:proposal for a monitoring protocol). *Quad. ICRAM* **2006**, *5*, 150.
85. APAT-ICRAM. Manuale di Movimentazione dei Sedimenti Marini 2007. (Handbook of Marine Sediment Transport: 77). Available online: http://www.isprambiente.gov.it/contentfiles/00006700/6770-manuale-apaticram-2007.pdf/view (accessed on 12 November 2022).
86. Targusi, M.; La Porta, B.; Lattanzi, L.; La Valle, P.; Loia, M.; Paganelli, D.; Pazzini, A.; Proietti, R.L.; Nicoletti, L. Beach nourishment using sediments from relict sand deposit: Effects onsubtidal macrobenthic communities in the Central Adriatic Sea (EasternMediterranean Sea-Italy). *Mar. Environ. Res.* **2018**, *144*, 186–193. [CrossRef]
87. Tortora, P. Failure of the nourishment intervention at Ladispoli Beach Central Latium coast, Italy, Part 1: An insignificant episode during the last 70 years of coastal evolution. *J. Mediterr. Earth Sci.* **2020**, *12*, 33–53.
88. Tortora, P. Failure of the nourishment intervention at Ladispoli Beach (Central Latium coast, Italy, Part 2: The causes. *J. Mediterr. Earth Sci.* **2020**, *12*, 55–75.
89. Gunn, D.A.; Pearson, S.G.; Chambers, J.E.; Nelder, L.M.; Lee, J.R.; Beamish, D.; Busby, J.P. An evaluation of combined geophysical and geotechnical methods to characterise beach thickness. *Q. J. Eng. Geol. Hydrogeol.* **2022**, *39*, 339. [CrossRef]
90. De Falco, G.; Budillon, F.; Conforti, A.; De Muro, S.; Di Martino, G.; Innangi, S.; Perilli, A.; Tonielli, R.; Simeone, S. Sandy beaches characterization and management of coastal erosion on western Sardinia island (Mediterranean sea). *J. Coast. Res.* **2014**, *70*, 395–400. [CrossRef]
91. De Muro, S.; Ibba, A.; Kalb, C. Morpho-sedimentology of a Mediterranean microtidal embayed wave dominated beach system and related inner shelf with Posidoniaoceanica meadows: The SE Sardinian coast. *J. Maps* **2016**, *12*, 558–572. [CrossRef]
92. Buosi, C.; Tecchiato, S.; Pusceddu, N.; Frongia, P.; Ibba, A.; De Muro, S. Geomorphology and sedimentology of Porto Pino, SW Sardinia, western Mediterranean. *J. Maps* **2017**, *13*, 470–485. [CrossRef]
93. Mastronuzzi, G.; Aringoli, D.; Aucelli, P.P.C.; Baldassarre, M.A.; Bellotti, P.; Bini, M.; Biolchi, S.; Bontempi, S.; Brandolini, P.; Chelli, A.; et al. Geomorphological map of the Italian coast: From a descriptive to a morphodynamic approach. *Geogr. Fis. E Din. Quat.* **2017**, *40*, 161–196.
94. Lapietra, I.; Lisco, S.; Mastronuzzi, G.; Milli, S.; Pierri, C.; Sabatier, F.; Scardino, G.; Moretti, M. The morpho-sedimentary dynamics of Torre Guaceto beach (Southern Adriatic Sea, Italy). *J. Earth Syst. Sci.* **2022**, *131*, 64. [CrossRef]
95. Pennetta, M. Caratteri granulometrici dei sedimenti del Golfo di Taranto (Alto Ionio). *Ann. Ist. Univ. Nav. Di Napoli* **1985**, *54*, 29–30.
96. Petrillo, A.F.; Bruno, M.F.; Nobile, B. *Supporto Scientifico per la Redazione del Piano Comunale delle Coste del Comune di Porto Cesareo (Le)*; Report; Politecnico di Bari: Bari, Italy, 2014.
97. Poppe, L.J.; Eliason, A.H.; Fredericks, J.J.; Rendigs, R.R.; Blackwood, D.; Polloni, C.F. *Grain-Size Analysis of Marine Sediments e Methodology and Data Processing*; Geological Survey Open File Report 00e358; U.S. Geological Survey: Woods Hole, MA, USA, 2000. Available online: https://pubs.usgs.gov/of/2000/of00-358/text/chapter1.htm (accessed on 11 November 2022).
98. Zuffa, G.G. Hybrid arenites: Their composition and classification. *J. Sediment. Petrol.* **1980**, *50*, 21–29.
99. Zuffa, G.G. Optical analysis of arenites: Influence of methodology on compositional results. In *Provenance of Arenites*; Zuffa, G.G., Ed.; Reidel: Dordrecht, The Netherlands, 1985; pp. 165–189.
100. Dahlin, T.; Zhou, B. A numerical comparison of 2D resistivity imaging with 10 electrode arrays. *Geophys. Prospect.* **2004**, *52*, 379–398. [CrossRef]

101. Loke, M.H.; Barker, R.D. Rapid least squares inversion of apparent resistivity pseudosections by a quasi-Newton method. *Geophys. Prospect.* **1996**, *44*, 131–152. [CrossRef]
102. Lapietra, I.; Lisco, S.N.; Milli, S.; Moretti, M. Sediment provenance of a bioclastic carbonate pocket beach—Le Dune (Ionian Sea, South Italy). *J. Palaeogeogr.* **2022**, *11*, 238–255. [CrossRef]

*Article*

# Effects of Mud Supply and Hydrodynamic Conditions on the Sedimentary Distribution of Estuaries: Insights from Sediment Dynamic Numerical Simulation

Qian Zhang [1], Mingming Tang [1,2,*], Shuangfang Lu [1,2], Xueping Liu [3] and Sichen Xiong [1]

1. School of Geosciences, China University of Petroleum (East China), Qingdao 266580, China
2. Key Laboratory of Deep Oil and Gas, China University of Petroleum (East China), Qingdao 266580, China
3. Engineering Technology Research Company Limited, China Petroleum and Chemical Corporation, Binzhou 256606, China
* Correspondence: tangmingming@upc.edu.cn

**Citation:** Zhang, Q.; Tang, M.; Lu, S.; Liu, X.; Xiong, S. Effects of Mud Supply and Hydrodynamic Conditions on the Sedimentary Distribution of Estuaries: Insights from Sediment Dynamic Numerical Simulation. *J. Mar. Sci. Eng.* **2023**, *11*, 174. https://doi.org/10.3390/jmse11010174

Academic Editors: George Kontakiotis, Angelos G. Maravelis and Avraam Zelilidis

Received: 26 November 2022
Revised: 29 December 2022
Accepted: 1 January 2023
Published: 10 January 2023

**Copyright:** © 2023 by the authors. Licensee MDPI, Basel, Switzerland. This article is an open access article distributed under the terms and conditions of the Creative Commons Attribution (CC BY) license (https://creativecommons.org/licenses/by/4.0/).

**Abstract:** Estuaries are important sediment facies in the fluvial-to-marine transition zone, are strongly controlled by dynamic interactions of tides, waves, and fluvial flows, and show various changes in depositional processes and sediment distribution. Deep investigations on the sediment dynamic processes of the sand component of estuaries have been conducted; however, the understanding of how mud supply affects estuaries' sedimentary characteristics and morphology is still in vague. Herein, the effects of mud concentration, mud transport properties, fluvial discharge, and tidal amplitude on the sedimentary characteristics of an estuary were systematically analyzed using sedimentary dynamic numerical simulation. The results show that the mud concentration has significant effects on the morphology of tidal channels in estuaries, which become more braided with a lower mud concentration, and straighter, with reduced channel migration, with a higher mud concentration. The mud transport properties, namely, setting velocity, critical bed shear stress for sedimentation, and erosion, mostly affect the ratio between the length and width (RLW) of the sand bar; a sheet-like sand bar with a lower RLW value develops in the lower settling velocity, while there are obvious strip shaped bars with a high RLW value in the higher settling velocity case. Moreover, the effects of hydrodynamic conditions on sedimentary distribution were analyzed by changing the tidal amplitudes and fluvial discharges. The results show that a higher tidal amplitude is often accompanied by a stronger tidal energy, which induces a more obvious seaward progradation, while a higher fluvial discharge usually yields a higher deposition rate and yields a greater deposition thickness. From the above numerical simulations, the statistical characteristics of tidal bars and mud interlayers were further obtained, which show good agreement with modern sedimentary characteristics. This study suggests that sedimentary dynamic numerical simulation can provide insights into an efficient quantitative method for analyzing the effects of mud components on the sediment processes of estuaries.

**Keywords:** estuary; numerical simulation; mud supply; hydrodynamic condition; tidal bar; mud interlayers

## 1. Introduction

The fluvial-to-marine transition zone is the area with the most complex sedimentary environments controlled by the continuous interaction of fluvial flows, tide currents, and waves [1–4]. Among which, estuaries are typical sediment systems [5,6], comprising predominantly sand components and minor mud components and salt marshes [7]. The interaction between mud and sand helps in promoting the long-term morphology of estuaries. Since the mud component has higher erosion resistance and lower sedimentary rate than the sand component [8,9], the mud layer can reduce the re-erosion of the underlying sand body and afford more stable sand bars and banks. Hence, the mud component might help in the dynamic balancing between sand body erosion and deposition in estuaries [10].

Recently, increasing attention has been paid to the sedimentary distribution and reservoir architecture of estuaries [11,12], and a series of works has been conducted on the main controlling factors of estuary sedimentation [13,14], such as fluvial discharge, slope angle, and wave. Winterwerp [15] believed that source supply direction is the main controlling factor in the location of mud deposits and hydrodynamic conditions are a secondary controlling factor. Verlaan [16] studied the mud distribution characteristics of different source supplies in estuaries and found that marine mud mainly settles in the entrance channels of lower estuaries, while relatively small amounts are deposited further upstream. Conversely, fluvial mud is mainly deposited in the inner estuary. Cleveringa and Dam [17] implied that hydrodynamic conditions have significant affects on the locations of mudflats, as mudflats exhibit a faster migration with rapid changes in flood-dominated peak velocities. Conversely, Kleinhans and M. [18] argued that the hydrodynamic conditions only change their scales instead of changing the location of mudflats. Schramkowski et al. [19] and Toffolon and Crosato [20] argued that hydrodynamic conditions are the major factor affecting the development of sand bars in an estuary. The length of sand bars is positively correlated with fluvial discharge and tidal range. Wave-generated currents are a third mechanism leading to sediment transport and sand bar morphodynamics [21]. The research show that waves mainly act on the shape of the estuary. Waves cause the estuary to widen and limit the deposition of mud. Although previous studies have provided a certain understanding on the sedimentary characteristics of estuaries and the factors affecting tidal bar development, there are several controversial points. Consequently, exploring the effects of mud supply and hydrodynamic conditions on the development mechanism of estuary sedimentation is necessary. Many research methods have been applied to estuary sediments, including sedimentary record analysis, modern sedimentary anatomy, physical simulation, and numerical simulation [22–26]. The lack of recorded information and the limitations of experimental operation have led to large errors in the experimental results of traditional research methods. Meanwhile, advances in computer technology have promoted the rapid development of sedimentary numerical simulation. Sedimentary dynamic numerical simulation has become the mainstream research method for investigating different hydrodynamic and sedimentary driving mechanisms [27,28], providing strong support for studying the morphological dynamics and stratigraphic patterns [7]. Weisscher et al. [29] published the use of the morphological dynamic model Nays2D to determine the impact of dynamic inflow disturbances on river patterns. Edmonds and Slingerland [30] argued the effect of flow velocity and sediment transport on the sedimentary body formation using sedimentary dynamic numerical simulation. van de Lageweg et al. [13] suggested that fluvial discharge and tidal amplitude have a functional relationship between mud-deposit coverage and mud-deposit thickness based on numerical simulation. Tang et al. [28] used numerical simulation to study the reservoir configuration of tidal-controlled estuaries. Hence, sedimentary numerical simulation is used to simulate the geomorphic evolution of estuaries, which provides a new idea for analyzing the characteristics of mud deposition [31].

Herein, sedimentary dynamic numerical simulation was used to set different mud supply and hydrodynamic conditions, perform the sedimentation simulation of the tidal bar and its interlayer, and explore the effects of the controlling factors on the sedimentary development and distribution of an estuary. Nine cases of an idealized estuary model were designed for identifying effect of the key factors, namely, mud concentration, mud transport properties, fluvial discharge, and tidal amplitude on the sedimentary characteristics of the estuary. This paper is arranged as follows. Section 2 introduces the simulation setting and numerical parameters, Section 3 shows the simulation results of nine scenarios, and Section 4 compares the numerical models and natural estuaries and discusses the length, thickness, and frequency distributions of mud interlayers in detail.

## 2. Method and Parameters

### 2.1. Numerical Simulation Method

The sedimentary processes in estuaries are simulated using the computational fluid dynamics software Delft3D [32], which uses numerical calculation methods to solve the Navier–Stokes equations and sediment transport based on an interleaved uniform finite-difference grid with the alternating direction implicit method. The flow module of Delft3D has been extensively applied to the study of topography and geomorphology in semi-enclosed coastal areas such as shallow seas, estuaries, and deltas [33–36]. The tidal–fluvial interaction of weakly forced stratified estuary systems is calculated using the momentum equations (Equations (1)–(3)).

$$\frac{\partial u}{\partial t} + \frac{u}{\sqrt{G_{\xi\xi}}}\frac{\partial u}{\partial \xi} + \frac{v}{\sqrt{G_{\eta\eta}}}\frac{\partial v}{\partial \eta} + \frac{w}{d+\zeta}\frac{\partial u}{\partial \sigma} - \frac{v^2}{\sqrt{G_{\xi\xi}}\sqrt{G_{\eta\eta}}}\frac{\partial \sqrt{G_{\eta\eta}}}{\partial \xi} + \frac{uv}{\sqrt{G_{\xi\xi}}\sqrt{G_{\eta\eta}}}\frac{\partial \sqrt{G_{\xi\xi}}}{\partial \eta} - fv$$
$$= -\frac{1}{\rho_0\sqrt{G_{\xi\xi}}}P_\xi + F_\xi + \frac{1}{(d+\zeta)^2}\frac{\partial}{\partial \sigma}\left(\nu_V \frac{\partial u}{\partial \sigma}\right) + M_\xi \quad (1)$$

$$\frac{\partial v}{\partial t} + \frac{u}{\sqrt{G_{\xi\xi}}}\frac{\partial v}{\partial \xi} + \frac{v}{\sqrt{G_{\eta\eta}}}\frac{\partial v}{\partial \eta} + \frac{w}{d+\zeta}\frac{\partial u}{\partial \sigma} + \frac{uv}{\sqrt{G_{\xi\xi}}\sqrt{G_{\eta\eta}}}\frac{\partial \sqrt{G_{\eta\eta}}}{\partial \xi} - \frac{u^2}{\sqrt{G_{\xi\xi}}\sqrt{G_{\eta\eta}}}\frac{\partial \sqrt{G_{\xi\xi}}}{\partial \eta} + fu$$
$$= -\frac{1}{\rho_0\sqrt{G_{\eta\eta}}}P_\eta + F_\eta + \frac{1}{(d+\zeta)^2}\frac{\partial}{\partial \sigma}\left(\nu_V \frac{\partial v}{\partial \sigma}\right) + M_\eta \quad (2)$$

$$\frac{\partial \zeta}{\partial t} + \frac{1}{\sqrt{G_{\xi\xi}}\sqrt{G_{\eta\eta}}}\frac{\partial\left((d+\zeta)u\sqrt{G_{\eta\eta}}\right)}{\partial \xi} \frac{1}{\sqrt{G_{\xi\xi}}\sqrt{G_{\eta\eta}}}\frac{\partial\left((d+\zeta)v\sqrt{G_{\eta\eta}}\right)}{\partial \eta} - \frac{\partial w}{\partial \sigma}$$
$$= (d+\zeta)(q_{in} - q_{out}) \quad (3)$$

where $t$ is time (s), $\xi$ and $\eta$ are horizontal coordinates, $\sigma$ is the scaled vertical coordinate, $u$ is the flow velocity in the $\xi$-direction (ms$^{-1}$), $v$ is the fluid velocity in the $\eta$-direction (ms$^{-1}$), $w$ is the fluid velocity in the z-direction (ms$^{-1}$), $\zeta$ is the water level above some horizontal plane of reference (datum), $F_\xi$ and $F_\eta$ are turbulent momentum fluxes in the $\xi$ and $\eta$ directions (ms$^{-2}$), $\sqrt{G_{\xi\xi}}$ and $\sqrt{G_{\eta\eta}}$ are coefficients used to transform the curvilinear to rectangular ones, $M_\xi$ and $M_\eta$ are the sources or sinks of momentum in the $\xi$ and $\eta$ directions (ms$^{-2}$), $P_\xi$ and $P_\eta$ are gradient hydrostatic pressures in the $\xi$ and $\eta$ directions (kgm$^{-2}$s$^{-1}$), $f$ is the Coriolis parameter (s$^{-1}$), $d$ is the depth below some horizontal plane of reference (datum), $\nu_V$ is the vertical eddy viscosity (m$^2$s$^{-1}$), and $q_{in}$ and $q_{out}$ are the local sources and sinks of water per unit of volume (s$^{-1}$), respectively.

The three-dimensional transport of suspended sediment is calculated by solving the three-dimensional advection–diffusion equation for the suspended sediment components (Equation (4)).

$$\frac{\partial c^{(l)}}{\partial t} + \frac{\partial u c^{(l)}}{\partial x} + \frac{\partial v c^{(l)}}{\partial y} + \frac{\partial \left(w - w_s^{(l)}\right)c^{(l)}}{\partial z} - \frac{\partial}{\partial x}\left(\varepsilon_{s,x}^{(l)}\frac{\partial c^{(l)}}{\partial t}\right) - \frac{\partial}{\partial y}\left(\varepsilon_{s,y}^{(l)}\frac{\partial c^{(l)}}{\partial x}\right)$$
$$-\frac{\partial}{\partial z}\left(\varepsilon_{s,z}^{(l)}\frac{\partial c^{(l)}}{\partial z}\right) = 0 \quad (4)$$

Here $x$, $y$, and $z$ are the Cartesian coordinates(m), $c^{(l)}$ is the mass concentration of sediment (kgm$^{-3}$), $\varepsilon_{s,x}^{(l)}$, $\varepsilon_{s,y}^{(l)}$, and $\varepsilon_{s,z}^{(l)}$ are the eddy diffusivities of sediment(m$^2$s$^{-1}$), and $w_s^{(l)}$ is the sediment settling velocity (ms$^{-1}$).

The sediment components are mainly of two types: cohesive and noncohesive components. The cohesive sediment component is controlled by the suspended-transport (Equation (4)), while the noncohesive sediment component is partly in suspension and partly through bed load [37]. For cohesive sediment components, the Partheniades–Krone formulations are used for calculating the fluxes between the water phase and bed [38]. Because the noncohesive sediment components in estuaries are mainly fine-grained mud, the Engelund–Hansen transport equation is selected [39] so that the sediment transport for bedload is calculated directly.

$$E^{(l)} = M^{(l)} S\left(\tau_{cw}, \tau_{cr,e}^{(l)}\right) \quad (5)$$

$$D^{(l)} = w^{(l)} c_b^l S\left(\tau_{cw}, \tau_{cr,d}^{(l)}\right) \quad (6)$$

$$c_b^{(l)} = c^{(l)}\left(z = \frac{\Delta Z_b}{2}\right), t \quad (7)$$

$$Q = \frac{0.05 \alpha q^5}{\sqrt{g} C^3 \Delta^2 D_{50}} \quad (8)$$

where $E^{(l)}$ is the erosion flux of mud (kgm$^{-2}$s$^{-1}$), $M^{(l)}$ is the erosion parameter (kgm$^{-2}$s$^{-1}$), $D^{(l)}$ is the deposition flux of mud (kgm$^{-2}$s$^{-1}$), $w^{(l)}$ is the fall velocity (ms$^{-1}$), $c_b^{(l)}$ is the average sediment concentration, $S$ is the erosion or deposition step function, $\tau_{cw}$ is the maximum bed shear stress due to currents and waves (Nm$^{-2}$), $\tau_{cr,e}^{(l)}$ is the critical shear stress for erosion (Nm$^{-2}$), $\tau_{cr,d}^{(l)}$ is the critical shear stress for deposition (Nm$^{-2}$), $Q$ is sediment transport (m$^3$m$^{-1}$s$^{-1}$), $q$ is the magnitude of flow velocity (ms$^{-1}$), $\alpha$ is calibration coefficient, $\Delta$ is the relative density $(\rho_s - \rho_w)/\rho_w$, and $D_{50}$ is the median grain size (m).

### 2.2. Numerical Simulation Parameters

The simulated estuary shape is characterized by an ideal funnel shape [40]. The model domain is 36 km × 100 km, comprising part of the fluvial zone, estuary area, and ocean area (Figure S1). The model comprises equal grids with a resolution of 200 m × 160 m, and the grid aspect ratios align consistent with the geometry of the funnel-shaped estuary. Following grid refinement, it facilitates the observation of more morphological details, such as smaller tidal channels and smaller-sized bar features [41]. The grid size is kept constant to improve the simulation convergence. The straight river is 15 km in length and 1.92 km in width, flowing into the inner estuary area (Figure 1). The length of the estuary area is set to be 45 km. The width of the estuary area increases from 2 km at the fluvial inlet to 30 km at the mouth of the estuary (Figure 1). The ocean area has a length of 40 km and width of 36 km. The bed level decreases linearly from the upstream fluvial boundary to the mouth of the estuary, and the overall slope of the model is set to 0.017 (Figure S2). The water depths are set to 8 m at the fluvial boundary and 28 m at the mouth of the estuary. This forms the estuary at the end of the rising sea-level cycle, and the sea level remains constant during the model runs [3]. The schematic of the model settings is shown in Figure 1.

Discharge boundary and open sea boundary are the open boundary conditions used. The average current, tidal frequency, and tidal range height of modern estuaries can provide reference for the model parameters [27,28,42,43]. We set the total fluvial discharge to a constant value of 3000 m$^3$s$^{-1}$ to keep a sufficient and stable source supply for the estuary (Table 1). The tidal boundary condition is the semidiurnal tide with a tidal amplitude of 6.7 m, providing continuous seaward transport power for estuary sediments (Table 1). For ensuring stability and accuracy, a time step of 0.5 min is used. The simulation time is set to 1 yr with a morphological scale factor of 100, which is comparable to a century of sedimentary evolution [41]. Although the large-scale time span of the simulation, the effects of the sea level change and organisms are ignored for generalization and simplicity. The stratigraphic record of the estuary sedimentation sequence is updated in each time step, including the bed level and stratigraphic sediment thickness (Table 1) [44,45]. The sediment subsurface is shown according to a multilayer concept [44,46]; hence, the virtual sedimentary successions of the estuary are set to 400 layers, each with a thickness of 0.1 m [41]. The underlayer fixed substrate tracks the sediment composition over time in the vertical direction. If the remaining sediment thickness is less than the sediment thickness threshold of 0.05 m and erosive conditions are expected. Above the underlayer, a transport layer of 0.2 m is used to exchange the sediment with the fluid layer (Table 1). To stabilize the complex hydrodynamic calculation, a factor for the erosion of adjacent dry cells is specified that determines the proportion of erosion evenly distributed to the adjacent cells. This factor is 0.5 in our simulation, meaning that half of the erosion that occurs in wet cells is distributed to adjacent dry cells [41].

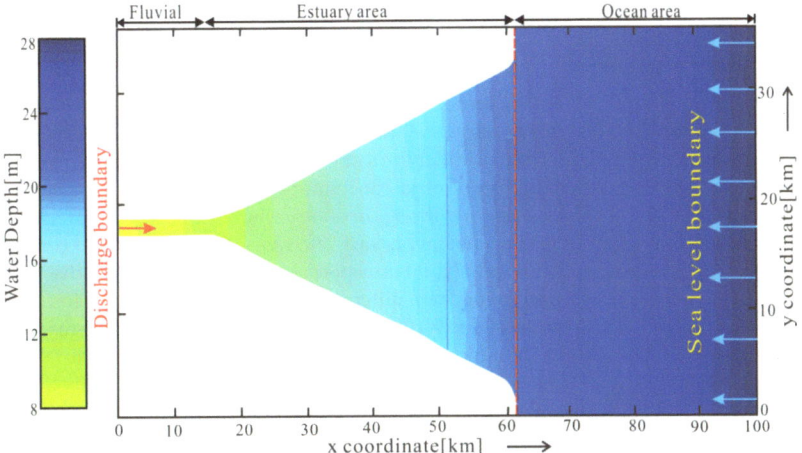

**Figure 1.** Schematic of the conceptual model of the estuary. The color indicates bathymetry, with an initial depth of 8 m in the fluvial boundary and 28 m in the ocean area.

**Table 1.** Initial parameter type, initial value, and range of the default model.

| Parameter | Symbol | Unit | Value | Range |
|---|---|---|---|---|
| Initial water depth | - | m | 28 | - |
| Discharge | - | $m^3 s^{-1}$ | 3000 | 1500–4500 |
| Tidal amplitude | - | m | 6.7 | 3.4–7.2 |
| Time step | dt | min | 0.5 | 0–999 |
| Threshold sediment thickness | - | m | 0.05 | 0.005–10 |
| Threshold depth | - | m | 0.1 | 0–10 |
| Min water depth for bed level change | SedThr | m | 0.1 | 0.1–10 |
| Morphological scale factor | H | - | 100 | 1–400 |
| Number of under layers | MxNULyr | - | 400 | - |
| Thickness of each under layer | ThUnLyr | m | 0.1 | - |
| Thickness of the transport layer | ThTrLyr | m | 0.2 | - |
| Erosion of adjacent dry cells | - | - | 0.5 | 0–1 |

The majority of the sediment supplied to estuaries consist of sand, with mud components and salt marshes [7]. Hence, noncohesive sand sediment and cohesive mud sediment are the sediment components used in the model [47]. For sediment supply, the flow carries sand and mud that supply the estuary area at a fixed concentration, meaning that sediment delivery to the model depends on hydrodynamic conditions. The total sediment supply is set to 7 $kgm^{-3}$ in the simulation. Two types of noncohesive sand sediment components and three types of cohesive mud sediment components are used. Table 2 lists each sediment component property. Cohesive sediment components are defined in terms of setting velocity and critical bed shear stress rather than grain size [48], wherein the default setting velocity is 0.25 $mms^{-1}$, critical shear stress for erosion is 0.5 $Nm^{-2}$, and critical shear stress for sedimentation is 1000 $Nm^{-2}$. If the bed shear stress for sedimentation of the cohesive sediment fractions is larger than the critical value, no sedimentation occurs; otherwise, mixed sediment fluxes are calculated following the Partheniades–Krone equations.

**Table 2.** Sediment fraction types and parameter settings for each sediment component in the models.

| Sediment Component | Type | Median Sediment Diameter (μm) | Setting Velocity (mms$^{-1}$) | Critical Bed Shear Stress for Sedimentation (Nm$^{-2}$) | Critical Bed Shear Stress for Erosion (Nm$^{-2}$) |
|---|---|---|---|---|---|
| Sand1 (S1) | NonCohesive | 125 | - | - | - |
| Sand2 (S2) | NonCohesive | 80 | - | - | - |
| Mud1 (M1) | Cohesive | - | 0.86 | 1000 | 0.3 |
| Mud2 (M2) | Cohesive | - | 0.25 | 1000 | 0.5 |
| Mud3 (M3) | Cohesive | - | 0.16 | 1000 | 0.6 |

A specific model parameter space is designed for investigating the effects of sediment composition and transport on estuary evolution (Table 3). The model is run in nine estuary scenarios under the same initial conditions. The model scenarios are analyzed by studying the effect of mud concentration, mud transport properties, tidal amplitude, and fluvial discharge on the sedimentary characteristics. Fluvial mud supply concentration at the upstream boundary is varied to assess the effect of mud concentration on estuary morphology. The effect of mud transport characteristics is further discussed by comparing scenarios with mud inputs with different setting velocities and erosion shear stress. Common factor analysis is used to further study the role of fluvial discharge and tidal amplitude on sedimentary distribution and to clarify the main factors that control the mud distribution characteristics in estuaries. These simulation results are quantitatively analyzed and compared to each other. Finally, the simulation results are compared with data from real estuaries.

**Table 3.** Parameter list of all model scenarios.

| Model Scenario | Type | Case ID | Fluvial Mud (kgm$^{-3}$) | Sediment Class | Tidal Amplitude (m) | Discharge (m$^3$s$^{-1}$) | Note |
|---|---|---|---|---|---|---|---|
| Base model | default | 01 | 1.75 | S1 S2 M2 | 6.7 | 3000 | Fluvial mud input |
| Mud supply | mud concentration | 02 | 3.5 | S1 S2 M2 | 6.7 | 3000 | Higher fluvial mud |
|  |  | 03 | 0 | S1 S2 | 6.7 | 3000 | No mud, only sand |
|  | mud transport properties | 04 | 1.75 | S1 S2 M1 | 6.7 | 3000 | Higher mud cohesive |
|  |  | 05 | 1.75 | S1 S2 M3 | 6.7 | 3000 | Lower mud cohesive |
| Hydrodynamic condition | tidal amplitude | 06 | 1.75 | S1 S2 M2 | 3.4 | 3000 | Lower tide |
|  |  | 07 | 1.75 | S1 S2 M2 | 7.2 | 3000 | Higher tide |
|  | fluvial discharge | 08 | 1.75 | S1 S2 M2 | 6.7 | 1500 | Lower discharge |
|  |  | 09 | 1.75 | S1 S2 M2 | 6.7 | 4500 | Higher discharge |

The sedimentary characteristics are presented in the form of a map, which assesses the sedimentary distribution and tidal bar morphology or cross-sectional view to enable the study of channel depth variation and sediment thickness evolution. From the three-dimensional sedimentary data, cumulative sedimentation and erosion are further calculated. The sedimentary components are tracked along with the vertical and horizontal directions, and each $x$ and $y$ coordinate point in the model space is recorded to represent the change in the corresponding bar elevation, i.e., decreasing elevation reflects the erosion process and increasing elevation reflects the deposition process [14].

## 3. Results

*3.1. Effect of Mud Concentration on the Sedimentary Characteristics*

For analyzing the effect of mud concentration on sediment characteristics, three sets of comparative scenarios are set, namely, no mud supply (zero mud concentration), lower mud concentration, and higher mud concentration (Table 3). In the simulation, mud supply concentration is changed by adjusting the sand–mud ratio and the total sediment supply is set constant. The sand–mud ratio of the smaller mud concentration model is 3:1, and the sand–mud ratio of the larger mud concentration model is 1:1 (Table 3).

The number of bar and tidal channel morphologies changes considerably with changing the supply of mud concentration (Figure 2). For the case with no mud supply (Case 03), the tidal channel in estuaries has a high degree of cutting and a large number of tidal bars (Figure 2a). Compared to Case 01, when the mud concentration is 1.75 kgm$^{-3}$, the degree of development of the tidal bar in the inner estuary is not high, but tidal bar thickness increases (Figure 2b). Tidal channels develop in the middle and outer estuaries, which become more braided owing to unhindered bank erosion. For the mud concentration of 3.5 kgm$^{-3}$ (Case 02), sediment diffusion becomes slight, and the tidal bar has a high degree of sedimentation in the inner estuary (Figure 2c). Compared with Cases 01 and 03, the tidal channels for higher mud concentration mainly develop in the middle of the river, which are straighter and less migrated. Figure 2 shows that the number of tidal bars decreases with increasing mud concentration and channel migration decreases greatly. Hence, it is concluded that the presence of mud components prevents sediment transport and increases erosion resistance, affording more stable tidal bars and banks.

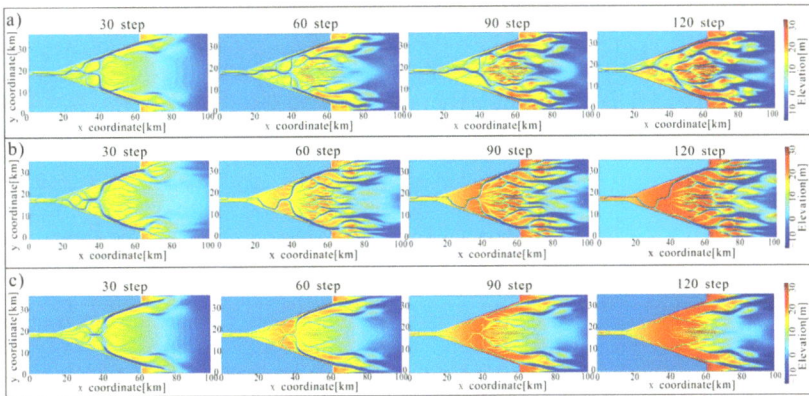

**Figure 2.** Sediment erosion changes in estuaries at different mud concentrations: (**a**) represents the mud concentration of 0 kgm$^{-3}$ (Case 03), (**b**) represents the mud concentration of 1.75 kgm$^{-3}$ (Case 01), and (**c**) represents the mud concentration of 3.5 kgm$^{-3}$ (Case 02). The records of the four stages (30, 60, 90 and 120) represent sediment erosion morphology after 25, 50, 75, and 100 years, respectively. Colors represent the elevation of accumulated erosion sediments, referred to as elevation.

In Figure 3, erosion is obvious in the channel without mud supply at the beginning of the simulation, with a mean channel depth of 3.23 m (Figure 3a). The number of active channels and the mean channel depth decrease when the mud concentrations are 1.75 and 3.5 kgm$^{-3}$, with mean channel depths of 2.58 m and 1.59 m, respectively (Figure 3a). At the end of the simulation, the sand-based estuary forms a deeper channel incision, and the mean depth of the channel is 4.21 m (Figure 3b). In the model with mud concentration of 3.5 kgm$^{-3}$, the mean channel depth is 2.02 m, and the erosion rate is 28% lower than that of the no mud supply model. In addition, the mean channel depth is 3.61 m in the inner estuary, and the mean channel depths of the middle and outer estuaries are 3.80 m and 4.04 m, respectively. The mean channel depth tends to increase as the distance increases from the supply source (Figure 3c). This is because the fluvial mud input enhances the mud deposition at the top of the inner estuary and causes the silting of mud components in the channel. The above simulation results indicate that mud components tend to deposit in the channel and the inner estuary with increasing mud concentration. These mud-dominated sediments are resistant to erosion, thus slowing down the erosion rate and reducing the tidal bar mobility.

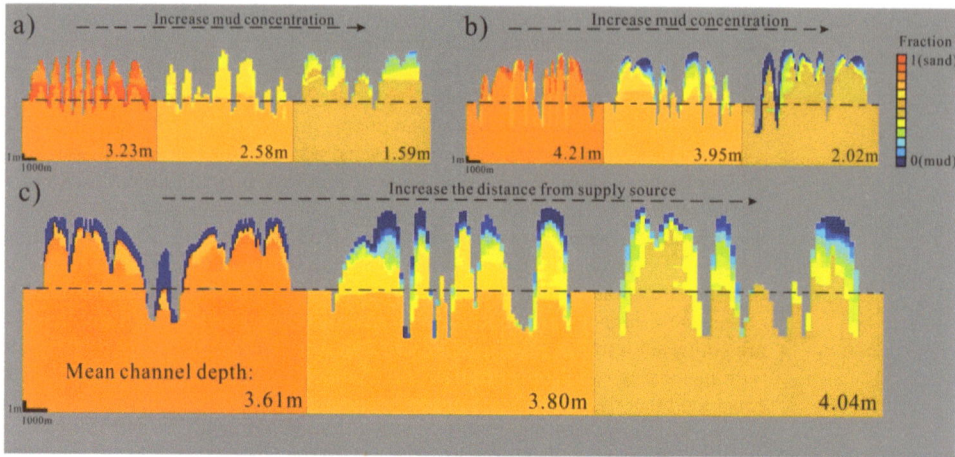

**Figure 3.** Simulated cross section of the channel in estuaries. From left to right in (**a**,**b**): mud concentrations of 0 kgm$^{-3}$ (Case 03), 1.75 kgm$^{-3}$ (default, Case 01), and 3.5 kgm$^{-3}$ (Case 02). (**a**) Cross section (x = 45 km) of the channel after two simulated months. (**b**) Cross section (x = 45 km) of the channel after one simulated year. (**c**) From left to right: the cross section of the inner (x = 30 km), middle (x = 45 km), and outer estuaries (x = 60 km) in the default model.

Mud concentration has a significant effect on the thickness and distribution of mud deposits in the estuary. When there is no mud supply, mud deposits derived from the initial stratigraphy are stirred up by the fluvial flows and tides and 90% of the mud deposits are thinner than 0.47 m. When the mud concentration is 1.75 kgm$^{-3}$, 50% of the mud-deposit thickness is less than 0.36 m and 90% of the mud-deposit thickness is less than 1.80 m in estuary sedimentation. For the mud concentration of 3.5 kgm$^{-3}$, 50% of the mud-deposit thickness is less than 0.36 m and 90% of the mud-deposit thickness is less than 1.92 m. Mud-dominated sediment aggradation occurs more rapidly with increasing mud concentration, yielding thicker mud deposits and higher bed levels. The simulation results herein indicate that the higher the mud concentration, the greater the mud-deposit thickness, and the stronger the self-confinement of the estuary. This self-confinement leads to a smaller surface area and narrower estuaries, eventually affecting the sedimentary characteristics of estuaries. In addition, there is an inevitable relationship between estuary morphology and sediment supply [49].

### 3.2. Effects of Mud Transport Properties on the Tidal Bar Characteristics

The mud transport properties cause complicated processes acting on sediment erosion and deposition under physicochemical forces [50]. This section focuses on the effects of mud transport properties, namely, settling velocity and erosion shear stress on mud deposition (Figures S3 and S4). Tables 2 and 3 list the parameter settings for each sediment component and the type of mud component for each scenario, respectively.

The simulation results show that the morphological characteristics are less affected by the change in mud transport properties at the early stage of simulation. However, significant changes occur in tidal channel development as the simulation continues. For higher and medium settling velocities (Cases 03 and 04; Case 01), the sediment deposition rate is faster in the inner estuary with an average sediment thickness of more than 20 m (Figure 4a,b). At the beginning of the simulation, multiple channels are developed in the inner estuary, the sediment is deposited along both sides of the bed and channel in the inner estuary, and the channels are developed on both sides simultaneously. For higher settling velocity, at the end of the simulation, these also happen (Figure 4a). For medium settling velocity, at the end of the simulation, the sediment accumulates in the inner estuary, which makes part of the channel fill with sediment and develop a single channel (Figure 4b). The estuary with a lower settling velocity shares many similarities in geomorphic morphology with the no mud model. For the lower settling velocity (Case 05), relatively large channel mobility and mean channel depth are observed (Figure 4c). From the perspective of the inner and outer estuaries, the channel is better developed in the inner estuary and the tidal channel is wide and deep in the outer estuary (Figure 4c). Comparing Cases 01, 04, and 05, as the settling velocity decreases, the degree of sediment diffusion increases and the thickness of sediment decreases.

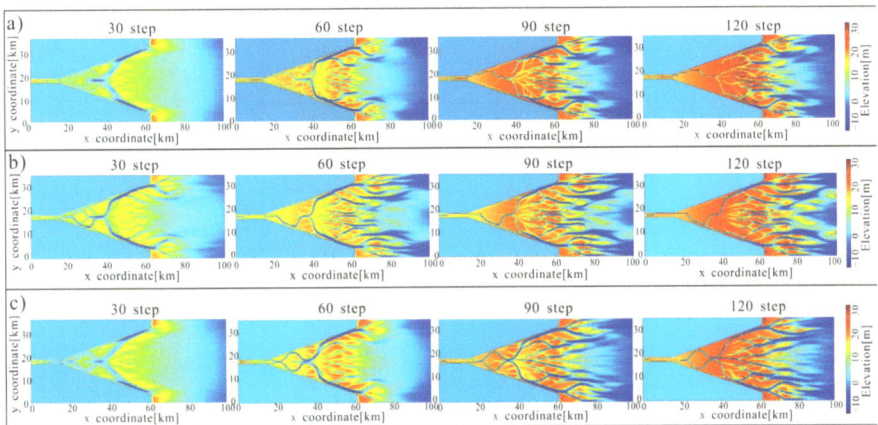

**Figure 4.** Sediment erosion changes in estuaries at different mud transport properties: (**a**) represents the higher settling velocity (Case 04), (**b**) represents the medium settling velocity (Case 01), and (**c**) represents the lower settling velocity (Case 05).

The length, width, and thickness distribution of the bar are further calculated based on the simulation results to quantitatively influence the mud sediment properties on the scale of the tidal bar [51]. The thickness of the tidal bar increases with increasing distance from the estuary mouth. In the higher settling velocity simulation scenario (Case 04), average tidal bar thicknesses in the inner and outer estuaries of 8.2 and up to 28 m are measured (Figure 5a). For lower settling velocity (Case 05), there is an increased average tidal bar thickness from 4 to 19.8 m from the inner estuary to the outer estuary (Figure 5a). The thickness of tidal bars gradually becomes thinner near the ocean area. The tidal bar width is microscopically affected by location and the setting velocity, and the distribution is concentrated between 1 and 4 km (Figure 5b). To be more specific, the tidal bar shape varies with the mud transport properties. In a sediment supply system with a lower setting velocity (Case 05), the RLW values of the tidal bar vary from 1 to 5 (Figure 5c). The higher setting velocity has a great influence on the RLW of the tidal bar, and the ratio is mainly concentrated from 4 to 15 (Case 04; Figure 5c). In addition, the tidal bar increasingly tends to become a long strip and the RLW is higher with increasing distance from the estuary mouth (Figure 5c). This is because the bar has a tendency of avulsion under a strong hydrodynamic disturbance while being protected by the strong cohesion of mud fractions, yielding a strip-shaped bar morphology. More specifically, the tidal bar shape varies with the mud transport properties. In a sediment supply system with a lower setting velocity (Case 05), the RLW value of the tidal bar mainly vary between 1 and 5 (Figure 5c). The higher setting velocity has great influence on the RLW of tidal bar, and the ratio is mainly concentrated between 4 and 15 (Case 04; Figure 5c). Furthermore, the tidal bar increasingly tends to become long strip shape and the RLW is higher as the distance from the estuary mouth increases (Figure 5c). This is because the bar has a tendency of avulsion under the strong hydrodynamic disturbance and, at the same time, it is protected by the strong cohesion of mud fractions. Hence, settling velocity is one of the key factors affecting the development of bar morphology.

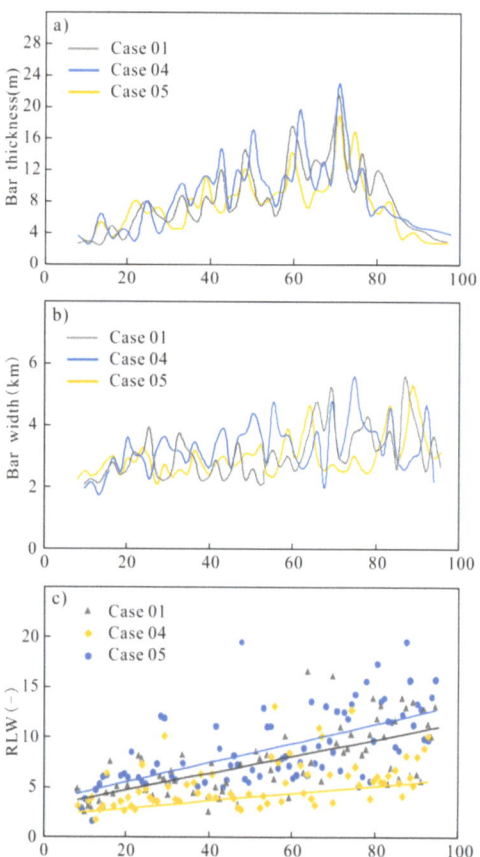

**Figure 5.** Statistics of (**a**) bar thickness, (**b**) bar width, and (**c**) the ratio between length and width of the bar along the estuary X-section.

*3.3. Effects of Hydrodynamic Conditions on the Sedimentary Distribution in Estuaries*

In estuaries, tides are the most important driving force for sediment transport, and fluvial discharge plays an important role in sediment source supply. The interaction of tidal amplitude and fluvial discharge in estuaries has led to a continuously evolving morphology with river channels and bars. In this section, the effects of hydrodynamic conditions on sedimentary distribution are analyzed by changing tidal amplitudes and fluvial discharges. Table 3 lists the specific parameter settings.

It is concluded that the development degree of the tidal bar changes obviously with tidal amplitude (Figure 6). With a lower tidal amplitude of 3.4 m (Case 06), the inner estuary is covered with mud deposition and shows slight progradation (Figure 6a). Although the sediment thickness is the highest among the case of tidal amplitudes (mean 25 m), the tidal bar has not developed very well, only with a mean thickness of 2.1 m. This is because the fluvial sand-carrying and sand-flushing plays a dominant role in estuaries, the tidal action is extremely weak, and hence, there is almost no sediment being reprocessed. For the estuary with a medium tidal amplitude of 6.8 m (Case 01) and higher tidal amplitude of 7.2 m (Case 07), the inner estuary develops multiple deeper channels. In the outer estuary, the tidal bar is well developed with a more complex shape owing to erosion and redeposition

(Figure 6b,c). The increased tidal amplitude makes the sediment deposit farther seaward, yielding a smaller mean sediment thickness in the entire estuary. In summary, the greater the tidal amplitude, the greater the degree of the seaward migration of the tidal bar and the deeper is the erosion of tidal channels.

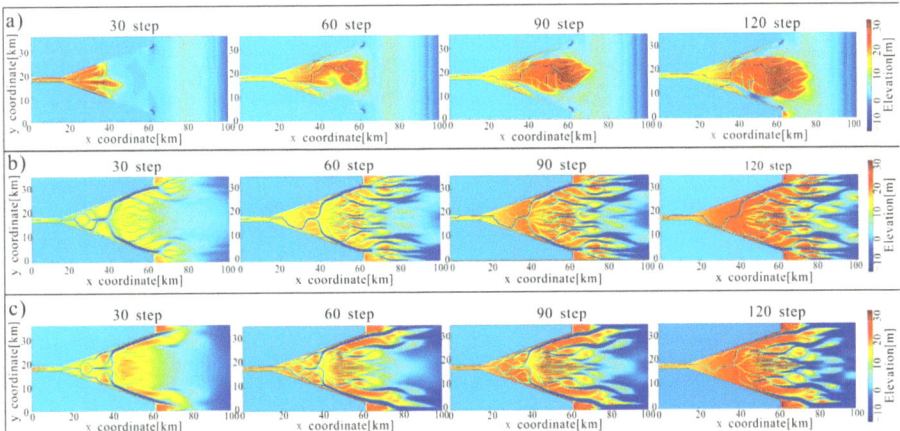

**Figure 6.** Sediment erosion changes in estuaries at different tidal amplitudes: (**a**) represents the tidal amplitude of 3.4 m (Case 06), (**b**) represents the tidal amplitude of 6.8 m (Case 01), and (**c**) represents the tidal amplitude of 7.2 m (Case 07).

The simulation results show that fluvial discharge considerably affects the development rate and sediment thickness of tidal bars. For lower fluvial discharge (Case 08), the tidal bar exhibits an elliptical shape in the inner estuary of ~4.5 m thickness (Figure 7a). For higher fluvial discharge (Case 09), the inner estuary bars appear to be an elongated shape, with a faster development rate (Figure 7c). In addition, the tidal bars gradually develop to a complex bar. The higher the fluvial discharge, the higher the deposition rate. In the later stage of the simulation, the sediments are concentrated in the middle and outer estuaries, where the sediment thickness is the largest. Based on the simulation results, this implies that river-dominated estuaries form larger and thicker deposits, which more easily cause the transition from filled estuaries to deltas [13,52].

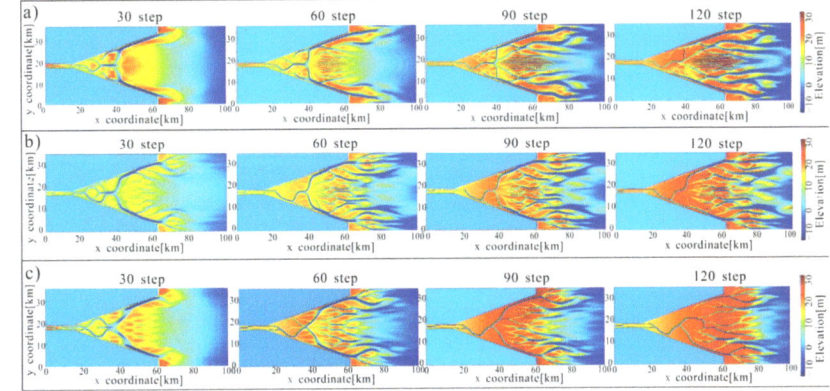

**Figure 7.** Sediment erosion changes in estuaries at different fluvial discharges: (**a**) represents the fluvial discharge of 1500 m$^3$s$^{-1}$ (Case 08), (**b**) represents the fluvial discharge of 3000 m$^3$s$^{-1}$ (Case 01), and (**c**) represents the fluvial discharge of 4500 m$^3$s$^{-1}$ (Case 09).

The evolution of sediment progradation in Figure 8 shows that the range of sediment progradation in estuaries is concentrated at 70 km under the three fluvial discharge conditions, indicating a slight effect of fluvial discharge on sediment transport distance. The increase in tidal amplitude brings an obvious prograde seaward. The progradation area in the lower tidal amplitude accounts for 53% of the estuary (Figure 8a), and the range of the progradation is ~58 km. The progradation area can reach the entire estuary in the higher tidal amplitude (Figure 8b). This indicates that tidal energy is the major factor that determines the range of sediment progradation in estuaries [15]. In addition, the higher fluvial discharge with higher tidal amplitude keeps the sediment in a suspended state, affording a more dynamic system. When the discharge is high and the tidal amplitude is low, the estuary fills and eventually evolves into a delta. This finding demonstrates that in the absence of a sea-level rise and fall, hydrodynamic conditions are enough to alter sediment retrogradation and progradation behavior [2,53].

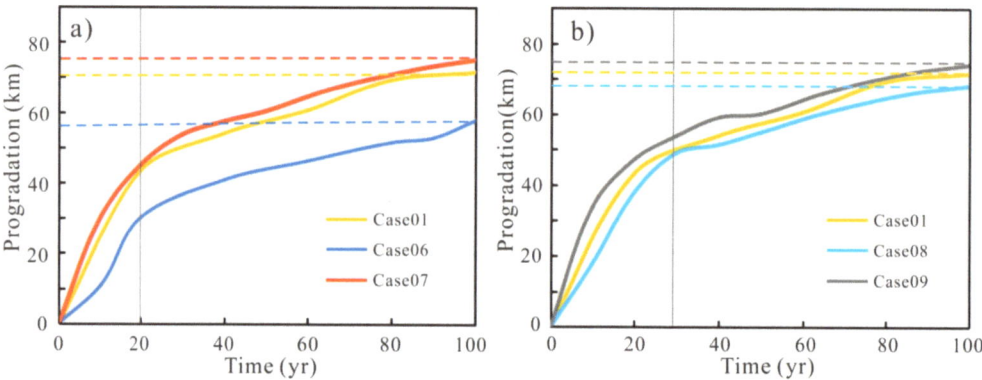

**Figure 8.** Evolution of sediment progradation over time. (**a**) Effects of tidal amplitude variation on sediment progradation. (**b**) Effects of fluvial discharge variation on sediment progradation.

## 4. Discussion

### 4.1. Comparison with Modern Sedimentation

The development of tidal bars and the sedimentary distribution characteristics in estuaries change with the changes of various factors. Based on sedimentary dynamic simulations, we quantified the effects of mud concentration, mud transport properties, tidal amplitude, and fluvial discharge on the sedimentary characteristics of the estuary. The numerical simulation results indicate that tidal amplitude plays a major role in tidal bar morphology. The greater the tidal amplitude, the higher the degree of development of tidal bars in estuaries, consistent with Schramkowski et al. [19]. Mud concentration and fluvial discharge also have a significant effect on estuary sediment thickness. With a higher mud concentration, relatively higher fluvial discharge, and lower tidal amplitude, the estuary sediment thickness becomes larger and the deposition area becomes wider, which is in good agreement with the simulation results of Hibma et al. [54].

The sediment in the models is similar to modern estuaries in terms of their distribution characteristics and behavior. There is a relatively large proportion of erosion sediments on both sides of the channel in the simulation results (Figure 4), forming mudflat deposition in agreement with the characteristics of modern estuary datasets [55]. There are fewer mud deposits in the center of the outer estuary compared to mudflats along the flank of the basin, and a similar pattern has been observed in the Western Scheldt [16,50]. The Scheldt estuary consists mainly of sand with a small amount of mud deposition [47]. The estuary has a freshwater discharge of 120 $m^3 s^{-1}$, and fluvial mud supply is 100 × $10^6$ kgyr$^{-1}$ [47]. Likewise, the Scheldt estuary supplied mud deposition from a single channel as in the simulated model.

Figure 9 demonstrates the cumulative probability distribution of mud-deposit thickness covering the top of the modeled estuary and Scheldt estuary. Average mud-deposit thicknesses of the modeled and Scheldt estuaries are 1.0 and 1.2 m, respectively (Figure 9). This is because the mud input in the Scheldt estuary is ~15% higher [13]. Here, the mud-deposit thickness of the Scheldt estuary increases from 1.3 m near the head to 2.7 m near the mouth [47]. About 60% of the mud-deposit thickness obtained from the models is less than 0.5 m, and 90% of the mud-deposit thickness is concentrated from 0.05 to 3.0 m (Figure 9), in general accordance with the Scheldt estuary of sediment distribution. From the above statistical characteristics of sediment thickness, the simulated estuary is in good agreement with modern sedimentation.

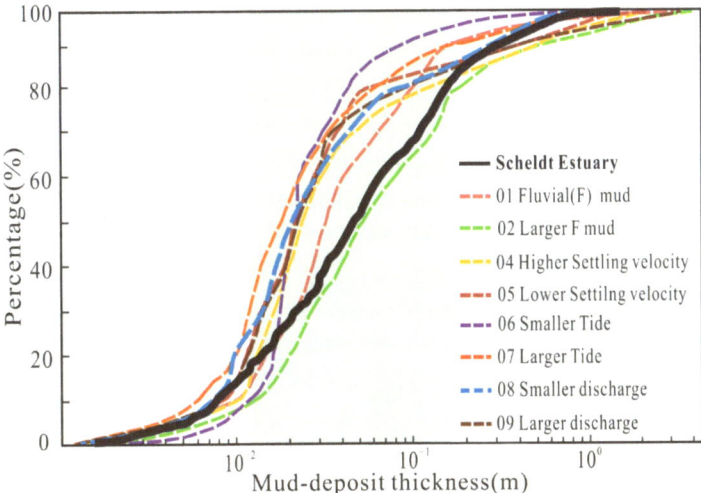

**Figure 9.** The cumulative probability distribution graphs of mud-deposit thickness for Scheldt estuary and all estuary models.

Comparing the dimensions of individual tidal bars from the datasets collected by Leuven et al. [55] with the tidal bars of the simulation results, it is found that the aspect ratios of the tidal bars are in the range of 3–10. The lengths of tidal bars in our simulations are concentrated at ~10 km, and the widths are distributed at 1.5 km and partitioned bar widths are distributed at 0.8 km (Figure 10), with a similar scatter to the datasets. Most of the simulated scatter points are distributed within the range of higher or lower confidence limit trends, and the tidal bars are close to the modern estuary bar size. Meanwhile, the bar morphology in the estuary models is consistent with the natural bar. A natural estuary is developed from the seaward direction with a sidebar, distributary bar, compound bar, U-shaped bar, and linear bar. The simulation herein shows that the linear bars are concentrated in the outer estuary, especially in the case with a tidal amplitude of 7.2 m (Figure 6). The U-shaped bars appear in the middle of the outer estuary, and the compound bars are stored in the inner estuary at the later stage of the simulation. The distributary bars are the most obvious in the case with a fluvial discharge of 1500 $m^3s^{-1}$ (Figure 7). From the comparative analysis of statistical data, the sedimentary characteristics of estuaries obtained using the sedimentary dynamic numerical simulation method are in good agreement with modern sedimentation characteristics.

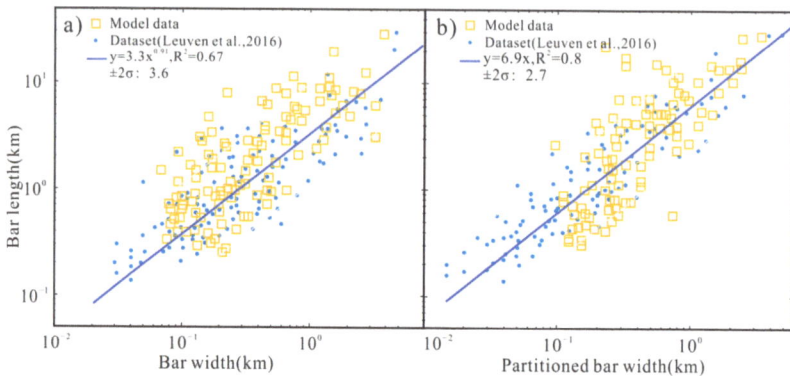

**Figure 10.** Comparison of simulation results with modern estuaries [55]. (**a**) Correlation of bar length and width. (**b**) Correlation of bar length and partitioned bar width.

### 4.2. Distribution of the Mud Deposits

Mud deposits are mainly stored in the estuary in two forms. Most of the mud is deposited at the top of the estuary in the form of mudflats, and a small amount is deposited in the middle of the bar as mud interlayers. The mud interlayer is formed in connection with the sedimentary environment, and the banded mudstone wall easily develops [56]. The estuary model reproduces the three-dimensional internal structure of the mud interlayer, enabling us to further study the quantified distribution of interlayers in modern estuaries. Based on the statistical data, the thickness of the mud interlayer is concentrated in 0.4–0.6 m, and a small part of the mud interlayer can reach 1.5 m thickness (Table 4). The length of mud interlayers varies considerably, with some of them concentrated in 2–4 km and others concentrated in 8–10 km, with the maximum length of the mud interlayer reaching 16 km (Table 4).

**Table 4.** Descriptive statistics of the sediment progradation and the mud interlayers in the estuary model.

| Case ID | Average Sediment Progradation (km) | Length of Mud Interlayer (km) | Thickness of Mud Interlayer (m) | Distribution Frequency (Pieces) |
|---|---|---|---|---|
| 01 | 67.3 | 4.93 | 0.47 | 1.23 |
| 02 | 60.8 | 5.41 | 0.52 | 0.98 |
| 03 | 70.1 | - | - | - |
| 04 | 58.8 | 5.59 | 0.50 | 0.92 |
| 05 | 62.0 | 7.03 | 0.32 | 1.20 |
| 06 | 57.5 | 2.47 | 0.28 | 0.48 |
| 07 | 75.4 | 7.19 | 0.39 | 0.57 |
| 08 | 65.3 | 3.89 | 0.41 | 0.68 |
| 09 | 69.6 | 5.43 | 0.72 | 1.05 |

The increased mud concentration has a positive feedback effect on the length and thickness of the mud interlayer (Table 4) [56]. Conversely, the alteration of mud concentration is opposite to the sediment progradation, in which the higher mud concentration cases restrict seaward progradation (Table 4). The average mud interlayer thickness and length are similar in most cases and no significant trend for changing mud properties or mud supply sources is observed. The effect of hydrodynamic conditions is intense for the internal structure of the mud interlayer [56]. The thickness of the interlayer increases with increasing fluvial discharge, but the length of the interlayer varies poorly with fluvial discharge. Under the smallest discharge, the interlayer length is the smallest (Table 4). With increasing tidal intensity, the thickness of the interlayer decreases, while the length of the interlayer increases and changes (Table 4). The above observations suggest that the tide is the main controlling factor and fluvial current is a secondary controlling factor for the length of the interlayer. For the thickness of the interlayer, the fluvial currents exerted the primary control and tides had a secondary effect. The quantitative statistics of the thickness and distribution of mud interlayer are of great significance for the classification and understanding of the internal architecture of estuaries [56].

## 5. Conclusions

A process-based numerical simulation method is proposed to investigate the effect of mud supply and hydrodynamic conditions on the sedimentary development and distribution in estuaries. Our model demonstrates the effect of mud supply, sediment transport, and hydrodynamic conditions on the long-term evolution of estuaries. The statistical data obtained using this the model provide a quantitative analysis of the process–product relationship in estuaries. A series of morphological maps and cross-sectional view results show that the estuary develops into dynamic channels and sandbars flanked by mudflats, and the thickness is in the range of 1–2 m. About 60% of the thickness of mud deposits is between 0.21 m and 0.36 m, and 90% of the mud deposits are more than 0.69 m thick. Meanwhile, a small amount of mud is deposited in the middle of the bar as mud interlayers. The thickness of the mud interlayer is concentrated from 0.4 to 0.6 m; however, the length varies considerably, up to 16 km. The study concludes that mud supply strengthens self-confinement, and a higher mud concentration yields stable tidal bars and banks as well as reduced channel migration. Conversely, the estuary without mud supply is less resistant to erosion and more highly incised; thus, there are more tidal bars. Mud transport properties considerably affect the tidal bar morphology. The estuary tidal bar with a higher settling velocity has a high degree of development and a larger length-to-width ratio, mainly forming a long strip tidal bar. The study of hydrological conditions focuses on rivers and tides, and fluvial discharge and tidal amplitude form negative feedback on the dynamic balance between deposition and erosion. When the fluvial discharge is low and the tidal amplitude is high, mud deposits are suspended in the estuary so that less mud settles. Meanwhile, with lower fluvial discharge and higher tidal amplitude, the estuary fills up and eventually becomes a tidal delta. The sediments in the models are in good agreement with sedimentation in modern estuaries in terms of their distribution characteristics and behavior. It is hoped that the sedimentary dynamic numerical simulation method used herein can help predict variations in stratigraphic structures and provide guidance for the further exploration and development of estuary sedimentary reservoirs.

**Supplementary Materials:** The following supporting information can be downloaded at: https://www.mdpi.com/article/10.3390/jmse11010174/s1, Figure S1: Sediment erosion changes in estuaries at different the shape of the funnel. Figure S2: Sediment erosion changes in estuaries at different the slope of the depth. Figure S3: Bed shear stress over time in default model (Case 01). Figure S4: Velocity field in default model (Case 01).

**Author Contributions:** Conceptualization, M.T. and Q.Z.; methodology, M.T.; software, M.T.; validation, M.T. and Q.Z.; formal analysis, Q.Z.; investigation, Q.Z. and X.L.; resources, Q.Z. and X.L.; data curation, S.L.; writing—original draft preparation, Q.Z.; writing—review and editing, M.T. and Q.Z.; visualization, M.T.; supervision, S.X.; project administration, M.T.; funding acquisition, M.T. All authors have read and agreed to the published version of the manuscript.

**Funding:** This research was funded by the National Natural Science Foundation (grant numbers 42072163, 41972250), and the Foundation of Shandong Province (grant number ZR2019MD006), and the Foundation of CNPC (grant number 2021DJ3302).

**Conflicts of Interest:** The authors declare no conflict of interest.

## References

1. Galloway, W.E. *Process Framework for Describing the Morphologic and Stratigraphic Evolution of Deltaic Depositional Systems*; Houston Geological Society: Houston, TX, USA, 1975; pp. 87–98.
2. Dalrymple, R.W.; Zaitlin, B.A.; Boyd, R. Estuarine Facies Models: Conceptual Basis and Stratigraphic Implications. *J. Sediment. Res.* **1992**, *62*, 1130–1146. [CrossRef]
3. Dalrymple, R.W.; Choi, K. Morphologic and facies trends through the fluvial–marine transition in tide-dominated depositional systems: A schematic framework for environmental and sequence-stratigraphic interpretation. *Earth-Sci. Rev.* **2007**, *81*, 135–174. [CrossRef]
4. Gugliotta, M.; Saito, Y.; Van Lap, N.; Thi Kim Oanh, T.; Nakashima, R.; Tamura, T.; Uehara, K.; Katsuki, K.; Yamamoto, S. Process regime, salinity, morphological, and sedimentary trends along the fluvial to marine transition zone of the mixed-energy Mekong River delta, Vietnam. *Cont. Shelf Res.* **2017**, *147*, 7–26. [CrossRef]
5. Van der Wegen, M.; Wang, Z.B.; Savenije, H.H.G.; Roelvink, J.A. Long-term morphodynamic evolution and energy dissipation in a coastal plain, tidal embayment. *J. Geophys. Res-Earth Surf.* **2008**, *113*, F03001. [CrossRef]
6. Gugliotta, M.; Saito, Y.; Nguyen, V.L.; Ta, T.K.O.; Tamura, T. Sediment distribution and depositional processes along the fluvial to marine transition zone of the Mekong River delta, Vietnam. *Sedimentology* **2019**, *66*, 146–164. [CrossRef]
7. Dalman, R.; Weltje, G.J.; Karamitopoulos, P. High-resolution sequence stratigraphy of fluvio-deltaic systems: Prospects of system-wide chronostratigraphic correlation. *Earth Planet. Sci. Lett.* **2015**, *412*, 10–17. [CrossRef]
8. Van Ledden, M.V.; Kesteren, W.; Winterwerp, J.C. A conceptual framework for the erosion behaviour of sand–mud mixtures. *Cont. Shelf Res.* **2004**, *24*, 1–11. [CrossRef]
9. Le Hir, P.; Cayocca, F.; Waeles, B. Dynamics of sand and mud mixtures: A multiprocess-based modelling strategy. *Cont. Shelf Res.* **2011**, *31*, S135–S149. [CrossRef]
10. Schuurman, F.; Shimizu, Y.; Iwasaki, T.; Kleinhans, M.G. Dynamic meandering in response to upstream perturbations and floodplain formation. *Geomorphology* **2016**, *253*, 94–109. [CrossRef]
11. Dalrymple, R.W.; Baker, E.K.; Harris, P.T.; Hughes, M.G. Sedimentology and Stratigraphy of a Tide-Dominated, Foreland-Basin Delta (Fly River, Papua New Guinea). In *Tropical Deltas of Southeast Asia*; SEPM (Society for Sedimentary Geology): Tulsa, OK, USA, 2003; pp. 143–173.
12. Fenies, H.; Tastet, J.P. Facies and architecture of an estuarine tidal bar (the Trompeloup bar, Gironde Estuary, SW France). *Mar. Geol.* **1998**, *150*, 149–169. [CrossRef]
13. Van de Lageweg, W.I.; Braat, L.; Parsons, D.R.; Kleinhans, M.G. Controls on mud distribution and architecture along the fluvial-to-marine transition. *Geology* **2018**, *46*, 971–974. [CrossRef]
14. Van de Lageweg, W.I.; Feldman, H. Process-based modelling of morphodynamics and bar architecture in confined basins with fluvial and tidal currents. *Mar. Geol.* **2018**, *398*, 35–47. [CrossRef]
15. Winterwerp, J.C. Fine sediment transport by tidal asymmetry in the high-concentrated Ems River: Indications for a regime shift in response to channel deepening. *Ocean Dyn.* **2011**, *61*, 203–215. [CrossRef]
16. Verlaan, P. Marine vs Fluvial Bottom Mud in the Scheldt Estuary. *Estuar. Coast. Shelf Sci.* **2000**, *50*, 627–638. [CrossRef]
17. Cleveringa, J.; Dam, G. *Slib in de Sedimentbalans van de Westerschelde*; Eindrapport G-3, 1630/U12376/C/GD; Vlaams Nederlandse Scheldecommissie: Bergen op Zoom, The Netherlands, 2013.
18. Kleinhans, M.G. Sorting out river channel patterns. *Prog. Phys. Geogr.* **2010**, *34*, 287–326. [CrossRef]
19. Schramkowski, G.P.; Schuttelaars, H.M.; de Swart, H.E. The Effect of Geometry and Bottom Friction on Local Bed Forms in a Tidal Embayment. *Cont. Shelf Res.* **2002**, *22*, 182–1833. [CrossRef]
20. Toffolon, M.; Crosato, A. Developing Macroscale Indicators for Estuarine Morphology: The Case of the Scheldt Estuary. *J. Coast. Res.* **2007**, *231*, 195–212. [CrossRef]
21. Tessier, B. Stratigraphy of tide-dominated estuaries. In *Principles of Tidal Sedimentology*; Davis, R.A., Jr., Dalrymple, R.W., Eds.; Springer: Dordrecht, The Netherlands, 2012; pp. 109–128.
22. Sprovieri, M.; Bonanno, A.; Mazzola, S.; Patti, B. Cyclostratigraphy: A methodological approach. *Riv. Ital. Paleontol. Stratigr.* **2002**, *108*, 179–182. [CrossRef]
23. Die Moran, A.D.; Abderrezzak, K.E.K.; Mosselman, E.; Habersack, H.; Lebert, F.; Aelbrecht, D.; Laperrousaz, E. Physical model experiments for sediment supply to the old Rhine through induced bank erosion. *Int. J. Sediment. Res.* **2013**, *28*, 431–447. [CrossRef]
24. Leuven, J.R.; Braat, L.; van Dijk, W.M.; de Haas, T.; Van Onselen, E.; Ruessink, B.; Kleinhans, M.G. Growing forced bars determine nonideal estuary planform. *J. Geophys. Res. Earth Surf.* **2018**, *123*, 2971–2992. [CrossRef]
25. Peng, Y.; Olariu, C.; Steel, R.J. Recognizing tide-and wave-dominated compound deltaic clinothems in the rock record. *Geology* **2020**, *48*, 1149–1153. [CrossRef]
26. Alshammari, B.; Mountney, N.P.; Colombera, L.; Al-Masrahy, M.A. Sedimentology and stratigraphic architecture of a fluvial to shallow-marine succession: The Jurassic Dhruma Formation, Saudi Arabia. *J. Sediment. Res.* **2021**, *91*, 773–794. [CrossRef]

27. Liu, X.; Lu, S.; Mingming, T.; Sun, D.; Tang, J.; Zhang, K.; He, T.; Qi, N.; Lu, M. Numerical Simulation of Sedimentary Dynamics to Estuarine Bar under the Coupled Fluvial-Tidal Control. *Diqiu Kexue-Zhongguo Dizhi Daxue Xuebao/Earth Sci.-J. China Univ. Geosci* **2020**, *46*, 2944–2957. [CrossRef]
28. Tang, M.; Lu, S.; Zhang, K.; Yin, X.; Ma, H.; Shi, X.; Liu, X.; Chu, C. A three dimensional high-resolution reservoir model of Napo Formation in Oriente Basin, Ecuador, integrating sediment dynamic simulation and geostatistics. *Mar. Pet. Geol.* **2019**, *110*, 240–253. [CrossRef]
29. Weisscher, S.A.H.; Shimizu, Y.; Kleinhans, M.G. Upstream perturbation and floodplain formation effects on chute-cutoff-dominated meandering river pattern and dynamics. *Earth Surf. Process. Landf.* **2019**, *44*, 2156–2169. [CrossRef]
30. Edmonds, D.A.; Slingerland, R.L. Mechanics of river mouth bar formation: Implications for the morphodynamics of delta distributary networks. *J. Geophys. Res.-Earth Surf.* **2007**, *112*, F02034. [CrossRef]
31. Burpee, A.P.; Slingerland, R.L.; Edmonds, D.A.; Parsons, D.; Best, J.; Cederberg, J.; McGuffin, A.; Caldwell, R.; Nijhuis, A.; Royce, J. Grain-size controls on the morphology and internal geometry of river-dominated deltas. *J. Sediment. Res.* **2015**, *85*, 699–714. [CrossRef]
32. Van Ledden, M.; Wang, Z.B.; Winterwerp, H.; de Vriend, H. Sand-mud morphodynamics in a short tidal basin. *Ocean. Dyn.* **2004**, *54*, 385–391. [CrossRef]
33. Lesser, G.R.; Roelvink, J.A.; van Kester, J.; Stelling, G.S. Development and validation of a three-dimensional morphological model. *Coast. Eng* **2004**, *51*, 883–915. [CrossRef]
34. Guo, L.; van der Wegen, M.; Wang, Z.B.; Roelvink, D.; He, Q. Exploring the impacts of multiple tidal constituents and varying river flow on long-term, large-scale estuarine morphodynamics by means of a 1-D model. *J. Geophys. Res.-Earth Surf.* **2016**, *121*, 1000–1022. [CrossRef]
35. Vona, I.; Palinkas, C.M.; Nardin, W. Sediment Exchange Between the Created Saltmarshes of Living Shorelines and Adjacent Submersed Aquatic Vegetation in the Chesapeake Bay. *Front. Mar. Sci.* **2021**, *8*, 727080. [CrossRef]
36. Zhu, Q.; Wiberg, P.L.; Reidenbach, M.A. Quantifying Seasonal Seagrass Effects on Flow and Sediment Dynamics in a Back-Barrier Bay. *J. Geophys. Res.-Oceans* **2021**, *126*, e2020JC016547. [CrossRef]
37. Caldwell, R.L.; Edmonds, D.A. The effects of sediment properties on deltaic processes and morphologies: A numerical modeling study. *J. Geophys. Res.-Earth Surf.* **2014**, *119*, 961–982. [CrossRef]
38. Partheniades, E. Erosion and Deposition of Cohesive Soils. *Am. Soc. Civ. Eng.* **1965**, *91*, 105–139. [CrossRef]
39. Engelund, F.A.; Hansen, E. *A Monograph on Sediment Transport in Alluvial Streams. Hydrotechnical Construction*; Tekniskforlag: Copenhagen, Denmark, 1967.
40. Davies, G.; Woodroffe, C.D. Tidal estuary width convergence: Theory and form in North Australian estuaries. *Earth Surf. Process. Landf.* **2010**, *35*, 737–749. [CrossRef]
41. Dam, G.; van der Wegen, M.; Labeur, R.J.; Roelvink, D. Modeling centuries of estuarine morphodynamics in the Western Scheldt estuary. *Geophys. Res. Lett.* **2016**, *43*, 3839–3847. [CrossRef]
42. Guo, L.; van der Wegen, M.; Roelvink, D.; He, Q. Exploration of the impact of seasonal river discharge variations on long-term estuarine morphodynamic behavior. *Coast. Eng.* **2015**, *95*, 105–116. [CrossRef]
43. Khojasteh, D.; Glamore, W.; Heimhuber, V.; Felder, S. Sea level rise impacts on estuarine dynamics: A review. *Sci. Total Environ* **2021**, *780*, 146470. [CrossRef]
44. Roelvink, J.A. Coastal morphodynamic evolution techniques. *Coast. Eng.* **2006**, *53*, 277–287. [CrossRef]
45. Braat, L.; van Kessel, T.; Leuven, J.R.F.W.; Kleinhans, M.G. Effects of mud supply on large-scale estuary morphology and development over centuries to millennia. *Earth Surf. Dyn.* **2017**, *5*, 617–652. [CrossRef]
46. Ribberink, J.S.; Blom, A.; van der Scheer, P.; van Straalen, M.P. Multi-fraction techniques for sediment transport and morphological modeling in sand-gravel rivers. *River Flow* **2002**, *2002*, 731–739.
47. Geleynse, N.; Storms, J.E.A.; Walstra, D.-J.R.; Jagers, H.R.A.; Wang, Z.B.; Stive, M.J.F. Controls on river delta formation; insights from numerical modelling. *Earth Planet. Sci. Lett* **2011**, *302*, 217–226. [CrossRef]
48. Van der Vegt, H.; Storms, J.E.A.; Walstra, D.J.R.; Howes, N.C. Can bed load transport drive varying depositional behaviour in river delta environments? *Sediment. Geol.* **2016**, *345*, 19–32. [CrossRef]
49. Gugliotta, M.; Saito, Y. Matching trends in channel width, sinuosity, and depth along the fluvial to marine transition zone of tide-dominated river deltas: The need for a revision of depositional and hydraulic models. *Earth-Sci. Rev.* **2019**, *191*, 93–113. [CrossRef]
50. Van Kessel, T.; Vanlede, J.; de Kok, J. Development of a mud transport model for the Scheldt estuary. *Cont. Shelf Res.* **2011**, *31*, S165–S181. [CrossRef]
51. Gugliotta, M.; Saito, Y.; Nguyen, V.L.; Ta, T.K.O.; Tamura, T.; Fukuda, S. Tide-and river-generated mud pebbles from the fluvial to marine transition zone of the Mekong River Delta, Vietnam. *J. Sediment. Res.* **2018**, *88*, 981–990. [CrossRef]
52. Feldman, H.; Demko, T. Recognition and prediction of petroleum reservoirs in the fluvial/tidal transition—ScienceDirect. *Dev. Sedimentol.* **2015**, *68*, 483–528. [CrossRef]
53. Boyd, R. Classification of clastic coastal depositional environments. *Sediment. Geol.* **1992**, *80*, 139–150. [CrossRef]
54. Hibma, A.; de Vrient, H.J.; Stive, M.J.F. Numerical modelling of shoal pattern formation in well-mixed elongated estuaries. *Estuar. Coast. Shelf Sci.* **2003**, *57*, 981–991. [CrossRef]

55. Leuven, J.R.F.W.; Kleinhans, M.G.; Weisscher, S.A.H.; van der Vegt, M. Tidal sand bar dimensions and shapes in estuaries. *Earth-Sci. Rev.* **2016**, *161*, 204–223. [CrossRef]
56. Yang, J.; Zhang, K.; Chen, H.; Lu, S.; Wan, X.; Tang, M.; Xiao, D.; Zhang, C. Genesis of mudstone dikes and their impact on oil accumulations in D-F oilfield of Oriente Basin, Ecuador. *Oil Gas Geol.* **2017**, *38*, 1156–1164. [CrossRef]

**Disclaimer/Publisher's Note:** The statements, opinions and data contained in all publications are solely those of the individual author(s) and contributor(s) and not of MDPI and/or the editor(s). MDPI and/or the editor(s) disclaim responsibility for any injury to people or property resulting from any ideas, methods, instructions or products referred to in the content.

*Article*

# Unraveling the Origin of the Messinian? Evaporites in Zakynthos Island, Ionian Sea: Implications for the Sealing Capacity in the Mediterranean Sea

Avraam Zelilidis [1,*], Nicolina Bourli [1], Konstantinos Andriopoulos [1], Eleftherios Georgoulas [1], Savvas Peridis [1], Dimitrios Asimakopoulos [1] and Angelos G. Maravelis [2]

1 Laboratory of Sedimentology, Department of Geology, University of Patras, 26504 Rion, Greece
2 Department of Geology, Aristotle University of Thessaloniki, 54124 Thessaloniki, Greece
* Correspondence: a.zelilidis@upatras.gr; Tel.: +30-697-203-4153

**Abstract:** The new approach on depositional conditions of the Messinian evaporites in Zakynthos Island indicates that the evaporites in the Kalamaki and Ag. Sostis areas were redeposited during the Early Pliocene. They accumulated either as turbiditic evaporites or as slumped blocks, as a response to Kalamaki thrust activity. Thrust activity developed a narrow and restricted Kalamaki foreland basin with the uplifted orogenic wedge consisting of Messinian evaporites. These evaporites eroded and redeposited in the foreland basin as submarine fans with turbiditic currents or slumped blocks (olistholiths) that consist of Messinian evaporites. These conditions occurred just before the inundation of the Mediterranean, during or prior to the Early Pliocene (Zanclean). Following the re-sedimentation of the Messinian evaporites, the inundation of the Mediterranean produced the "Lago Mare" fine-grained sediments that rest unconformably over the resedimented evaporites. The "Trubi" limestones were deposited later. It is critical to understand the origin of the "Messinian" Evaporites because they can serve as an effective seal rock for the oil and gas industry. It is thus important to evaluate their thickness and distribution into the SE Mediterranean Sea.

**Keywords:** slide; messinian evaporites; turbiditic evaporites; Ionian thrust; Ionian foreland; Kalamaki foreland

**Citation:** Zelilidis, A.; Bourli, N.; Andriopoulos, K.; Georgoulas, E.; Peridis, S.; Asimakopoulos, D.; Maravelis, A.G. Unraveling the Origin of the Messinian? Evaporites in Zakynthos Island, Ionian Sea: Implications for the Sealing Capacity in the Mediterranean Sea. *J. Mar. Sci. Eng.* **2023**, *11*, 271. https://doi.org/10.3390/jmse11020271

Academic Editor: Antoni Calafat

Received: 22 December 2022
Revised: 11 January 2023
Accepted: 19 January 2023
Published: 25 January 2023

**Copyright:** © 2023 by the authors. Licensee MDPI, Basel, Switzerland. This article is an open access article distributed under the terms and conditions of the Creative Commons Attribution (CC BY) license (https://creativecommons.org/licenses/by/4.0/).

## 1. Introduction

Western Greece has been the focus of long-standing academic and industry interest regarding the regional hydrocarbon generative potential [1–4]. Despite the significant oil and gas reserves that were discovered in nearby regions with similar geological characteristics, such as Albania, Croatia, and Italy [5–8], the exploration activities in western Greece have proved unsatisfactory. The lack of sufficient data (e.g., 3D seismic lines, exploration wells, geochronological data) and the complex tectonic setting are considered the principal reasons for these results [2,3]. However, recent studies that suggest the presence of working petroleum systems rejuvenated the exploration activities [1–8], and nowadays, the Governments of Greece try to explore the offshore areas (Ionian foreland basins) for oil and gas fields. Greek authorities run licensing rounds and offer offshore blocks (more than 74 blocks in total; in Greece about 65 sq km each in the Ionian Sea and south of Crete) [1]. Despite the latest encouraging findings, a major issue that needs to be addressed is the existence of suitable seal rocks that would keep the oil and/or gas within the potential reservoirs. The principal type of seal rock in the eastern Mediterranean is the Messinian evaporites [9–11], and thus, the determination of their characteristics (e.g., origin, thickness and thickness variations, aerial distribution) in the study area is of paramount importance for defining future exploration strategies. In the Mediterranean Sea, the accumulation of the Messinian evaporites was triggered by the reduction in water supply from the Atlantic Ocean to the Mediterranean Sea and occurred in two stages (lower Evaporites and upper

Evaporites). These stages are divided by the Messinian Erosional Surface [12,13]. The Messinian Erosional Surface reflects a significant relative sea-level fall that followed the accumulation of the lower Evaporites. The upper evaporites consist of redeposited material that belong to lower Evaporites, carbonates, evaporites, and/or brackish to freshwater deposits (Lago-Mare facies) [12–15]. The extend and lateral distribution of such deposits in the Mediterranean is critical and enigmatic because they display significant lithological and thickness variations [14]. For instance, the Lower Evaporites are only preserved in deep-water settings (e.g., Apennines, Sicily, Calabria, Tuscany, Cyprus) where they are considered as suitable seal rock candidates [16].

Similar evaporitic deposits outcrop in the external part of the Hellenic fold and thrust belt (Zakynthos Island). This system was established because of Alpine orogenic processes related to plate convergence between Apulia and Eurasia and the closure of the Mesozoic Neo-Tethyan Ocean [3,4]. The Hellenides FTB developed during the Tertiary after closure of the Pindos Ocean and following a continent–continent collision between Apulia and Eurasia [4,5]. These evaporitic deposits were studied about their origin, and the results are controversial [17]. The discrimination between the in situ or resedimented origin of the evaporitic succession has very important paleogeographic implications because the different types of evaporites can be ascribed to different parts of the fold and thrust belt system (e.g., the more proximal regions often contain in situ evaporites in contrast to the more distal and deeper parts). Therefore, revealing their origin has also economic implications because it can add constraints to the regional geotectonic setting and assist future exploration activities.

The aim of this study is to define the depositional conditions (in situ vs. slump and/or turbiditic in origin), thickness, and spatial distribution of the Messinian deposits in the southwestern part of the Hellenides FTB (Zakynthos Island). Further, this work will add constraints on their relationship with the regional tectonic activity that could serve as an analogue for other parts of the foreland's basins in the Balkan Peninsula.

## 2. Geological Setting

The Hellenides FTB includes both platforms (Pre-Apulian and Gavrovo zones) and deep basins (Ionian and Pindos zones) that exhibit a NNW–SSE direction [18,19]. Zakynthos Island is located in the Pre-Apulian and Ionian zones and was influenced by the major orogenic processes that are related to the Hellenides FTB. The principal tectonic features are the Ionian and the Kalamaki Thrusts (Figure 1). The Pre-Apulian zone lies to the east of the Apulian platform and to the west of the Ionian zone and corresponds to an edge-slope facies belt that is largely covered by the Ionian Thrust [20,21]. The Pre-Apulian zone was renamed to Apulian Platform Margins (APM) [11]. The west directed Hellenides FTB as indicated from deep seismic profiles that acquired north [22] and south of Zakynthos Island [2,10] that influenced the sedimentation from Triassic to Pliocene. Triassic evaporites served as major decollement zones and separated the overlying sediments from the basement rocks [23]. Listric faults that are steeper than the major thrusts occur internally in the uplifted wedge [23]. Further, diapiric intrusions are related either to thrust (during Pliocene) or to back-thrust and strike-slip (during Pleistocene) faults [10]. These intrusions influenced the basin geometry, bathymetry, and the processes that controlled the sedimentation of the basin [10]. The uplift that resulted from these intrusions led to the uplift of the Skopos Mountain. Pleistocene post-depositional deformation took place when the Kalamaki thrust evolved into a back-thrust, eroding the pre-existing deposits [10]. The diapiric intrusions modified the sea-floor topography causing steeper slopes and, in conjunction with the syn-sedimentary strike-slip faults, controlled the depositional processes for the lower Pleistocene deposits.

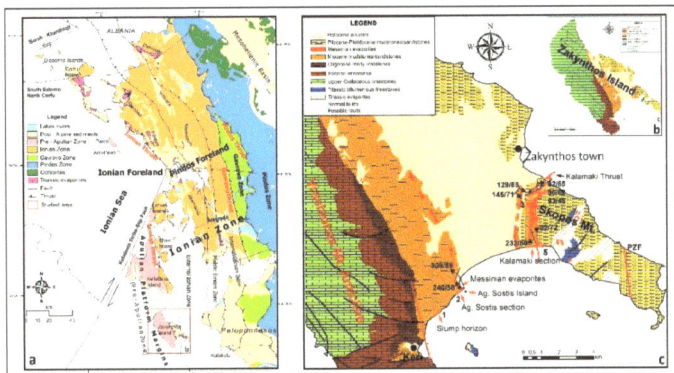

**Figure 1.** (**a**) Geological map of western Greece, where the characteristic Ionian thrust and Ionian foreland are marked, respectively. Red box shows Figure 1b of the Zakynthos Island. (**b**) The geological map of Zakynthos Island. Red box shows the location of the studied area of Figure 1c. (**c**) Detailed geological map of the studied area where bed direction and angle of inclination are marked. Moreover, studied sections and studied places are marked.

## 3. Previous Works and Their Results

Two sections along the south coast of Zakynthos were selected (Figure 1) for re-assessment of the depositional conditions of the Messinian Evaporites, Ag. Sostis to the west and Kalamaki to the east. In both sections at the base, there are upper Miocene clastic sediments (Figure 2), deposited in a shelf environment, evaporites in the middle part and in the upper part white carbonates (Trubi formation) interbedded with marls of the early Pliocene age (Zanclean). The subject of this work is to check and to present details in order for re-assessment, if it is necessary, the depositional conditions of the middle part with evaporites. There are many theories and many published results where these evaporites are a debate between teams of researchers. The major question or difference between these research teams is if the total outcrops with evaporites correspond to in situ Messinian evaporites or a major part of them resedimented as mass flow deposits (gypsum turbidites) in a marine environment after erosion of pre-existing Messinian evaporites during the early Pliocene time.

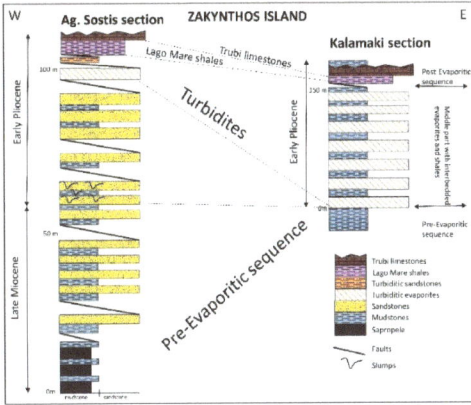

**Figure 2.** Synthetic stratigraphic column of the studied deposits, modified from [24,25], showing the two studied sections of Ag. Sostis and Kalamaki. With blue stars in Kalamaki section the places from where the Early Pliocene age was determined [24] were marked.

The problem or the different approach between the two research groups has an impact on the basin evolution and the time of the tectonic activity. Accepting the theory of the Messinian age in total, then the inundation of the basin could relate with the general inundation of the Mediterranean during the early Pliocene. In this case, these Messinian in origin evaporites were redeposited later during the early Pliocene, but before the Mediterranean inundation, as the inundation was related with the sedimentation of Trubi formation with an early Pliocene age. Probably in the second case, the total thickness must be greater from the previous in situ Messinian evaporites, and additionally the fault activity was started earlier than the until now knowledge.

*3.1. Publications That Accept the Theory of Resedimented Turbiditic Deposits That Took Place during Early Pliocene*

Kontopoulos et al. [17] mostly worked on the Ag. Sostis section with a minor mention on the Kalamaki section. They interpreted the Ag. Sostis section with Messinian evaporites as gypsum turbidites (the section along the coast) (Figure 3b–d) intercalated in a terrigenous turbiditic succession (the sandstones in Ag. Sostis Island) (Figure 4a). These turbidites mostly were transported by dense briny underflows and in minor by erosion in shallow water during sea-level fall. The section of Ag. Sostis harbor was interpreted as shelf deposits due to the presence of hummocky cross-stratification (Figure 4b,c). They mentioned that Kalamaki section (Figure 5b) as consisting at the lower part of 55 m marls with minor calcarinites and gypsiferous partings, probably belonging to the pre-evaporitic sequence of the late Miocene age. The middle part with 113 m thick comprises six cycles, 9–14 m thick each with evaporites at the base and marls on the top. These deposits appear to be shallow water deposits with nodular and banded gypsum but no clear turbiditic deposits. In the upper three cycles, there are nodular gypsum and gypsum conglomerates interbedded with laminated and crystalline gypsum. The upper part rests unconformably, with an erosional contact, over the middle part and corresponds to the Pliocene Trubi limestones (2 m thick) and overlying calcarenites (10 m thick). Moreover, authors agree with the interpretation of Braune and Heimann [26] that the gypsum in the eastern part of the Island is of shallow water origin and is not resedimented. Shallow water Pliocene sediments, suggesting sea level fall during the Messinian, unconformably overlie it.

**Figure 3.** (**a**) Messinian evaporites very close to Ag. Sostis section. (**b**) Part of Ag. Sostis section with the fault-controlled contact between pre-evaporitic sections with the turbiditic evaporites. (**c**) The turbiditic evaporites and (**d**) from close the turbidites mostly with Tb Bouma sub-interval.

**Figure 4.** (**a**) The Island of Ag. Sostis where the turbiditic sandstones were outcropped. See the marked faults due to which there is no continuity from turbiditic evaporites to the post-evaporitic sequence in the harbor where the Trubi limestones (**b**) were outcropped. (**c**) Hummocky cross-stratification within the post-evaporitic sequence introducing a shelf environment.

Zelilidis et al. [27] focused mostly on the Pliocene deposits mentioning that Trubi limestones, at the Ag. Sostis harbor (Figure 4b,c), representing the inundation of the Mediterranean during the Zanclean with the sedimentation in a shelf environment.

Mpotziolis et al. [24] gave the age of sediments in the Ag. Sostis and Kalamaki sections of previous [17–27] and later [28] works. They found that beneath Ag. Sostis evaporites there are also early Pliocene deposits and a characteristic slumped Pliocene horizon with resedimented fossils from the Late Oligocene, Tortonian, and Messinian age (their synthetic stratigraphic column). The age of turbiditic succession is also early Pliocene in age with many reworked fossils of the late Oligocene age. The Kalamaki section showed that the shales on the top of each cycle with evaporites, in the middle part of the Kalamaki section, were deposited during the Early Pliocene with many reworked fossils of the late Oligocene (see the position of the samples in Figure 5a).

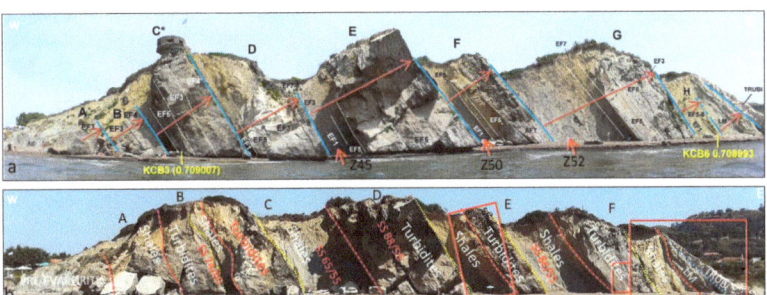

**Figure 5.** The Kalamaki section into two interpretations. (**a**) [29] where we add the position of the three samples from Mpotziolis et al. [24] with which the Early Pliocene age was determined for this part of the Kalamaki section. (**b**) The new interpretation of the same section with additional data. Bed directions were added and the cycles with evaporites and shales re-organized.

Maravelis et al. [25] in Figure 5 present the middle part of the Kalamaki section as early Pliocene in age (Zanclean) based on [24] work. Moreover, they gave a stratigraphy from the western part of the Island, including the Ag. Sostis section where they present a characteristic presence of a slump horizon internally to the lower Pliocene deposits.

Zelilidis et al. [10] questioned the position of the Ionian Thrust in Zakynthos Island (Kalamaki area), and based on sedimentological and structural analyses, reassigned the N–S directed thrust fault in the Kalamaki area (formerly assigned to the Ionian Thrust) as the Kalamaki fault (Figure 1). Zelilidis et al. [10] suggested that this fault commenced during the Pliocene as a thrust fault (in agreement with [30]) and gradually evolved into a back-thrust fault during the Pleistocene. Zelilidis et al. [10] placed the Ionian thrust further westwards at the contact between the clastic deposits and the Vrachionas mountain (Pre-Apulian limestones). This conclusion modifies the boundary between Ionian and Pre-Apulian zones (see [10]), and the two studied locations belong to the same geotectonic zone (Ionian Zone).

*3.2. Publications That Accept the Theory of an Evaporitic Sequence That Accumulated during the Messinian Time*

Kontakiotis et al. [29] focused on the uppermost part of the Kalamaki section (Trubi Formation, Figure 5a), using the same samples from the work of [28], published first online in 2015. They propose a lowery Pliocene (Zanclean age) for these sediments, after the MSC (after 5.33 Ma), during the inundation of the Mediterranean.

Karakitsios et al. [28] used detailed micropaleontological, magnetostratigraphic, and geochemical data on 176 samples and divided the Kalamaki section into three parts (Figure 5a). One hundred and ten (110) samples were tested from the lower part; no samples were analyzed from the middle part, and sixty-three samples were taken from the upper part. The middle part corresponds to the evaporites, and strontium isotope analyses on two samples were conducted indicating a Messinian age (KCB 3 and KCB6 in [29]). The stratigraphic evolution that was proposed by Karakitsios et al. [30] involves a pre-evaporitic succession (two intervals) that corresponds to the lower part and formed between 6.45 and 5.97 Ma (late Miocene). The middle part (two intervals) was formed between 5.971 and 5.55 Ma, during the MSC (latest Miocene–Messinian). The upper part (two intervals) formed the lower interval during 5.6–5.33 Ma (Lago Mare formation) and the upper interval since 5.33 Ma (Trubi formation). This part corresponds to the Mediterranean inundation (latest Miocene to early Pliocene-Zanclean) (Figure 5a). Despite the lack of paleontological data, the evaporites were assigned to the Messinian evaporites. Karakitsios et al. [28] proposed the same subdivision for the Ag. Sostis section. The lower part corresponds to the pre-evaporitic succession, and the middle part (especially the lower interval) belongs to Messinian stage (5.60–5.55 Ma). This part includes redeposited Messinian evaporites that accumulated through gravity flows. These evaporites slide from the west to east (lower interval). The upper part contains sandstone and marl intercalations and is outcropped in Ag. Sostis Island. This part reflects the inundation of the Mediterranean and is represented by the Trubi formation (1.5 m thick, Ag. Sostis harbor).

Despite the consensus that the evaporites in Ag. Sostis are turbiditic in origin [24,26,28], the tectonic contact (normal fault) between these deposits and the pre-evaporitic mudstones has yet to be explained.

*3.3. Additional Contribution to the Debate*

Duermeijer et al. [31] from paleomagnetic measurements in Zakynthos showed that, between 0.77 Ma and recent times, Zakynthos underwent a $21.6° \pm 7.4°$ clockwise shift, whereas there is no significant rotation between 8.11 and 0.77 Ma. In contrast, only in one site, on Kalamaki beach, sampling from evaporites, the results showed a rotation $11°$ anticlockwise, indicating an angle about $32°$ between the rest deposits and these of evaporites. Although authors suggest omitting this result, from the Kalamaki section, they cannot give an alternative explanation or suggestion of what accounted for this difference. Moreover, authors [31] according to their paleontological analysis suggest a Messinian age for these deposits (5.95–5.21 Ma) without mention on what they based their results on or from which part of the section were the selected/analyzed samples.

The evaporites in the middle part of the Ag. Sostis and Kalamaki sections are of Messinian origin (5.971–5.60 Ma). The lower part of the Ag. Sostis and Kalamaki sections corresponds to the pre-evaporitic succession and was deposited before the MSC (late Miocene, before 5.971 Ma).

The upper part is represented by the post-Messinian evaporitic succession that commences with the Lago Mare Formation (5.36 Ma) and is followed by the lower Pliocene (Zanclean, 5.33–5.08 Ma) Trubi limestones that are interbedded with marls.

The Disagreements

One research group accepts that in the middle part of the Ag. Sostis and Kalamaki sections is of the early Pliocene age (based on a paleontological data set from the evaporites and especially from the mudstone beds, internally to the evaporites, that document an early Pliocene age (Zanclean)). The second research group introduces sedimentation during the Messinian, based mostly on sedimentological data.

## 4. New Additional Data

### 4.1. With Re-Interpretation of the Previous Published Papers

Slump horizons and sediments with Bouma subdivisions (turbidites) that were documented in the Ag. Sostis area and are of an early Pliocene age occur under the middle evaporitic part (west of the Ag. Sostis section and towards the Keri village, Figure 6) also introducing an early Pliocene age for the middle part of the Ag. Sostis evaporitic sequence [24]. The presence of Bouma sub-divisions introduce turbiditic currents that were developed in deep environments within an unstable basin floor. Additionally, the fact of the slump's development introduces that, from the uplifted western part of the Vrachionas mountain (Figure 1c), the pre-existing Miocene deposits slumped eastwards, within the Kalamaki foreland basin, during the early Pliocene, explaining the reason of many re-deposited fossils in the upper Miocene age together with the early Pliocene in situ fossils.

**Figure 6.** (**a**)The section along the beach, between Keri and Ag. Sostis villages, where an Early Pliocene slump horizon (**b**) was recognized with Bouma sub-divisions (**c**). For the position, see Figure 1c.

The above two factors introduce that, due to strong tectonic activity, the basin floor was tilted, uplift of the Vrachionas wedge-top, and, due to the instability and the presence of unconsolidated deposits, these deposits slumped eastwards to the Kalamaki foreland basin.

Additionally, the evaporites in the middle part of the Kalamaki section display a large discrepancy in rotation (over 30°) compared to the rest of the deposits [31] probably introducing different depositional conditions. As the [31] showed that there is not any rotation between 8.11 and 0.77 Ma, then the same age deposits must present the same rotation that took place after 0.77 Ma.

Therefore, the middle part with evaporites in the Kalamaki section, if they were deposited during the Messinian time, must present the same rotation with the same age Messinian evaporites. The fact that this part presents a discrepancy in rotation with more than 30° could be explained only if these evaporitic deposits were redeposited from other locations, such as the wedge top of the Kalamaki thrust fault, showing in this case the early Pliocene activity of the Kalamaki thrust fault.

*4.2. Adding New Thoughts with New Data*

In the Ag. Sostis area (western part of the studied area), the Messinian evaporites exist and outcrop on the top of the pre-evaporitic succession, just west of the turbiditic evaporites (Figure 3a), showing the exact primary location of Messinian evaporites.

In the eastern part of the studied area, close and parallel to the Kalamaki thrust fault, there are four additional facts that were analyzed:

a. Within the lower Pliocene shelf deposits, in the Zakynthos town, a slump horizon was recognized (Figure 7a,b), just west of the Kalamaki back-thrust fault, showing the activity of the thrust during lower Pliocene sedimentation and the so on produced instability of the basin floor. It is obvious that during the lower Pliocene sedimentation, pre-existing deposits were slumped within the Kalamaki foreland from the wedge top of the Kalamaki thrust fault.

b. In the Panagoula section, the contact between the Miocene and Pliocene is characterized by the absence of Messinian evaporites, and this conduct is an unconformity (Figure 7c). The facts of the Messinian evaporites' absence and the unconformity boundary, although it is in places with a fault-controlled contact (Figure 7d), could suggest either the erosion of the Messinian pre-existing evaporites and/or that these Messinian evaporites never deposited. The unconformity between Pliocene and Miocene deposits could indicate a shifting of the basin floor, before the sedimentation of the lower Pliocene, probably due to Kalamaki thrust activity. Due to the thrust fault activity, additional faults were created influencing the depositional conditions.

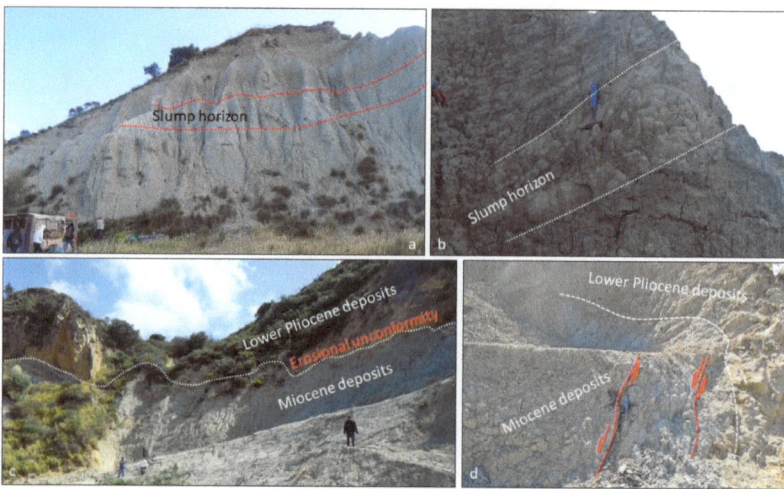

**Figure 7.** (**a**,**b**) The slump horizon within the lower Pliocene deposits close to the Zakynthos town. (**c**,**d**). The contact between upper Miocene and lower Pliocene deposits with either erosional contact or fault-controlled. See the absence of Messinian evaporites. For the position, see Figure 1c.

c.  The Kalamaki section, specifically the middle part, is organized into six cycles that contain evaporites at the base and mudstone at the top (Figure 5b). The detailed measurements of bedding planes showed that bed directions indicate three blocks that display different dip directions but the same angle of bedding (74/60, 65/55, 88/55) (Figure 5b). In the same middle part of the Kalamaki section, a deformation structure (slump) was recognized at the top of the thin-bedded mudstone (Figure 8a). The folding is tight at the one edge but laterally diminishes (Figure 8c). Sedimentary structures similar to load casts occur at the lower bed of evaporites (Figure 8b). Evaporites are composed of cycles with a normally graded, fining upward trend (Figures 8d and 9a,b,d). Amalgamation surfaces exist in the Ta Bouma subdivisions (Figure 9a). Water escape structures, associated with high sedimentation rates, occur in this middle part of the evaporites (Figure 9c).

**Figure 8.** (**a**) Thinly bedded shales between turbiditic evaporites. Arrow shows b, the base of evaporites. (**b**) Evaporites at their base formed structures like load-casts. (**c**) On the top of shales, there are strong deformed shales (white arrow) and this deformation gradually finished. (**d**) Cycles internally to the evaporites; white line marks the contact between cycles, which probably represents an amalgamation surface between Ta Bouma sub-divisions.

**Figure 9.** (**a**) The upper part of turbiditic evaporites between shales. See amalgamation surfaces internally to the evaporites, marked with white dashed lines. Red line shows the contact between two cycles. (**b**) The lower part of the above evaporitic horizon with characteristic fining and thinning upward trend. Blue arrow shows the base of these evaporites. (**c**) Water-escaped structures clearly showed at the base of the above evaporitic horizon. (**d**) Strong deformation within this lower cycle of the evaporitic turbidites.

Explaining the new findings in the middle part of the Kalamaki section, it seems that there are at least three different depositional packages, due to internal unconformities, as this suggested different bedding planes' directions. It seems that sedimentation took place during the thrust fault activity, and as the Kalamaki foreland basin was subsided, there was a shifting of the preexisting beds.

d.  In this Kalamaki section and in the eastern side, Lago Mare and Trubi limestones rest unconformably over the middle evaporitic part, but in some places this contact seems fault-controlled (Figure 10c). This type of contact indicates that tectonic activity controlled the accumulation of these deposits. Mudstone beds that occur at the top of the Trubi limestones rest unconformably over the Trubi limestones (Figure 10a). The mudstone beds underneath the Lago Mare formation contain several slump horizons (Figure 10d), suggesting slope instability during the sedimentation. Trubi formation is cross-cut by normal faults that are directed parallel to the coast (Figure 10b). These faults are probably responsible for the outcrop of the Kalamaki section, which rests on the footwall of these normal faults.

**Figure 10.** (**a**) View of Trubi limestones with bed direction and the contact with the overlying shales. See the great difference between them in bed directions. (**b**) Lago Mare and Trubi limestones in contact. See the faults that cross-cut the Trubi limestones. In this figure, it is marked the position of Kalamaki fault and the Skopos Mountain with the Triassic evaporites. (**c**) See both the-fault controlled and unconformity contact between the upper part of turbiditic evaporites and Lago Mare and Trubi limestones. (**d**) Strong deformation internally to the beds under the Lago Mare deposits.

The fault-controlled contact and unconformity between the middle and upper parts (turbiditic evaporites with Lago Mare shales and Trubi limestones), in the Kalamaki section, indicate that sedimentation during the early Pliocene took place in a restricted area close to the thrust fault.

The presence of the Bouma sequence both in the Ag. Sostis section and in the Kalamaki section supports the idea of a foreland basin, this of the Kalamaki foreland where submarine fans were accumulated during the early Pliocene.

## 5. Discussion—Conclusions

The new presented evidence can led to a new approach for the paleogeographic evolution of Zakynthos Island during late Miocene–Messinian to early Pliocene time and the Messinian evaporites' development:

The Ionian thrust (in its revised position) and the Kalamaki Thrust (old position of the Ionian Thrust) commenced just prior the Mediterranean inundation (during or before the early Pliocene, Zanclean). The uplifted areas (wedge top) of the two thrust faults exposed the pre-existing upper Miocene shelf deposits and especially the Messinian evaporites that were eroded. This process triggered gravity flows and slumping, leading to the deposition of turbiditic gypsum or slumped blocks that are composed of Messinian evaporites (olisthostroms).

The pre-evaporitic succession (upper Miocene shelf deposits) belongs to the western end of the Pindos foreland basin and has gentle slopes (Figures 11 and 12) that were accumulated before the Kalamaki thrust fault activity.

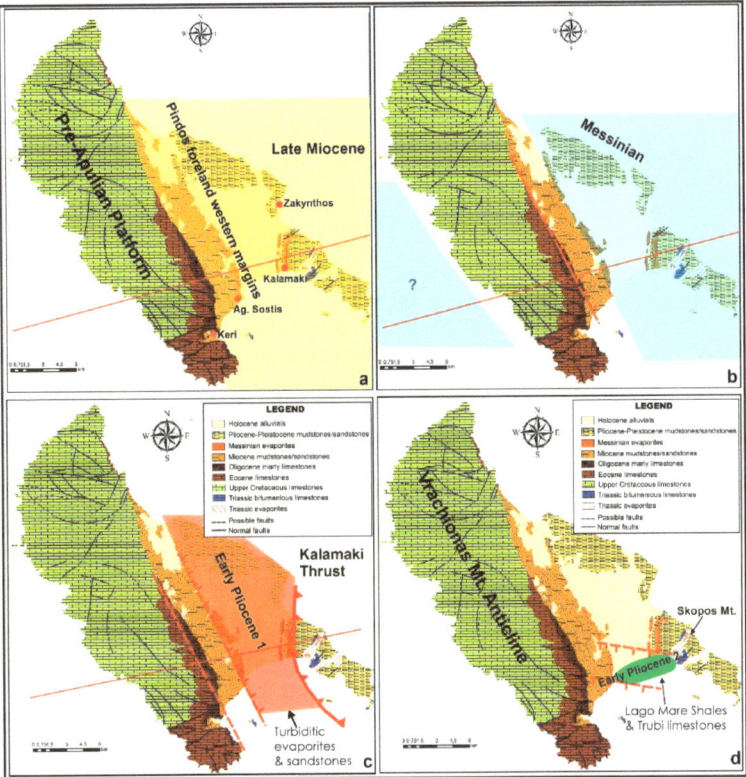

**Figure 11.** Paleogeographic maps showing in four stages the distribution of the sediments into the existing basins. (**a**) During Late Miocene, as the western part of the Pindos foreland, (**b**) during the Messinian, when the whole basin was desiccated, (**c**) during Early Pliocene, when the uplifted Messinian evaporites slid or eroded and resedimented into the Kalamaki Foreland (stage 1), (**d**) during Early Pliocene (stage 2), when the basin was restricted southwards, and Lago Mare and Trubi limestones were deposited.

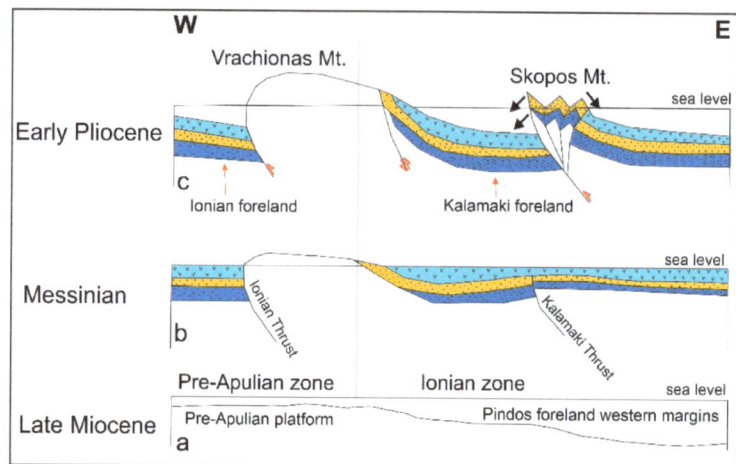

**Figure 12.** Evolutionary models showing in three stages (Late Miocene, Messinian, and Early Pliocene) the studied area that influenced from the two major thrust faults. (**a**) During Late Miocene, Vrachionas Mountain represents the Apulian Platform Margins, passing gradually eastwards to the Ionian Basin. (**b**) During Messinian, both thrusts were active, producing the Kalamaki and Ionian foreland basins, respectively. (**c**) During Early Pliocene, Kalamaki thrust acted as back-thrust producing on its wedge top three small-restricted sub-basins (Argasi, Mellas, and Xirakastello [10]). Red line in Figure 11 shows the position of cross-sections.

The Vrachionas anticline, representing the Pre-Apulian zone [28] or the Apulian Platform Margins [11], formed because of the Ionian Thrust activity and represents the wedge top of the Ionian thrust that took place/was activated during the middle to upper Miocene. The new Ionian Foreland is situated to the west of Zakynthos Island (in the Ionian Sea, Figures 11 and 12). During the early Pliocene, this wedge top (Vrachionas anticline) was sourced, with slumping processes, eastwards towards the Kalamaki foreland basin.

Turbiditic evaporites (of Messinian age evaporites in origin) in Ag. Sostis were sourced from the uplifted western area (Vrachionas anticline—wedge top of the new Ionian thrust). The uplift triggered slumping in the lower Pliocene mudstone beds, and subsequently the deposition of the evaporitic turbidites indicate an early Pliocene age.

Slumps that consist of Messinian evaporites were transported and redeposited in the Kalamaki Foreland Basin from the uplifted Skopos Mountain. In the new position, these evaporites present a 32° deficiency in rotation, compared to the in situ clastic deposits (the pre-evaporitic upper Miocene or post-evaporitic lower Pliocene deposits, Figures 11 and 12).

The new Kalamaki Foreland Basin was a narrow and restricted basin and was formed during the Messinian time and was still active during the early Pliocene ( Figures 11b,c and 12b,c).

The evaporites display their greater in total thickness in the Kalamaki foreland basin, and especially close to the thrust front, because of the presence of both in situ Messinian evaporites that act as a blanket in the whole basin and lower Pliocene resedimented turbiditic evaporites sourced from the uplifted areas of Skopos mountain (Figure 12b,c).

The Lago Mare shales and Trubi limestones accumulated south of the Kalamaki and Ag. Sostis sections (now are situated offshore, Figure 11d) in a fault-controlled basin. The northern part was filled up with sediments from the eroded uplifted Kalamaki (Skopos Mountain) and Ionian (Vrachionas Mountain) thrust faults.

It is critical to examine thoroughly such outcrops in order to determine the thickness, origin, and aerial distribution of the Messinian evaporites or their re-sedimentation during the early Pliocene. Further, it is important to understand the driving mechanisms and timing of their erosion and re-deposition since they add on to the pre-existing thickness of evaporites. It is very important to understand the timing of fault activity, in regions adjacent to the Messinian evaporites, in order to understand their behavior.

As the Ionian Thrust is critical in hydrocarbon exploration in the Ionian and Adriatic Seas, because many exploration targets are active, the oil companies would like to know if there is and how it is the thickness of the in situ Messinian evaporites internally to the Ionian foreland.

Unravelling the depositional conditions of the Messinian in origin evaporites, in Zakynthos Island, the implication could be established for the sealing capacity in the Mediterranean Sea, as there are many areas with limited detailed measurements of the sedimentological and stratigraphic analysis of the Messinian in origin evaporites that accumulated into the restricted and confined basin with quite different total thicknesses.

**Author Contributions:** Conceptualization, A.Z., N.B., A.G.M.; data collection, A.Z., K.A., E.G., S.P., D.A.; methodology, A.Z., N.B., A.G.M.; software, N.B.; writing—original draft preparation, A.Z., N.B., A.G.M.; supervision; A.Z.; writing—review and editing; A.Z., N.B., A.G.M. All authors have read and agreed to the published version of the manuscript.

**Funding:** This research received no external funding.

**Institutional Review Board Statement:** Not applicable.

**Informed Consent Statement:** Not applicable.

**Data Availability Statement:** Not applicable.

**Acknowledgments:** The authors would like to thank reviewers for their improvements on the manuscript.

**Conflicts of Interest:** The authors declare no conflict of interest.

# References

1. Zelilidis, A.; Maravelis, A.G. Introduction to the thematic issue: Adriatic and Ionian Seas: Proven petroleum systems and future prospects. *J. Pet. Geol.* **2015**, *38*, 247–253. [CrossRef]
2. Zelilidis, A.; Maravelis, A.G.; Tserolas, P.; Konstantopoulos, P.A. An overview of the petroleum systems in the Ionian zone, onshore NW Greece and Albania. *J. Pet. Geol.* **2015**, *38*, 331–348. [CrossRef]
3. De Graciansky, P.; De Dardeau, G.; Lemoine, M.; Tricart, P. The inverted margin of the French Alps. In *Inversion Tectonics*; Cooper, M.A., Williams, G.D., Eds.; Geological Society of London Special Publications: London, UK, 1989; Volume 44, pp. 105–122.
4. Doutsos, T.; Koukouvelas, I.; Xypolias, P. A new orogenic model for the External Hellenides. In *Tectonic Evolution of the Eastern Mediterranean Regions*; Robertson, A.H.F., Mountrakis, D., Brun, J.-P., Eds.; Geological Society of London Special Publications: London, UK, 2006; Volume 260, pp. 507–520.
5. Kokkalas, S.; Kamberis, E.; Xypolias, P.; Sotiropoulos, S.; Koukouvelas, I. Coexistence of thin- and thick-skinned tectonics in Zakynthos area (western Greece): Insights from seismic sections and regional seismicity. *Tectonophysics* **2013**, *598*, 73–84. [CrossRef]
6. Avramidis, P.; Zelilidis, A. The nature of deep-marine sedimentation and palaeocurrent directions as evidence of Pindos foreland basin fill conditions. *Episodes* **2001**, *24*, 252–256. [CrossRef]
7. Underhill, J.R. Late Cenozoic deformation of the Hellenide foreland, Western Greece. *Geol. Soc. Am. Bull.* **1989**, *101*, 613–634. [CrossRef]
8. Maravelis, A.; Makrodimitras, G.; Zelilidis, A. Hydrocarbon prospectivity in Western Greece. *Oil Gas Eur. J.* **2012**, *38*, 84–89.
9. Maravelis, A.; Makrodimitras, G.; Pasadakis, N.; Zelilidis, A. Stratigraphic evolution and source rock potential of a Lower Oligocene to Lower-Middle Miocene continental slope system, Hellenic fold and thrust belt, Ionian Sea, northwest Greece. *Geol. Mag.* **2014**, *151*, 394–413. [CrossRef]
10. Zelilidis, A.; Papatheodorou, G.; Maravelis, A.; Christodoulou, D.; Tserolas, P.; Fakiris, E.; Dimas, X.; Georgiou, N.; Ferentinos, G. Interplay of thrust, back-thrust, strike-slip and salt tectonics in a Fold and Thrust Belt system: An example from Zakynthos Island, Greece. *Int. J. Earth Sci.* **2016**, *105*, 2111–2132. [CrossRef]
11. Bourli, N.; Iliopoulos, G.; Zelilidis, A. Reassessing Depositional Conditions of the Pre-Apulian Zone Based on Synsedimentary Deformation Structures during Upper Paleocene to Lower Miocene Carbonate Sedimentation, From Paxoi and Anti-Paxoi islands, Northwestern end of Greece. *Minerals* **2022**, *12*, 201. [CrossRef]
12. CIESM (Commission Internationale pour l'Exploration Scientifi que de la mer Méditerranée). The Messinian salinity crisis from mega-deposits to microbiology. In *A consensus Report: CIESM Workshop Monographs*; Briand, F., Ed.; CIESM Workshop Monographs: Menton, France, 2008; Volume 33, 168p.

13. Iadanza, A.; Sampalmieri, G.; Cipollari, P.; Cosentino, D.; Mola, M. Hydrocarbon-seep Scenarios in the Western Mediterranean Margin during the Messinian Salinity Crisis: Onland Case Studies from the Italian Peninsula. *AAPG Eur. Reg. Conf. Exhib. Barc. Sp* **2013**. Search and Discovery Article 41176.
14. Manzi, V.; Gennari, R.; Hilgen, F.; Krijgsman, W.; Lugli, S.; Roveri, M.; Sierro, F.J. Age refinement of the Messinian salinity crisis onset in the Mediterranean. *Terra Nova* **2013**, *25*, 315–322. [CrossRef]
15. Roveri, M.; Bertini, A.; Cosentino, D.; Di Stefano, A.; Gennari, R.; Gliozzi, E.; Grossi, F.; Iaccarino, S.M.; Lugli, S.; Manzi, V.; et al. A high-resolution stratigraphic framework for the latest Messinian events in the Mediterranean area. *Stratigraphy* **2008**, *5*, 323–342.
16. Lugli, S.; Gennari, R.; Gvirtzman, Z.; Manzi, V.; Roveri, M.; Schreiber, C. Evidence of clastic evaporites in the canyons of the Levant Basin (Israel): Implications for the Messinian salinity crisis. *J. Sediment. Res.* **2013**, *83*, 942–954. [CrossRef]
17. Kontopoulos, N.; Zelilidis, A.; Piper, D.J.W.; Mudie, P.J. Messinian evaporites in Zakynthos, Greece. *Palaeogeogr. Palaeoclimatol. Palaeoecol.* **1997**, *129*, 361–367. [CrossRef]
18. Robertson, A.H.F.; Clift, P.D.; Degnan, P.J.; Jones, G. Palaeogeographic and palaeotectonic evolution of the eastern Mediterranean Neotethys. *Palaeogeogr. Palaeoclimatol. Palaeoecol.* **1991**, *87*, 289–343. [CrossRef]
19. Papanikolaou, D. Timing of tectonic emplacement of the ophiolites and terrane paleogeography in the Hellenides. *Lithos* **2009**, *108*, 262–280. [CrossRef]
20. BP Co. Ltd. The Geological Results of Petroleum Exploration in Western Greece. *Inst. Geol. Subsurf. Res.* **1971**, *10*, 1971.
21. Jenkins, D.A.L. Structural development of western Greece. *AAPG Bull.* **1972**, *56*, 128–149.
22. Kamberis, E.; Marnelis, F.; Loucoyannakis, M.; Maltezou, F.; Hirn, A.; Group Streamers. Structure and deformation of the External Hellenides based on seismic data from offshore western Greece. In *Oil and Gas in Alpidic Thrust Belts and Basins of Central and Eastern Europe*; Wessely, G., Liebl, W., Eds.; EAGE Special Publication: Amsterdam, The Netherlands, 1996; pp. 207–214.
23. Kokinou, E.; Kamberis, E.; Vafidis, A.; Monopolis, D.; Ananiadis, G.; Zelilidis, A. Deep seismic reflection data from offshore western Greece: A new crustal model for the Ionian Sea. *J. Pet. Geol.* **2005**, *28*, 185–202. [CrossRef]
24. Mpotziolis, C.; Kostopoulou, S.; Triantaphyllou, M.; Zelilidis, A. Depositional environments and hydrocarbon potential of the Miocene deposits of Zakynthos Island. *Bull. Geol. Soc. Greece* **2013**, *XLVII*, 2101–2110. [CrossRef]
25. Maravelis, A.; Koukounya, A.; Tserolas, P.; Pasadakis, N.; Zelilidis, A. Geochemistry of Upper Miocene-Lower Pliocene source rocks in the Hellenic fold and thrust belt, Zakynthos Island, Ionian Sea, western Greece. *Mar. Pet. Geol.* **2015**, *66*, 217–230. [CrossRef]
26. Braune, K.; Heimann, K.O. Miocene evaporites on the Ionian Islands. *Bull. Geol. Soc. Greece* **1973**, *10*, 25–30.
27. Zelilidis, A.; Kontopoulos, N.; Avramidis, P.; Piper, D.J.W. Tectonic and sedimentological evolution of the Pliocene-Quaternary basins of Zakynthos Island, Greece: Case study of the transitions from compressional to extensional tectonics. *Basin Res.* **1998**, *10*, 393–408. [CrossRef]
28. Karakitsios, V.; Roveri, M.; Lugli, S.; Manzi, V.; Gennari, R.; Antonarakou, A.; Triantaphyllou, M.; Agiadi, K.; Kontakiotis, G.; Kafousia, N.; et al. A record of the Messinian salinity crisis in the eastern Ionian tectonically active domain (Greece, eastern Mediterranean). *Basin Res.* **2017**, *29*, 203–233. [CrossRef]
29. Kontakiotis, G.; Karakitsios, V.; Mortyn, P.G.; Antonarakou, A.; Drinia, H.; Anastasakis, G.; Agiadi, K.; Kafousia, N.; De Rafelis, M. New insights into the early Pliocene hydrographic dynamics and their relationship to the climatic evolution of the Mediterranean Sea. *Palaeogeogr. Palaeoclimatol. Palaeoecol.* **2016**, *459*, 348–364. [CrossRef]
30. Underhill, J.R. Triassic evaporites and Plio-Quarternary diapirism in Western Greece. *J. Geol. Soc. Lond.* **1998**, *145*, 269–282. [CrossRef]
31. Duermeijer, C.E.; Krijgsman, W.; Langereis, C.G.; Meulenkamp, J.E.; Triantafyllou, M.V.; Zachariasse, W.J. A late Pleistocene clockwise rotation phase of Zakynthos (Greece) and implications for the evolution of the western Aegean arc. *Earth Planet. Sci. Lett.* **1999**, *173*, 315–331. [CrossRef]

**Disclaimer/Publisher's Note:** The statements, opinions and data contained in all publications are solely those of the individual author(s) and contributor(s) and not of MDPI and/or the editor(s). MDPI and/or the editor(s) disclaim responsibility for any injury to people or property resulting from any ideas, methods, instructions or products referred to in the content.

*Article*

# Sedimentological and Petrographical Characterization of the Cambrian Abbottabad Formation in Kamsar Section, Muzaffarabad Area: Implications for Proto-Tethys Ocean Evolution

Syed Kamran Ali [1,*], Rafiq Ahmad Lashari [2], Ali Ghulam Sahito [2], George Kontakiotis [3,*], Hammad Tariq Janjuhah [4], Muhammad Saleem Mughal [1], Ahmer Bilal [5], Tariq Mehmood [6] and Khawaja Umair Majeed [1]

[1] Institute of Geology, Azad Jammu and Kashmir University, Muzaffarabad 13100, Pakistan
[2] Center for Pure and Applied Geology, University of Sindh, Jamshoro 76060, Pakistan
[3] Department of Historical Geology and Paleontology, Faculty of Geology and Geoenvironment, National and Kapodistrian University of Athens, Panepistimiopolis, 15784 Athens, Greece
[4] Department of Geology, Shaheed Benazir Bhutto University, Sheringal 18050, Pakistan
[5] Shandong Provincial Key Laboratory of Depositional Mineralization & Sedimentary Minerals, Shandong University of Science and Technology, Qingdao 266590, China
[6] Oil and Gas Development Corporation Limited (OGDCL), Islamabad 44000, Pakistan
* Correspondence: kamran.ali@ajku.edu.pk (S.K.A); gkontak@geol.uoa.gr (G.K.)

**Citation:** Ali, S.K.; Lashari, R.A.; Sahito, A.G.; Kontakiotis, G.; Janjuhah, H.T.; Mughal, M.S.; Bilal, A.; Mehmood, T.; Majeed, K.U. Sedimentological and Petrographical Characterization of the Cambrian Abbottabad Formation in Kamsar Section, Muzaffarabad Area: Implications for Proto-Tethys Ocean Evolution. *J. Mar. Sci. Eng.* **2023**, *11*, 526. https://doi.org/10.3390/jmse11030526

Academic Editor: Dimitris Sakellariou

Received: 7 January 2023
Revised: 23 February 2023
Accepted: 27 February 2023
Published: 28 February 2023

**Copyright:** © 2023 by the authors. Licensee MDPI, Basel, Switzerland. This article is an open access article distributed under the terms and conditions of the Creative Commons Attribution (CC BY) license (https://creativecommons.org/licenses/by/4.0/).

**Abstract:** The current sedimentological and petrographical research of the Abbottabad Formation has been carried out in order to understand the formation and evolution of the Proto-Tethys Ocean during the Cambrian on the northern margin of the Indian Plate. The Muzaffarabad region is located east of the Upper Indus Basin and the southern part of the Hazara Kashmir Syntaxis. The geological history of the region varies from the Precambrian to the recent period. The Cambrian Abbottabad Formation is well exposed along the Hazara Kashmir Syntaxis at the core of the 500-m-thick Muzaffarabad anticline. The Abbottabad Formation is an unconformity-bounded allo-stratigraphic unit. It has an unconformable lower contact with the Late Precambrian Dogra Formation and an unconformable upper contact with the Paleocene Hangu Formation. The Abbottabad Formation has been divided into four lithofacies, from bottom to top, namely, thinly interbedded dolomite and shale, cherty-stromatolitic dolomite, oxidized limonitic-brecciated zone, and quartzite, with significant lithological changes. Petrographic studies revealed four types of dolomites: fine crystalline dolomite (Dol. I), dolomitic cryptocrystalline chert (Dol. II), algal mat-stromatolitic dolomite (Dol. III), and intraclastic-dolo-grain stone (Dol. IV). The mineral composition of dolostone was analyzed using X-ray diffraction (XRD) and found to be consistent with previous petrographic studies. The dolomite mineral content decreased from base to top, while chert increased towards the top. Elemental weight percentages through energy dispersive X-ray (EDX) analysis show different elements constitute the minerals found in the dolostone, as confirmed by petrographic and XRD analysis. Using outcrop data, facies information, and geochemical data, a modified depositional model of the Abbottabad Formation was developed. During the Early Cambrian period, the formation was deposited in a shallow subtidal to supratidal setting of the Proto-Tethys Ocean. The top of this deposit marks the Cambrian–Paleocene boundary. Because of the progressively coarsening outcrop sequences, this formation seems to be at the very top of the Proto-Tethys Ocean's shallow marine system.

**Keywords:** stromatolites; dolomite microfacies; Proto-Tethys Ocean; shallow marine carbonates; subtidal-supratidal depositional environments; sequence stratigraphic development; sedimentary basin dynamics; stratigraphic correlations; diagenetic processes; paleoenvironmental reconstruction

## 1. Introduction

The Himalayas are classified into four subdivisions, i.e., sub-Himalaya, lesser Himalaya, higher Himalaya, and trans-Himalaya [1]. The Muzaffarabad area situated in the east of the Upper Indus Basin (UIB) in Pakistan [2] and is the southern portion of the Hazara Kashmir Syntaxis (HKS), sub-Himalaya (Figure 1). The HKS of the Northern Pakistan has been the focus of geological study because of its sedimentology, complicated structure, and seismicity. The area is highly deformed by the folds and faults owing to tectonic activity [3]. The sedimentological sequence ranges in age from Cambrian to the most recent era. The inner core of the HKS mainly consists of the Cambrian Abbottabad Formation, which is thrust over the Miocene Murree Formation [3]. The study area forms the core and the northern limb of a Muzaffarabad overturned anticline, which is part of regional HKS in the Muzaffarabad city along Neelum valley road, Azad Jammu and Kashmir, Pakistan (AJKP).

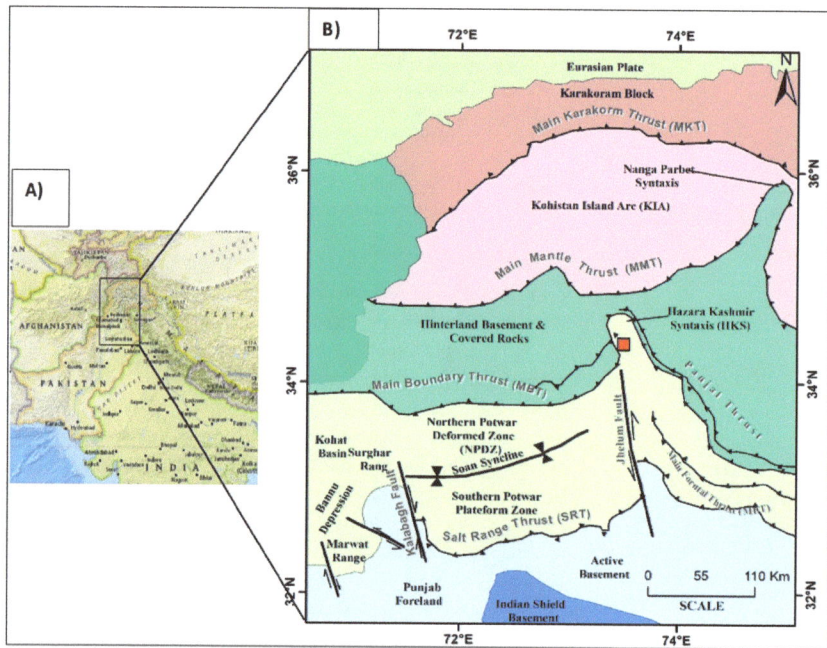

**Figure 1.** (**A**) Regional location map of the northern Pakistan; (**B**) tectonic map of northern Pakistan. Red square showing the study area (compiled and adopted after [4–6]).

Various researchers have worked on the stratigraphy and tectonics of HKS deposits in different parts of Azad Jammu and Kashmir, Pakistan (AJKP) [3–5,7–18]. The geology and stratigraphy of this region was originally described by Medlicott [19]. The Swiss geologist Bossart, et al. [20]) described the lithological, stratigraphic and structural features of HKS. Greco [8] described the stratigraphic and metamorphic features of the rocks of the Hazara-Kashmir Syntaxis. Baig and Iqbal Siddiqi [9] and Baig, et al. [21] worked on the Muzaffarabad fault and discussed the active tectonic evidence.

Although general information about the stratigraphic and sedimentological aspects of the various formations in the HKS exists, detailed study regarding sedimentology and stratigraphy for the dolomites of the Abbottabad Formation is lacking, which is crucial for understanding the Cambrian deposits of Proto-Tethys Ocean in the northern Pakistan. The Abbottabad Formation was selected for sedimentological and petrographic analysis. The Abbottabad Formation consists of stromatolitic, thinly to thickly bedded dolomite of the Cambrian age. Different researchers presented various models for the genesis of dolomite and the associated diagenetic environment [22–27].

This study aims to determine the litho- and micro-facies of dolostone, its depositional environment, and stromatolite development in the Abbottabad Formation. The depositional model proposed here will provide additional insights into the paleogeographic evolution of the Proto-Tethys basin and a better correlation with nearby tectonostratigraphic successions.

## 2. Geological Framework

The study area lies in the Kamsar to Yadgar region of the Muzaffarabad, a division of AJKP and is part of HKS (Figure 1). The HKS is one of the most conspicuous northwest Himalayan structures and is a complex tectonic zone formed by a regional shift in the northeast-to-northwest strike of orogeny. The syntaxial zone and its surroundings exhibit sedimentary, volcanic, and metamorphic rocks from the Precambrian to the Neogene [5]. The thickness of the Precambrian and Cambrian rock strata increases eastward from the UIB while it diminishes westward (Figure 2) [28]. The HKS comprises numerous thrust sheets that overlap and are composed of Precambrian to Mesozoic formations that were thrust onto molasses deposits [13,29,30].

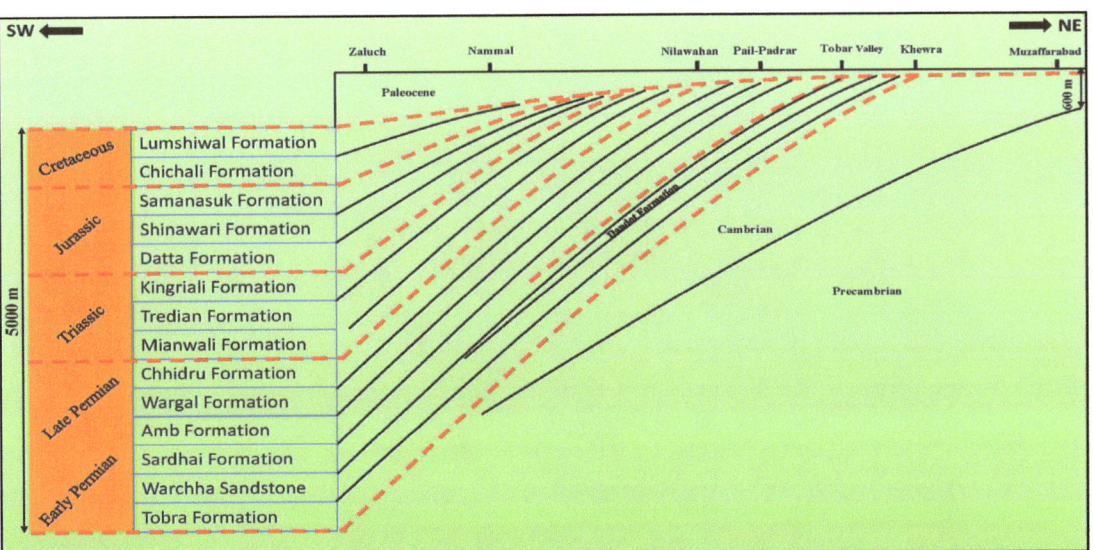

**Figure 2.** Surface and subsurface deposition of different formations and unconformities in UIB and HKS.

In the Cambrian period, the study region was a part of Gondwanaland characterized by generally warm, shallow marine environments [29,30]. After a period of non-deposition during the Lower Cambrian, the region once again experienced a transgression of the sea and the deposition of the Abbottabad Formation took place [28]. The Ordovician to Cretaceous rock sequence was not deposited during tectonic uplift and erosion related to earlier orogenic events on the Gondwana of the Indian Plate (Figure 2) [31]. In the early Paleocene, the Hangu Formation was deposited at the base of the Late Paleocene–Early Eocene Neo-Tethys limestone and shale sequence [32]. After the collision of the Indo-Eurasian plates in the Cretaceous, the regional thrust systems evolved, including the Main Karakoram Thrust (MKT), the Main Mantle Thrust (MMT), the Main Boundary Thrust (MBT), and the Salt Range Thrust (SRT) [1,33,34]. Currently, the research area is in the sub-Himalayas. The Main Boundary Thrust (MBT) restrains the sub-Himalayan lithosphere from the north. The MBT divides the rocky regions of the sub-Himalayan area from the lesser Himalayan region (Figure 1).

*Stratigraphy of Kamsar Section Abbottabad Formation*

The study area contains rocks from the Precambrian Dogra Formation to the Miocene Murree Formation (Figure 3). The majority of the Dogra Formation is composed of graphitic schist, phyllite, and marble. The uppermost contact between the Dogra Formation and the Abbottabad Formation is nonconformable, and the Abbottabad Formation marks the beginning of the Cambrian period.

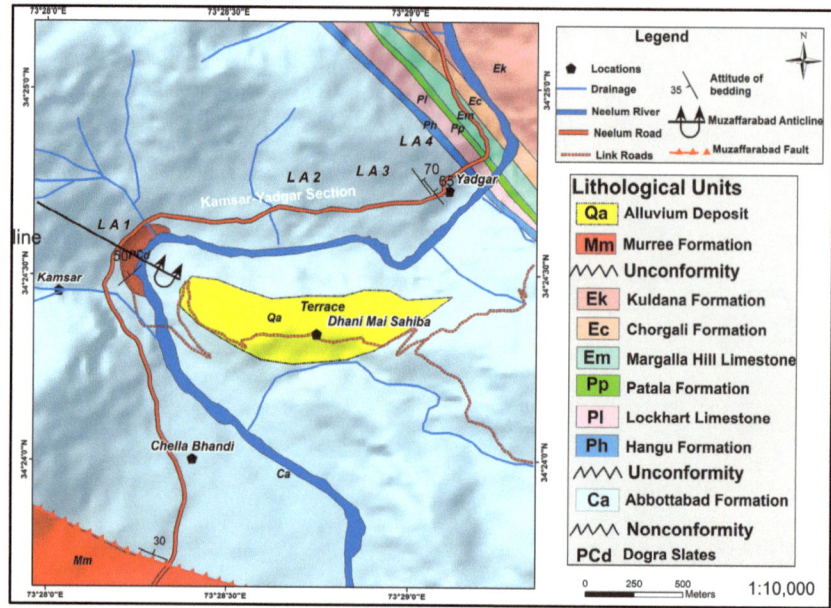

**Figure 3.** Simplified geological map of the study area.

The Stratigraphic Committee of Pakistan has classified the primarily dolomite, quartzite, and limestone rocks of the Abbottabad district as the Abbottabad Formation [35]. This nomenclature has been expanded to encompass the similar and correlative rocks previously referred to as "great limestone" [36], "Upper Dolomite Member" [7], "Muzaffarabad Formation" [4], and the Sirban Formation [8]. In the study area, the Abbottabad Formation is exposed under the MBT in the sub-Himalaya region and the center of the Muzaffarabad anticlinal system. This formation consists of cliffs and very steep slopes and have average thickness of 200–833 m [37].

The Abbottabad Formation is composed mostly of cherty dolomite, with subordinate shales, stromatolitic dolomite, and quartzite facies. Latif [14] identified a few Cambrian-aged fossils in the Abbottabad Formation, which is mostly fossil-free. The microfossils in the Abbottabad Formation are interpreted to reflect the Cambrian Period. About 500 mya, after the Cambrian deposit ended, the Paleocene deposit began. Overlying the Abbottabad Formation, the Hangu Formation's bauxite, coal seams, and sandstone deposits with average thickness of 15–35 m [37] indicate the primary unconformity at the erosional breccia surface of dolomite [13].

## 3. Methodology

### 3.1. Field Study

Field observations were conducted in Muzaffarabad's Kamsar-Yadgar region (Figure 3). This investigation shows a variety of outcrop lithofacies. Using Jacob's staff technique, the outcrops of the Abbottabad Formation were recorded, sampled, and measured. Based on sedimentological observations, such as lithological variations, color, and depositional texture, detailed investigations were performed to identify distinct lithofacies (Table 1). A total of 100 samples from the Abbottabad Formation were obtained from different places.

**Table 1.** Litho facies of the Abbottabad Formation, Kamsar-Yadgar section, Muzaffarabad.

| Lithofacies | Description | Bedding and Structures | Environment of Deposition |
|---|---|---|---|
| LA-1 Dolomite with subordinate shale | The basal zone is fine grained, light grey dolomite. The upper zone consists of dark grey, thinly bedded dolomite with blackish shales. | Lenticular layers with minor erosional surfaces against shale; small-scale cross-lamination; rare normal graded bedding. | Shallow subtidal |
| LA-2 Stromatolitic dolomite member | This zone is typical dolomite showing growth of thinly laminated algal mats to thick domal shape stromatolites. The colour of the dolomite is light grey with some chert bands. | Convex to undulatory laminae; abundant stromatolites of varying shapes and size; large domal and columnar stromatolites up to 10 to 15 cm, 1–2 m long and grading upwards into microbial laminae. | Shallow subtidal to supratidal |
| LA-3 Oxidized, ferruginous, dolomite | This zone is greenish and reddish brown, rusty stained, brecciated, and limonitic. The thickness of this zone at Yadgar is 16.2 m and extends more than 100 m along dip and strike. | Monomict with lithology, such as host rocks as well as dolomite and chert clasts; poorly sorted, sub-angular to angular clasts of varying sizes and coated by oxidized limonite. | Surface karstification products; karstic depression fills |
| LA-4 Quartzite member | Quartzite is fine grained, snow white to white, with brownish to yellowish encrustations on the joints. | The thickness of the quartzite varies from 9 to 25 m and brecciated at the top. Calcrete deposits. | Subaerial exposure |

*3.2. Petrography*

Thin sections of dolomitic rock were prepared at the Hydrocarbon Development Institute of Pakistan (HDIP) in Islamabad. A petrographic microscope (Leica DM-750P) was used to examine thin sections for mineral identification and diagenetic markers using reflected and transmitted light. Tables 2 and 3 show the results of the modal mineralogical composition using the point-counting technique. Photomicrographs were taken and used to characterize microfacies. The basic terminology of dolomite textures employed in thin sections considers crystal structure, crystal mutual relationships, and uniform or non-uniform size distribution. Commonly, dolomite has an idiotopic (planar) to a xenotopic (non-planar) texture. The idiotopic texture is further divided into idiotopic e, idiotopic s, idiotopic c, and idiotopic p, representing euhedral, subhedral, cement, and porphyrotopic, respectively. The xenotopic texture is further classified into xenotopic a, xenotopic c, and xenotopic p, indicating anhedral, cement with saddle shaped dolomite crystals, and porphyrotopic, respectively. Saddle dolomite derived from hydrothermal fluids is a late diagenetic product that often overprints earlier diagenetic phases. The carbonate classification methods of Dunham [33], Sibley and Gregg [34] and Flügel and Munnecke [37] were used to analyze the texture and microfacies of the carbonate rocks, which comprised both biogenic and inorganic dominant components. The classification technique of Logan, et al. [38] was used to classify stromatolites. Kalkowsky [39] used the term stromatolith to describe "limestone masses with a thin, planar layered structure", composed of a collection of laminae ("stromatoid") thought to be of vegetal origin. Logan, Rezak and Ginsburg [38] developed a classification to characterize the growth types of intertidal and shallow subtidal stromatolites. Laminated mats are frequent in quiet-water supratidal and intertidal environments, whereas domal growth forms are widespread in subtidal conditions. Several petrographic observations and diagenetic phases were documented and quantified to analyze the depositional environment using lithofacies analysis. Figure 4 presents a summary of the results based on the above sedimentological and petrographic observations.

**Table 2.** Modal mineralogical composition of the Abbottabad Formation, Kamsar section, Muzaffarabad.

| Sr. No. | Sample No. | Dol | Chert | Cr. Silica | Chl | Calcite | Pr | CM | St | Hm | BC | Sc | IC | Mc | Pl | Microfacies |
|---|---|---|---|---|---|---|---|---|---|---|---|---|---|---|---|---|
| 1 | KA-1 | 70 | – | – | – | – | – | 30 | – | – | – | – | – | – | – | Dol. I |
| 2 | KA-2 | 90 | 5 | – | – | – | 5 | – | – | – | – | – | – | – | – | Dol. I |
| 3 | KA-3 | 75 | 5 | – | 1 | – | 4 | – | 15 | – | – | – | – | – | – | Dol. I |
| 4 | KA-4 | 90 | 1 | – | – | 4 | 2 | – | – | 3 | 2 | – | – | – | – | Dol. I |
| 5 | KA-5 | 30 | 5 | 40 | 20 | 3 | 2 | – | – | – | – | – | – | – | – | Dol. II |
| 6 | KA-8 | 75 | 5 | – | – | 9 | 5 | 15 | – | – | – | – | – | – | – | Dol. III |
| 7 | KA-19 | 90 | 1 | – | – | – | – | – | – | – | – | – | – | – | – | Dol. I |
| 8 | KA-22 | 85 | 5 | – | 4 | – | 2 | 8 | – | – | – | – | – | – | – | Dol. I |
| 9 | KA-27 | 75 | 20 | – | – | – | 1 | – | – | – | 1 | 1 | – | – | – | Dol. I |
| 10 | KA-31 | 95 | 1 | – | – | – | – | – | – | 2 | 1 | – | – | – | – | Dol. III |
| 11 | KA-35 | 78 | 10 | – | 2 | – | 2 | 10 | – | – | – | – | – | – | – | Dol. I |
| 12 | KA-37 | 65 | 15 | – | – | 15 | – | – | – | 3 | – | – | – | – | – | Dol. II |
| 13 | KA-41 | – | 25 | 65 | 8 | – | – | – | – | 2 | – | – | – | – | – | Dol. I |
| 14 | KA-44 | 70 | 5 | – | 3 | 20 | – | – | – | 2 | – | – | – | – | – | Dol. I |
| 15 | KA-50 | 20 | 5 | – | – | – | 1 | – | – | – | – | – | 60 | 12 | 2 | Dol. IV |

**Abbreviations**: Dol = dolomite, Cr. Silica = cryptocrystalline silica, Chl = chalcedony, Pr = pyrite, CM = carbonaceous material, St = stylolite features, Hm = hematite, BC = bioclast, Sc = sericite, IC = intraclasts, Mc = micrite, Pl = peloids.

Table 3. Modal mineralogical composition of the Abbottabad Formation, Kamsar section, Muzaffarabad.

| Sr. No | S. No. | Dol | Chert | Cr. Sil. | Chl | CM | Calcite | Pr | Pl | IC | Qtz | Sp | Mc | Hm | BC | Oo | Zr | Tr | Apt | Opm | Microfacies |
|---|---|---|---|---|---|---|---|---|---|---|---|---|---|---|---|---|---|---|---|---|---|
| 1 | KA-56 | 30 | 20 | – | 5 | 5 | – | – | – | 10 | – | 15 | 15 | 5 | – | – | – | – | – | – | Dol. III |
| 2 | KA-59 | – | 20 | 60 | – | – | – | 15 | – | – | – | – | – | – | 5 | – | – | – | – | – | Dol. II |
| 3 | KA-60 | 30 | 60 | – | – | – | – | – | – | – | – | – | – | 5 | – | – | – | – | – | – | Dol. II |
| 4 | KA-61 | 5 | 50 | 30 | 5 | – | – | 10 | – | – | – | – | – | – | – | – | – | – | – | – | Dol. III |
| 5 | KA-63 | 18 | 30 | – | 5 | 7 | – | 5 | 25 | – | – | – | – | 2 | – | 5 | 1 | 1 | 1 | 1 | Dol. IV |
| 6 | KA-65 | 25 | – | – | 10 | 20 | – | – | – | 25 | 25 | 13 | – | – | – | – | – | 1 | – | – | Dol. IV |
| 7 | KA-66 | 65 | – | – | 5 | 15 | 5 | 2 | – | – | – | – | – | 3 | 1 | – | – | – | – | – | Dol. III |
| 8 | KA-67 | 50 | 4 | – | 5 | 15 | 7 | – | 15 | – | – | – | – | 3 | – | – | – | 1 | – | – | Dol. III |
| 9 | KA-68 | – | 20 | 60 | 10 | – | 10 | – | – | – | – | – | – | 10 | – | – | – | – | – | – | Dol. II |
| 10 | KA-69 | 73 | – | – | 5 | 7 | – | 7 | – | – | – | – | – | 5 | – | – | – | – | – | – | Dol. III |
| 11 | KA-70 | 75 | 5 | – | – | 5 | – | – | – | – | – | – | – | 3 | 5 | – | – | – | – | – | Dol. III |
| 12 | KA-76 | 20 | 20 | – | – | – | – | – | – | 5 | 55 | – | – | – | – | – | – | 1 | – | 1 | Dol. IV |
| 13 | KA-82 | – | 3 | – | – | – | – | – | – | – | 95 | – | – | – | – | – | – | – | – | – | Arenite |
| 14 | KA-84 | 15 | 15 | 45 | 24 | – | – | – | – | – | – | – | – | 1 | – | – | – | – | – | – | Dol. II |
| 15 | KA-87 | 3 | – | – | – | – | – | – | – | – | 95 | – | – | – | – | – | – | 1 | – | 1 | Arenite |

**Abbreviations:** Dol = dolomite, Cr. Sil. = cryptocrystalline silica, Chl = chalcedony, CM = carbonaceous material, Pr = pyrite, Pl. = peloids, IC = intraclasts, Qtz. = quartz, Sp = spar, Mc = micrite, Hm = hematite, BC = bioclast, Oo = ooid, Zr = zircon, Tr = tourmaline, Apt = apatite, Opm = opaque minerals.

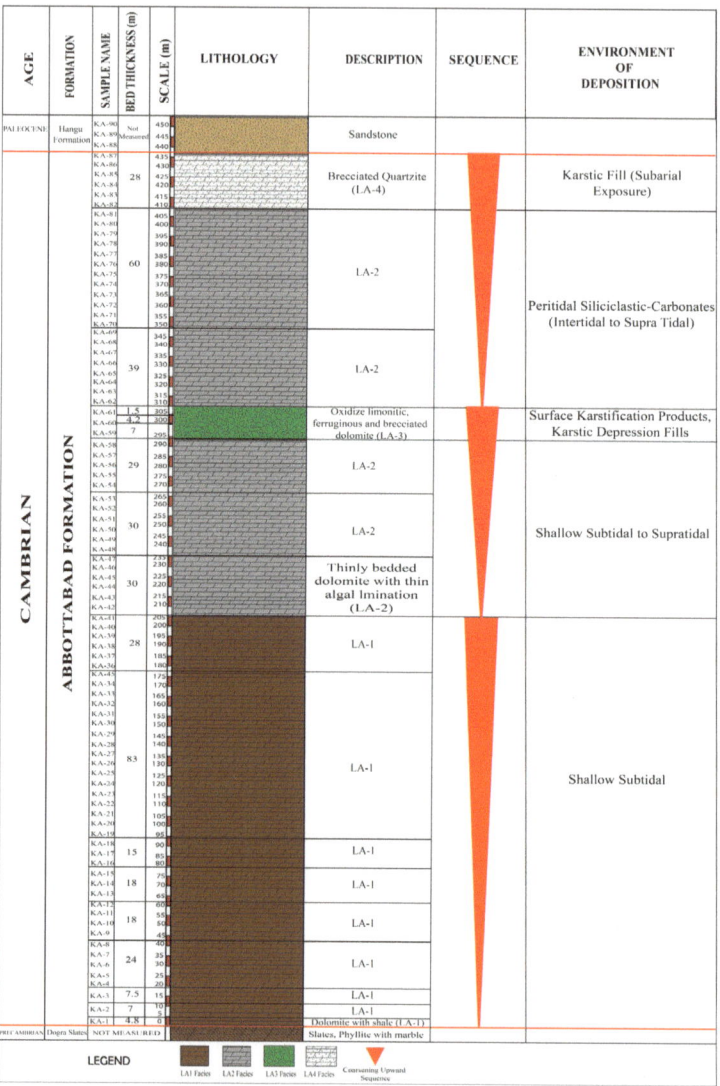

**Figure 4.** Litho-log of the Kamsar-Yadgar section, Muzaffarabad, Pakistan.

### 3.3. X-ray Diffraction (XRD) Energy Dispersive X-ray (EDX)

For the current research work, four samples of dolomite from the Abbottabad Formation were taken for XRD and eight samples for EDX analysis. The samples for XRD analysis were sent to Centralized Resource Laboratory (CRL), Peshawar University, and samples for EDX analysis were sent to Laboratory of Center for Pure and Applied Geology, University of Sindh, Pakistan, where they underwent detailed analysis.

The X-ray diffraction (XRD) instrument utilized at CRL Peshawar was a JDX-3532 Diffractometer from Japan with the capacity to generate voltages ranging from 20 to 40 kV and currents ranging from 2.5 to 30 mA. The instrument employed a copper (CuKa) X-ray source with a wavelength of 1.5418 nm and a 2Theta-range of 0–160°. Copper is a widely used X-ray source, and its K peak at 1.5418 nm offers advantages for analyzing minerals due to its higher-order lattice spacing, which is 10–15 times greater than the wavelength. Finally, the XRD data were analyze by Jade 6.5 software.

To perform EDX analyses, the samples were first coated with gold using the DII-29030SCTR smart coater, and then analyzed using a JEOL EDS system with Bruker software.

## 4. Results

### 4.1. Litho- and Petro-Facies Characterization of the Abbottabad Formation

The Abbottabad Formation has a non-conformable lower contact with the Precambrian Dogra slates in the study area (Figures 4 and 5A), while the Palaeocene Hangu Formation unconformably overlies the Abbottabad Formation (Figure 5I).

**Figure 5.** Detailed field characteristics and lithofacies of the Abbottabad Formation at Kamsar-Yadgar section, Muzaffarabad: (**A**) non-conformable contact between the Dogra slate and thin interbedded dolomite of the Abbottabad Formation (LA-1); (**B**) thinly bedded dolomite with blackish shale (LA-1); (**C**) thinly bedded greyish dolomite (LA-1); (**D**) thinly laminated convex algal mats (LA-2); (**E**) thick dome-shaped stromatolites (LA-2); (**F**) reddish brown, rusty stained, brecciated, and highly limonitic dolomitic zone (LA-3); (**G**) contact between the limonitic, brecciated zone, and medium bedded algal mat dolomite; (**H**) snow white, brecciated quartzite (LA-4); (**I**) composite unconformity between the brecciated quartzite of the Abbottabad Formation and the Hangu Formation.

The following four lithofacies were observed in the field (Table 1).

#### 4.1.1. Dolomite Member (LA-1)

The dolomite member can be separated into two zones. Figure 5B shows fine-grained, light grey to cream grey dolomite with cherty bands in the basal zone. Dolomite is thin to medium-bedded, highly jointed and fractured, hard, compact, fine grained, and cherty in general. Chert is found as grey to dark grey layers, patches, and lenses. Dolomite is brittle and fractures into small, angular fragments due to the presence of silica. The upper zone is composed of thinly bedded grey to dark grey dolomite with blackish shale (Figure 5C). This zone extends from 0 to 205 m above the ground (Figure 4).

#### 4.1.2. Stromatolitic Dolomite Member (LA-2)

This is a fine-grained dolomite zone, with thinly laminated algal mats (Figure 5D) to thick, dome-shaped growing stromatolites (Figure 5E).

The color of dolomite is light gray with chert bands. Stromatolites are most common in the study area as stacked hemispheroids (SH) that form columns separated by sediment (Figure 5E) [38]. The laminae domes have varying widths (subtypes SH-C and SH-V). Even though the region experienced periodic storm activity, low energy conditions prevailed. Tidal flat conditions occurred in general throughout the formation of the investigated sequence. This zone is around 200 m thick and displays a gradual increase in water depth (Figure 4).

#### 4.1.3. Oxidized, Ferruginous Dolomite Member (LA-3)

An oxidized zone can be seen in the central part of the stromatolitic dolomite facies (Figure 5F). This zone is greenish-reddish brown, rusty stained, brecciated, and highly limonitic. This zone is 16.2 m thick at Yadgar section (Monument Point) and extends more than 100 m along the dip and strike (Figure 5G). The zone is characterized by shearing and considerable silicification. Quartz lenses are found in phyllitic-looking altered rocks (Figure 4).

#### 4.1.4. Quartzite (Quartzose Sandstone) Member (LA-4)

In the study area, quartzite rests on top of stromatolitic dolomite. This quartzite is fine-grained, white to snow white, with brownish to yellowish encrustations on the joints (Figure 5H). The quartzite ranges in thickness from 9 to 25 m, is brecciated at the top, and is unconformably overlain by bauxite from the Hangu Formation (Figure 5I). This zone extends from 410 to 438 m above the LA-2 facies (Figure 4).

#### 4.1.5. Microfacies of the Abbottabad Formation

The petrographic analysis of 30 dolomite samples revealed different grain types, such as ooids, pellets, intraclasts, and the remains of unidentified fossils (Figures 6–9). The lower part of the Abbottabad Formation has fine crystalline dolomite, but the upper part is entirely quarzitic and brecciated. Dominant microfacies include those composed of dolomite, chert, chalcedony, various types of bioclasts, and quartz, with textures ranging from crystalline to grain stone. Five microfacies have been identified and are listed below.

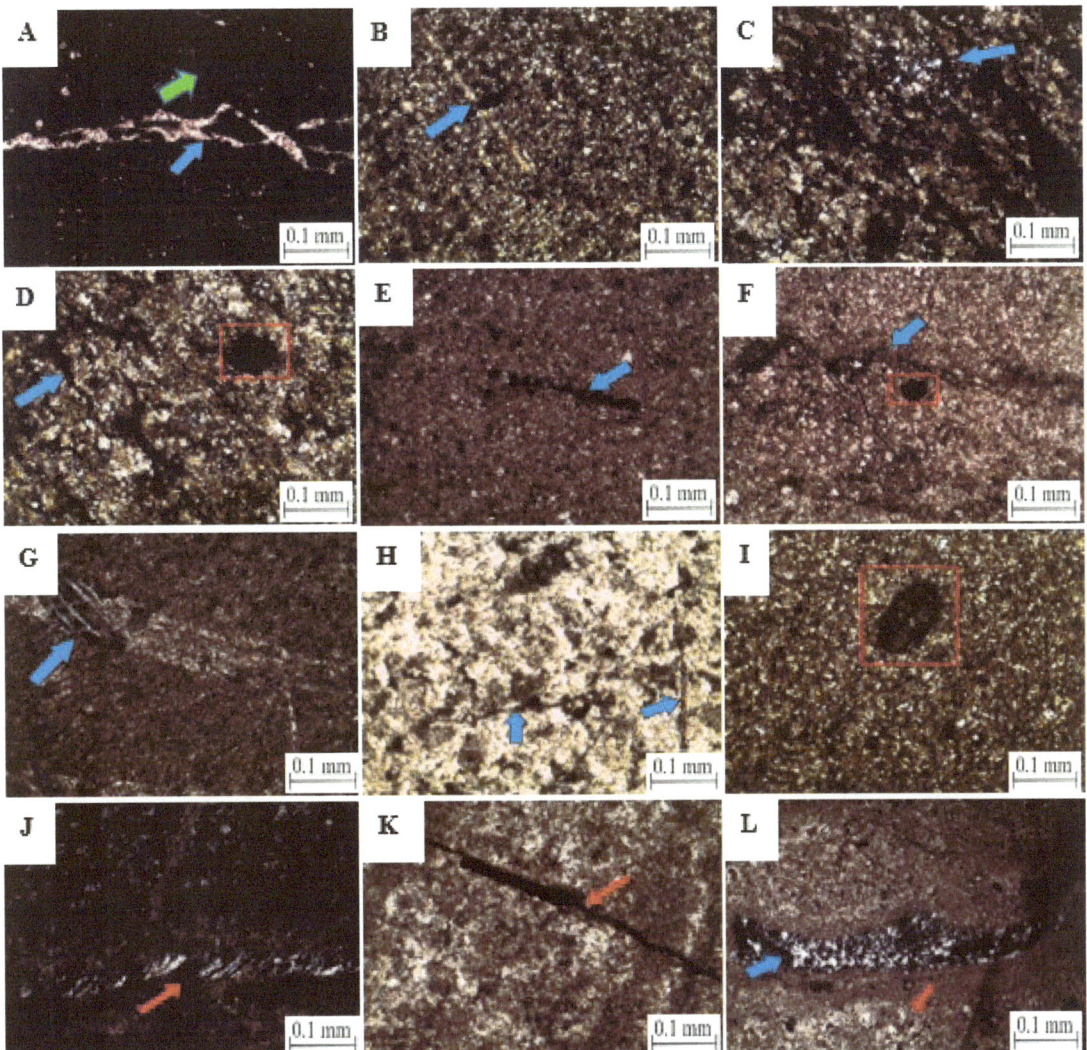

**Figure 6.** Photomicrographs of the Dol. I microfacies of the Abbottabad Formation. (**A**) Partially dolomitized vein (blue arrow) and carbonaceous material in fine dolomite rhombs (green arrow) (KA-1; PPL. 10×). (**B**) Fine equigranular, planar-s dolomite facie and pyrite grains upon dolomite vein (blue arrow) (KA-2; XPL. 10×). (**C**) Fine crystalline dolomite with chalcedony (blue arrow) (KA-3; XPL. 10×). (**D**) Stylolite veins highlighted by hematite (blue arrow) and rectangle showing cubic pyrite in planar dolomite facie (KA-3; XPL. 10×). (**E**) Bioclast in fine crystalline dolomite (blue arrow) (KA-4; PPL. 10×). (**F**) Stylolite vein (blue arrow) and rectangle showing hematite grain (KA-4; PPL. 10×). (**G**) Calcite and dolomite vein (blue arrow) in fine crystalline dolomite (KA-19; XPL. 10×). (**H**) Bioclasts in fine crystalline dolomite (blue arrows) (KA-27; XPL. 10×). (**I**) Bioclast in fine crystalline dolomite (KA-31; PPL. 10×). (**J**) Chalcedony vein (red arrow) and chert in dolomite (KA-37; XPL. 10×). (**K**) Hematite vein in dolomite (KA-37; XPL. 10×). (**L**) Chalcedony patches (blue arrow), and chert (red arrow) in fine planar-s dolomite (KA-44; XPL. 4×).

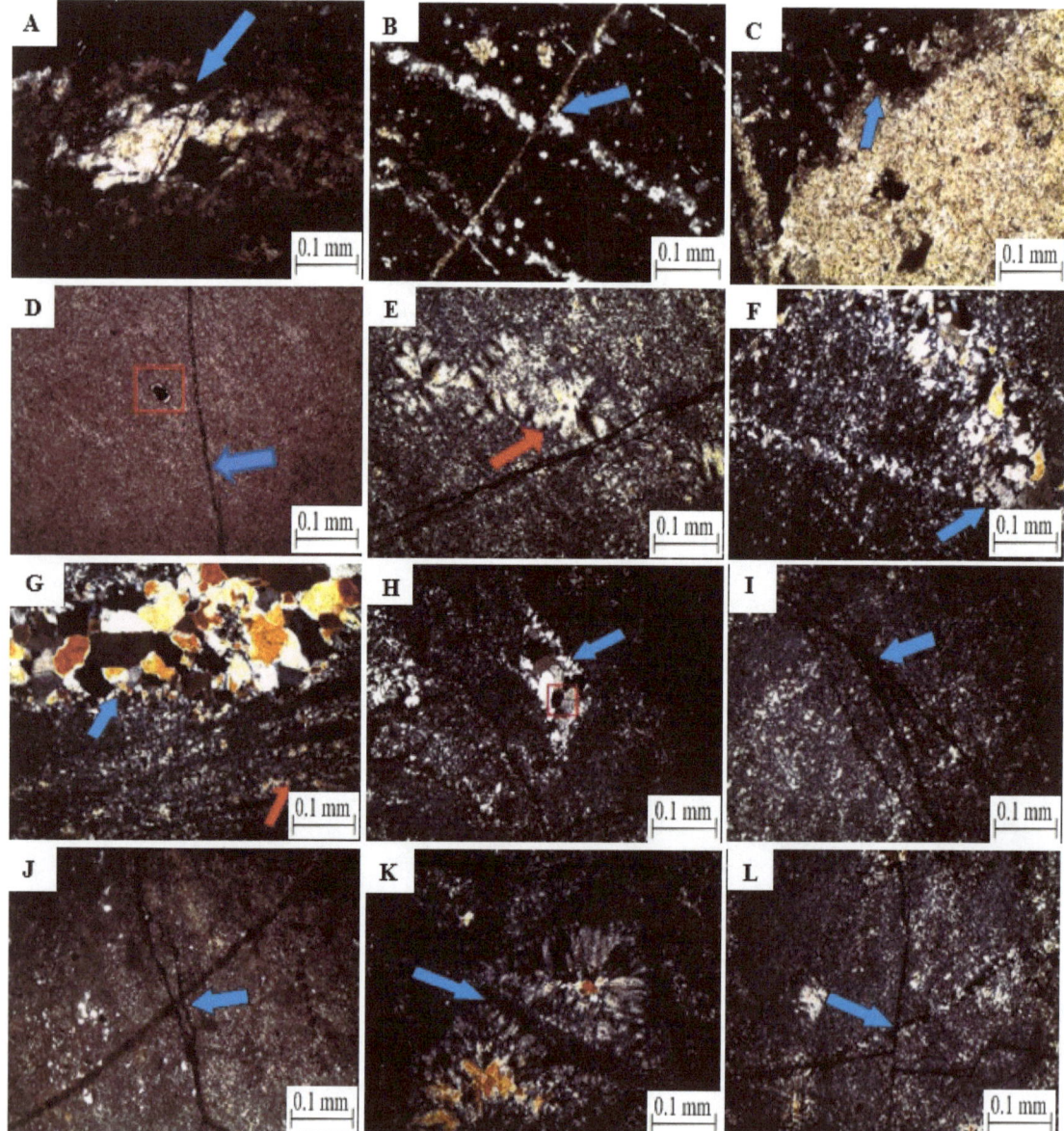

**Figure 7.** Photomicrographs of the Dol. II microfacies of the Abbottabad Formation. (**A**) Chert inclusion in dolomite vein (blue arrow) of cherty dolomite (KA-5; PPL. 10×). (**B**) Dolomite vein cutting chalcedony (blue arrow) (KA-5; XPL. 10×). (**C**) Irregular contact of dolomite and chert (blue arrow) (KA-5; XPL. 10×). (**D**) Dolomitized veins highlighted by hematite (blue arrow) and rectangle showing apatite (KA-41, PPL. 4×). (**E**) Chalcedony (red arrow) and hematite vein (KA-41; XPL. 4×). (**F**) cross cutting cherty veins (blue arrow) in cryptocrystalline silica (KA-41; XPL. 4×). (**G**) Microcrystalline (red arrow) and cryptocrystalline chert (blue arrow) (KA-41; XPL. 4×). (**H**). Chert nodule (blue arrow) containing pyrite (KA-59; XPL. 4×). (**I**) Hematite veins (blue arrow) in microcrystalline silica (KA-59; XPL. 4×). (**J**) Hematite veins cut by bioclast (KA-60; XPL. 4×). (**K**) Chalcedony (blue arrow) (KA-61; XPL. 4×). (**L**) Unfilled veins showing micro faults (blue arrow) (KA-61; XPL. 4×).

**Figure 8.** Photomicrographs of the Dol. III microfacies of the Abbottabad Formation. (**A**) Microbial lamination (blue arrow) and organic matter (KA-8; PPL. 4×). (**B**) Chert patches with stylolite vein (red arrow) (KA-35; XPL. 4×). (**C**) Algae bioclast (blue arrow) (KA-56; XPL. 4×). (**D**) Dolomitized algal mats (KA-56; XPL. 4×). (**E**) Dolomitized algae bioclast (red arrow) (KA-56; XPL. 4×) (**F**) Spary calcite (blue arrow) (KA-56; XPL. 4×). (**G**) Algal mats (blue arrow) partially dolomitized (KA-63; XPL. 10×). (**H**). Chalcedony vein (blue color) and hematite vein (red arrow) along algal mats (KA-63; XPL. 10×). (**I**) Rectangle showing pellets (KA-63; XPL. 4×). (**J**) Rectangle showing zircon and polygon showing tourmaline crystals (KA-63; XPL. 10×). (**K**) Algal mat (blue arrow) (KA-67; XPL. 4×). (**L**) Rectangle showing cavity filled by calcite and then dolomitized (KA-67; XPL. 4×).

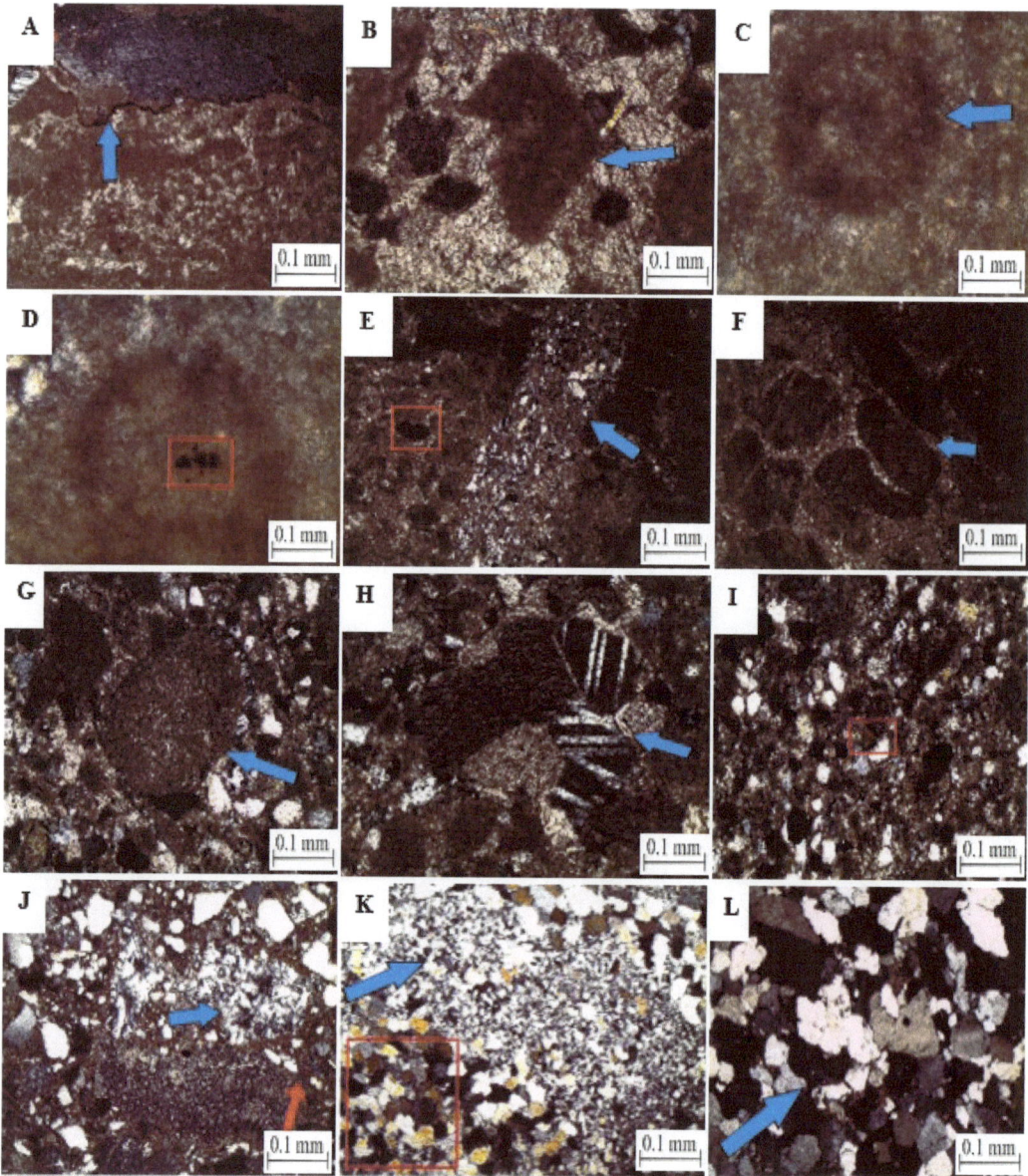

**Figure 9.** Photomicrographs of the Dol. III, IV and arenite microfacies of the Abbottabad Formation. (**A**) Chert (red arrow) and stylolite vein (blue arrow) along algal mats in Dol. III facies (KA-67; XPL. 4×). (**B**) Bioclast (blue arrow) in Dol. III facies (KA-67; XPL. 10×). (**C**) Dolomitized ooid (blue arrow) in Dol. III facies (KA-70; XPL. 40×). (**D**) Pyrite grains in bioclast in Dol. III facies (KA-70; XPL. 40×). (**E**) Cherty and dolomitic vein and rectangle showing pellets in Dol IV facies (KA-50; XPL. 4×). (**F**) Intraclasts (blue arrow) in Dol IV facies (KA-50; XPL. 4×). (**G**) Dolomite clast (blue arrow) in Dol IV facies (KA-65; XPL. 10×). (**H**). Spary calcite (blue arrow) in Dol IV facies (KA-65; XPL. 10×). (**I**). Tourmaline crystal (red rectangle) in Dol IV facies (KA-65; XPL. 10×). (**J**). Chalcedony (red arrow) and chert (blue arrow) in Dol IV facies (KA-76; XPL. 4×). (**K**). Chert (blue arrow) and quartz (red rectangle) in arenite facies (KA-82; XPL. 4×). (**L**). Sub-angular, sutured grains of quartz in arenite facies (KA-87; XPL. 4×).

### 4.1.6. Fine Crystalline Dolomite (Dol. I)

In the Kamsar section, ten samples (KA-1, KA-2, KA-3, KA-4, KA-19, KA-22, KA-27, KA-31, KA-37, and KA-44; Table 2) are microcrystalline to fine-crystalline dolomite (Dol. I). Around 70% to 95% of this microfacies are composed of dolomite, which has a nonplanar to planar texture (Figure 6A–C).

Dolomite crystals are compacted, subhedral, and are of the same size. Chert, which is composed of small fibers of silica, is between 1% and 5% (Figure 6C,J,L), and pyrite, which is a Fe-rich residue, is also between 1% and 5%. Pyrite is observed in both cubic (Figure 6D) and framboidal (Figure 6F) shapes, which formed during deposition and diagenesis, respectively. Bioclasts from the Cambrian period are between 1% and 2% (Figure 6E,H). Dolomitized bioclast in fine crystalline dolomite is present (Figure 6I). This microfacies also has stylolite veins (Figure 6D,F), chalcedony veins (Figure 6I,J), a hematite vein (Figure 6K), and calcite and dolomite veins (Figure 6G). Chemical compaction occurs in many sediments. This is shown by pressure solutions and the formation of stylolites [40]. Load and/or tectonic stress can cause pressure to build up. Based on the Sibley and Gregg [34] textural classification, the Dol. I microfacies of the study area are crystalline planar-s to planar-e type dolomite, most of which is dolomite replacing other sediments.

### 4.1.7. Dolomitic Cryptocrystalline Chert (Dol. II)

*Dolomitic cryptocrystalline chert* was observed in seven samples (KA-5, KA-41, KA-59, KA-60, KA-61, KA-68, and KA-84; Tables 2 and 3) (Dol. II). Petrographically, this microfacies has between 5% and 30% dolomite with planar-c texture (Figure 7A). In this microfacies, the amount of silica is very high, and it is visible in the form of chert, which is a fine-grained sedimentary rock made of microcrystalline or cryptocrystalline quartz ($SiO_2$) and chalcedony. Chert is between 5% and 60% (Figure 7C,F,H), cryptocrystalline silica is between 30% and 65% (Figure 7F,G), and chalcedony is between 8% and 24% (Figure 7B,E,K). The lower part of this facies (KA-5) is in a transitional phase, and the middle and upper parts have more silica. Grains of apatite (Figure 7D) and pyrite (Figure 7H) are also observed.

### 4.1.8. Dolomite with Algal Mats (Dol. III)

These facies were observed in eight samples (KA-8, KA-35, KA-56, KA-63, KA-66, KA-67, KA-69, and KA-70; Tables 2 and 3) (Dol. III). The quantity of dolomite in this microfacies ranges from 18% to 78% and shows a non-planar-c type texture. Algal mats are presented as a sedimentary structure along carbonaceous material, ranging from 5–55%. The thickness of the micritic laminae is variable (Figure 8A,D,G). These crusts, which are several centimeters thick, are thought to be carbonate precipitation due to micro-organisms. Thin sections of studied stromatolites show that they are composed of a fine-grained sediment that is well-laminated. This is because fine-grained sediment was deposited in small amounts at different times and was then trapped (Figure 8B,C,K). The iron minerals, such as hematite, in the dark microbial laminae of stromatolites (Figure 8A,H) are present. Chert is between 4% and 30% (Figure 9A). There are also partially dolomitized bioclast (Figure 8B) and peloids (Figure 9C,D) in this microfacies.

### 4.1.9. Intraclastic–Dolo-Grain Stone (Dol. IV)

Three samples (KA-50, KA-65, and KA-76; Tables 2 and 3) are intraclast dolo-grain stone (Dol. IV). Dolomite makes up between 20 and 25% of this microfacies, revealing planar idiotopic (Planar-p). In this microfacies, intraclasts of different types of rock, such as dolomite, chert, and carbonaceous material, are common and constitute 5–60% (Figure 9F,G), and chert makes up 5–25% (Figure 9E,J).

### 4.1.10. Fine Quarzitic Arenite

Two samples. (KA-82 and KA-87; Table 3) are arenites. The quartz amount in this microfacies is around 95%. Quartz grains are moderately sorted and sutured (Figure 9K,L).

### 4.1.11. XRD and EDX Analysis of Dolostone

The XRD analysis of dolomite indicates similar mineral composition as determined through petrographic studies. The main minerals found in dolostone are dolomite (26–68%), calcite (5–47%), quartz/chert/moganite (13–66%), pyrite (up to 3%), and carbonaceous material (up to 5%). The XRD analysis of dolostone further reveals that dolomite mineral decreases from base to top and chert increases towards top (Figure 10A–D).

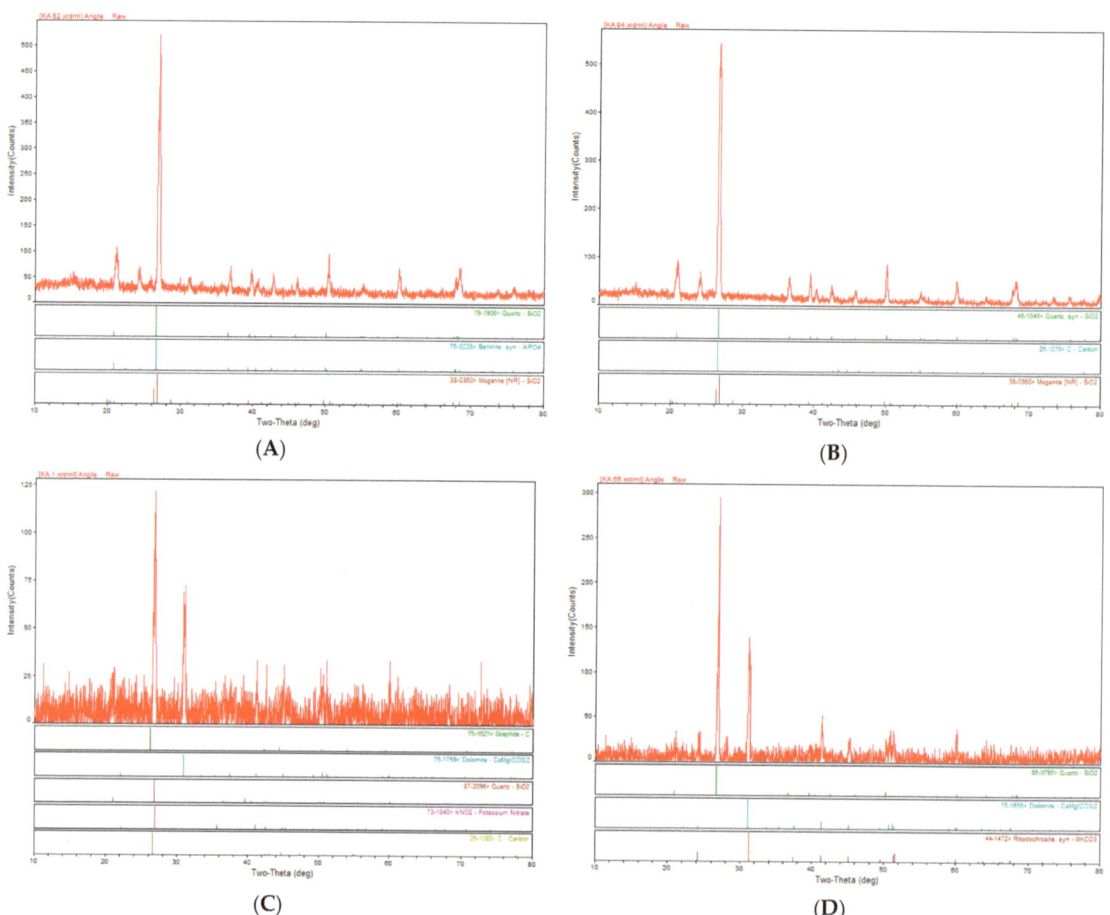

**Figure 10.** XRD pattern of the Abbottabad Formation. (**A**) Fine crystalline dolomite (LA-1; Dol. I; sample no. KA-1); (**B**) dolomite (LA-3; Dol. IV; sample no. KA-65); (**C**) quartzite (LA-4; arenite microfacies; sample no. KA-82); (**D**) dolomite (LA-1; Dol. II; sample no. KA-84).

The diffractogram A of Dol. I microfacies (sample no. KA-1) demonstrates the presence of dolomite, carbonaceous material, potassium nitrate ($KNO_3$), and quartz (Figure 10A). The occurrence of potassium nitrate in the Dol. I microfacies probably indicates that the rocks have been altered by the actions of microorganisms, such as bacteria or fungi, which produce nitrate as a byproduct of their metabolic processes. Diffractogram B of Dol. IV microfacies (sample no. KA-65) shows the presence of dolomite, quartz, and rhodochrosite (Figure 10B). The presence of rhodochrosite ($MnCO_3$) in dolomite indicates a hydrothermal alteration or replacement of the original dolomite. Diffractogram C of arenite microfacies (sample no. KA-82) shows the presence of quartz; berlinite ($AlPO_4$), which is a high-temperature hydrothermal or metasomatic phosphate mineral; and moganite (Figure 10C).

Diffractogram D of Dol II shows the presence of quartz ($SiO_2$), carbonaceous material (C), and moganite ($SiO_2$) (Figure 10D). However, no dolomite mineral was observed in this diffractogram. The dolomite mineral was replaced by quartz and moganite during metasomatism. In the Dol. II microfacies, hydrothermal solutions played vital role in replacing dolomite during metasomatism. In the field, this Dol. II microfacies was undifferentiated due to the preservation of the original texture of the dolomite. Texturally, Dol. II was similar to LA-1 lithofacies.

The elemental analysis of dolostone through EDX determined the major elements in weight percent and atomic percent. The weight percent of elements found in different facies of dolostone are carbon (up to 12%), oxygen (31–56%), magnesium (up to 9%), aluminum (up to 4%), silicon (2–64%), calcium (up to 53%), potassium (up to 6%) iron (up to 7%), and sulfur (up to 3%) (Table 4). These different elements constitute the minerals that found in the dolostone as confirmed by petrographic and XRD analysis.

Table 4. Energy dispersive X-ray (EDX) percentage of elements present in samples of the Abbottabad Formation.

| Sample No. | Elements Weight% | | | | | | | | | Elements Atomic% | | | | | | | | |
|---|---|---|---|---|---|---|---|---|---|---|---|---|---|---|---|---|---|---|
| | C | O | Mg | Al | Si | Ca | K | Fe | S | C | O | Mg | Al | Si | Ca | K | Fe | S |
| KA-3 | 0.94 | 52.30 | 4.12 | 4.13 | 24.93 | 5.90 | 5.55 | 2.17 | – | 0.85 | 67.21 | 3.49 | 3.45 | 18.26 | 3.03 | 2.92 | 0.8 | – |
| KA-8 | 0.39 | 47.05 | 0.65 | 0.37 | 2.42 | 47.72 | 1.40 | – | – | 0.76 | 67.97 | 0.62 | 0.31 | 1.99 | 27.52 | 0.83 | – | – |
| KA-22a | 0.33 | 55.80 | 1.37 | 1.33 | 5.36 | 33.73 | 2.08 | – | – | 0.58 | 74.10 | 1.20 | 1.05 | 4.06 | 17.88 | 1.03 | – | – |
| KA-22b | 0.81 | 54.07 | 4.80 | 4.13 | 13.31 | 14.53 | 4.25 | – | – | 1.40 | 70.17 | 4.10 | 3.18 | 9.84 | 7.53 | 2.26 | – | – |
| KA-27 | 12.36 | 31.02 | 0.55 | 0.27 | 2.70 | 52.71 | 0.39 | – | – | 23.27 | 43.85 | 0.51 | 0.22 | 2.17 | 29.74 | 0.23 | – | – |
| KA 60 | – | 31.91 | – | 2.99 | 51.32 | – | 3.41 | 7.40 | 2.98 | – | 46.98 | – | 2.61 | 43.05 | – | 2.05 | 3.12 | 2.2 |
| KA 61 | | 33.48 | | | 64.29 | | 2.23 | | | | 47.32 | | | 51.77 | | 0.90 | | |
| KA 67 | 14.66 | 32.13 | 9.26 | 2.21 | 9.94 | 28.60 | 2.15 | 1.06 | | 25.25 | 41.55 | 7.88 | 1.70 | 7.32 | 14.76 | 1.14 | 0.39 | |

Overall, the Cambrian dolostone of the Abbottabad Formation mainly comprised dolomite mineral ($CaMg(CO_3)_2$), with chert ($SiO_2$), calcite ($CaCO_3$), Pyrite ($FeS_2$), and carbonaceous material (C). Trace amounts of zircon ($ZrSiO_4$) and muscovite ($KAl_2(AlSi_3O_{10})(F,OH)_2$,) were also found in the dolostone of the Abbottabad Formation (Figure 8; Table 3).

## 5. Discussion

*5.1. Regional Stratigraphic Correlations*

Early Cambrian rocks are exposed in various locations in the northern Pakistan, including the Salt Range [28,41], Peshawar Basin [42], Abbottabad [43,44], Muzaffarabad, and Kotli [15,45,46]. The Cambrian Ambar Formation in the Peshawar basin is a possible equivalent of the Abbottabad Formation [47]. Latif [48] classifies the Abbottabad Group into Sirban and Kakul formations. The Kakul Formation is divided into four members: the Tanakki Conglomerate, the Sangargali Member, the Mahmdagali Member, and the Mirpur Sandstone. In the Abbottabad region, Qasim, Khan and Haneef [44] divided the Abbottabad Formation into three distinct units: (1) the base arenaceous unit, (2) the intermediate dolomite unit, and (3) the upper quartzite unit. In contrast to the Hazara region, the Abbottabad Formation in the studied area lacks the lower arenaceous member and stromatolitic-dolomitic unit, which are exclusively visible in the Muzaffarabad region. Hazara's Abbottabad Formation is associated with two lithofacies, LA-1 and LA-4, while the LA-1 of the study area is correlated with the Cambrian Jutana Formation of Potwar/Salt Range only [22]. The main difference is the absence of stromatolitic unit in the Salt Range and the Abbottabad area compared to the Muzaffarabad area. It is worth mentioning that in the Salt Range, the Cambrian rocks are present in the Precambrian Salt Range Formation instead of the Hazara Formation (Abbottabad area) or the Dogra Formation (Muzaffarabad area).

## 5.2. Dolomite Texture and Mechanism of Dolomitization

Heterogeneous textures were observed in the Cambrian dolomite. The texture of the dolomite varies from base to top and correspond to four microfacies of the dolomite, i.e., fine crystalline dolomite (Dol. I), dolomitic cryptocrystalline chert (Dol. II), dolomite with algal mats (Dol. III), and intraclastic-dolo-grain stone (Dol. IV). The texture of the Dol. I microfacies is non-planar xenotopic to planar e and planar s, where dolomite crystals are tightly packed anhedral to euhedral. Planar subhedral to euhedral replacive dolomite crystals formed in the early diagenesis stage (Figure 11). The Dol-I microfacies is present as fine-grained rhombohedral crystals and indicates that the dolomitization process was slow and controlled by kinetic constraints. The texture of the Dol. II microfacies is planar c type, in which pores are filled by micro crystalline to crypto crystalline chert. The addition of this silica occurred through the replacement of dolomite crystal during the syndiagenetic–epigenetic period. The large amounts of silica were provided in the Proto-Tethys Ocean by the hydrothermal fluids, and this silica replaced the dolomite selectively. Selective metasomatism was probably caused by the weak acidic environment due to the decomposition of organic matter, which is beneficial for the deposition and metasomatism of silica (Figure 11). The texture of the Dol. III microfacies is the non-planar c type, where a scimitar-like termination has been observed between algal mats dolomite crystals and non-algal mats dolomite. Non-planar dolomite can form at low temperatures, under conditions of high supersaturation, although such occurrences are fairly rare (Figure 11). The texture of the Dol. IV microfacies is the planar p (porphyrotopic) type, where dolomite crystals are embedded in a calcite matrix. The porphyrotopic planar dolomite probably formed in peritidal environments through reflux dolomitization. Based on the petrography it seems as though Dol. IV was dolomitized than eroded and deposited. This suggests early dolomitization. Additionally, dolomitization occurred prior to the regression at least in some places on the dip of this site. During the regression, it was subsequently eroded and transported, at which point dolomitization could have also been enhanced. The presence of detrital grains (tourmaline, quartz, zircon, and apatite) with intraclasts of dolomite also support this phenomenon.

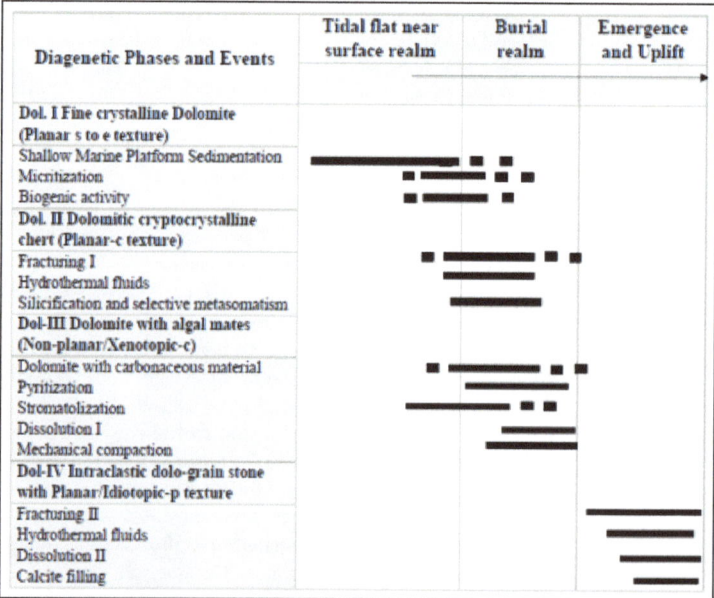

**Figure 11.** Paragenetic history of the dolomite of the Abbottabad Formation, representing early tidal flat period followed by s burial period and uplift.

Dolomitization is a replacement reaction, whereby Mg replaces Ca in the crystal lattice. This is dictated by hydrogeological mechanisms as well as thermodynamic and kinetic factors [45–47]. Dolomitization is often the product of early diagenetic processes in carbonate-rich mud or limestone, while it is the result of a higher grade of dolomitization in solid limestone [49–51]. In the Upper Indus Basin (UIB), above the Precambrian crystalline basement, the Salt Range Formation comprises thick reserves of salt. The thickness of these deposits provides a thread of a lagoonal environment [41]. The Cambrian-aged Abbottabad Formation in the Proto-Tethys Ocean overlies the older salt-related paleolatitudes. The sedimentological hierarchy of facies presents an intricate picture of the depositional history of geological strata. A careful analysis of this hierarchy reveals that the episodic regressive events have played a significant role in shaping the depositional environment. Specifically, these events have given rise to a horizon where limestone was deposited in a shallow intertidal to subtidal environment. Furthermore, this limestone was subsequently subjected to dolomitization, either via penecontemporaneously or through reflux mechanisms (Figure 12). The presence of a significant amount of silica and silicate in the Dol I and Dol II suggests that there were high concentrations of silica in the solution during dolomite formation. Recent studies have shown that dissolved silica could promote primary dolomite formation with silica adsorbed onto carbonate surfaces, which transforms either to authigenic clay or chert/chalcedony during burial [52,53].

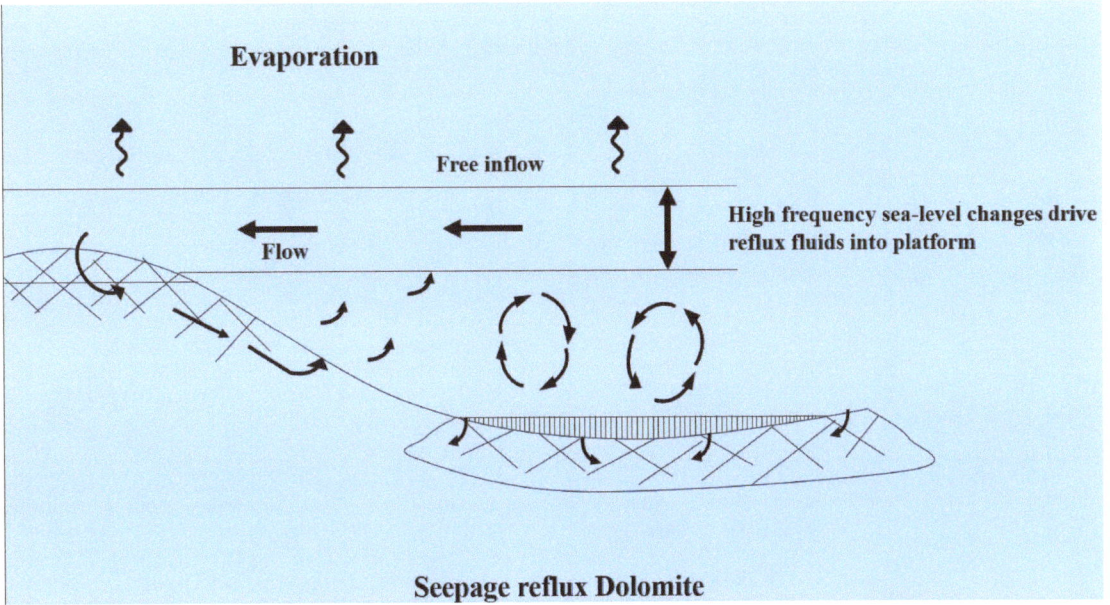

**Figure 12.** Mechanism of dolomitization (modified after [54,55]).

*5.3. Depositional Environment and Proto-Tethys Evolution*

In the studied area, a non-conformity indicates the top of the Precambrian Dogra slates and the fine crystalline dolomite is followed by shale. The overall succession of this lithofacies (LA-1) from base to top demonstrates an upward coarsening trend: from shallow subtidal dominated units to supratidal dominated units (Figure 4). The coarsening upward sequence in carbonate rocks can be recognized in the sediment layers as it become thicker as it rises and displays diagenetic features, such as cementation and dissolution, which are more common in the coarser upper part of the sequence. Above the LA-1, stromatolitic dolomites (LA-2) contain fenestrae, microbial laminae, and chert lenses, which are characteristic of shallow subtidal to supratidal settings [56]. This stage is characterized by the deposition of microbial dolomite before a major base-level decrease (Figure 13). Stromatolites vary in size and form in these facies. Stromatolites are laminated, arched organic sedimentary formations produced by blue-green algae that adhere to sediment (cyanobacteria). Bacteria precipitate, trap, and bind sediment layers to create accretionary structures. This appears as domes, conical, or complexly branched structures ranging in size from a little finger to a house. In this lithofacies, both cubic and framboidal pyrite ($FeS_2$) were found as syngenetic to diagenetic grains, respectively [13,57]. These grains can form in low-temperature diagenetic environments or precipitate in anoxic waters. They are typical minerals found in organic-rich sediments and indicate euxinic depositional conditions [58]. These euxinic conditions occurred when water was both anoxic and sulfidic. Oxidized, ferruginous, and limonitic dolomite (LA-3) characterized surface karstification products as well as karstic depression fills, revealing a hundred million years of exposure. Numerous studies of modern and ancient karst features indicate that the most favorable conditions for prominent karstification are moderate to high rainfall [59] and relatively pure, dense, and thick carbonate rocks with appropriate conduits, such as fractures, joints, faults, or selective solutional pipes [60–62]. In the Kamsar section, the quartzite unit (LA-4) is composed of medium- to fine-grained sandstone and brecciated sandstone at the top; these characteristics indicate the subaerial exposure (Figure 13).

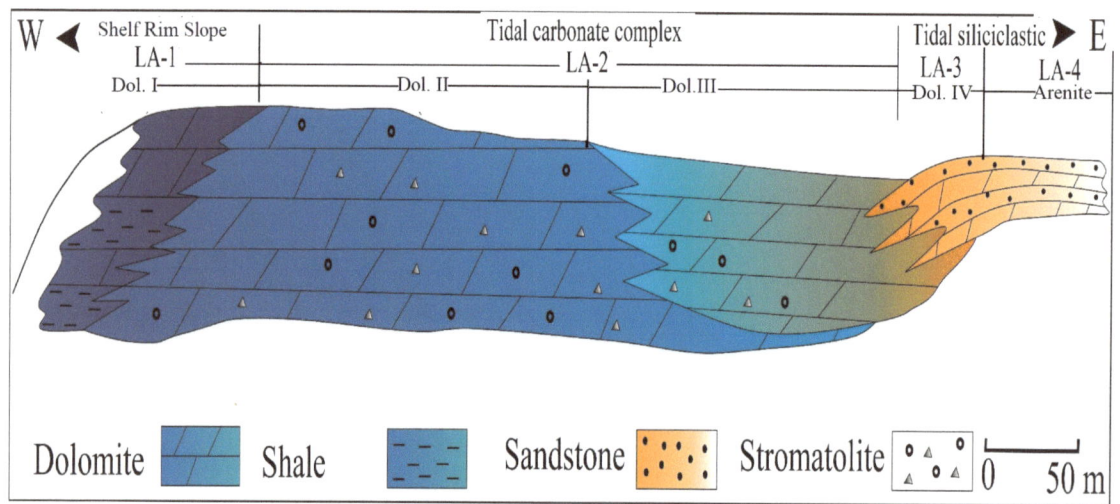

**Figure 13.** Depositional model showing the lateral facies distribution of the Abbottabad Formation from west (**left**) to east (**right**).

The presence of various types of dolomites, chert, and quartz in thin sections also aided in the petrographical investigation of the depositional phenomenon. During the process of diagenesis, the silica in sediments change from opal to microcrystalline quartz in mature chert [63]. Silica mobilized from hydrothermal fluids and clay minerals might potentially contribute to silicification [64]. Petrographic studies helped in delineating four phases of distinct characteristic features. These include fine crystalline dolomite (Dol. I), dolomitic crystalline chert (Dol. II), dolomite with algal mats (Dol. III), and intraclastic dolo-grain stone (Dol. IV) (Figure 11).

An initial phase of dolomite (Dol. I) resulted from interaction with Mg-rich fluids originated during earliest diagenetic phase, representing penecontemporaneous or reflux mechanism during early Cambrian. This was followed by an increase in different types of silica (Dol. II) from hydrothermal fluids and the growth of stromatolites/algal mats (Dol. III). Dol. II and Dol. III are associated with subtidal to supratidal conditions, whereas Dol. IV formation is associated with sub-aerial exposure or orogenic events. In general, the available data indicate that dolomitization resulting from reflux mechanisms can be identified in thin sections by the occurrence of saddle-shaped dolomite crystals, a coarse-crystalline texture, the replacement of previously formed minerals, the selective dolomitization of certain rock fabrics, and atypical crystallographic orientations.

The Proto-Tethys Ocean existed on the Earth between 550–330 mya [65]. The Proto-Tethys Ocean was highly susceptible to evaporitic conditions during the Late Precambrian, which resulted in the deposition of salt basement over a broad region in the Upper Indus Basin (UIB) [37]. These sediments comprise the Precambrian Salt Range Formation. The Muzaffarabad region of the Kamsar section is located on the Proto-Tethys Ocean's passive margin. The sedimentological hierarchy of facies indicates that the episodic regressive events provide a horizon where limestone was deposited in shallow intertidal to subtidal environment and later subjected to dolomitization penecontemporaneously or via a reflux mechanism (Figure 12). Based on the observed data, it can be stated that during regression events and increasing restriction, both temperature and salinity tend to increase. Overall, the combination of kinetic, geochemical, and biological factors initiates dolomitization depending on the specific geological setting and local conditions. Additionally, if evaporites precipitate, there is a likelihood of an increase in the Mg:Ca ratio [66]. These kinetic conditions collectively facilitate penecontemporaneous dolomitization [67].

## 6. Conclusions

The following results were obtained based on field observations, facies analysis, petrographic, and geochemical studies:

1. The Abbottabad Formation lithologically consists of fine crystalline dolomite (LA-1), microbially laminated dolomite (LA-2), oxidized, ferruginous dolomite (LA-3), and quartzite (LA-4), corresponding to four microfacies, i.e., fine crystalline dolomite (Dol. I), dolomitic cryptocrystalline chert (Dol. II), dolomite with algal mats (Dol. III), and intraclastic dolo-grain stone (Dol. IV).
2. Stromatolites are often composed of vertically stacked hemispheroids that are subordinated by laterally linked hemispheroids. In general, conditions suggestive of tidal flats existed throughout the formation of this sequence.
3. The overall succession of the Abbottabad Formation from base to top demonstrates a coarsening trend succession, indicated by the sediment layers becoming thicker as they rise and cementation and dissolution being more common in the upper part of the sequence. This succession represents shallow subtidal units becoming supratidal-dominated units. Oxidized, limonitic dolomite (LA-3) represents surface karstification products, whereas brecciated quartzite (LA-4) denotes subaerial exposure.
4. The X-ray diffraction (XRD) analysis of dolomite revealed a mineral composition consistent with that determined by petrographic studies, with dolostone containing primarily dolomite, calcite, quartz/chert/moganite, potassium nitrate ($KNO_3$), rhodochrosite ($MnCO_3$), berlinite ($AlPO_4$), and minor amounts of pyrite and carbonaceous material and shows the dolomite mineral decreasing from base to top, while chert increases towards the top.
5. The EDX analysis determined the weight percent of different elements, such as carbon, oxygen, magnesium, aluminum, silicon, calcium, potassium, iron, and sulfur. These elements constitute the minerals that found in the dolostone as confirmed by petrographic and XRD analysis.
6. The Dol. I microfacies is characterized by non-planar xenotopic texture, with tightly packed anhedral to euhedral dolomite crystals formed in the early diagenesis stage, indicating slow and controlled dolomitization. Dol. II microfacies has a planar-c type texture with pores filled by micro to crypto crystalline chert, formed by the selective replacement of dolomite by silica. The abundance of silica in the Proto-Tethys Ocean was due to hydrothermal fluids. The Dol. III microfacies shows non-planar-c type texture, where scimitar-like termination is observed between algal mats and non-algal mats dolomite. Dol. IV microfacies exhibits a planar-p (porphyrotopic) texture,

where dolomite crystals embedded in a calcite matrix probably formed in peritidal environments through reflux dolomitization.
7. The sedimentological hierarchy of facies indicates that the episodic regressive events provide a horizon where limestone was deposited in the shallow intertidal to subtidal environment and later subjected to dolomitization penecontemporaneously or via a reflux mechanism.
8. Overall, the evidence of reflux dolomitization in thin sections includes the presence of saddle-shaped dolomite crystals, coarse-crystalline texture, the replacement of earlier-formed minerals, fabric-selective dolomitization, and unusual crystallographic orientations.

**Author Contributions:** Conceptualization, S.K.A., R.A.L. and A.G.S.; methodology, S.K.A., T.M. and M.S.M.; software, S.K.A., H.T.J. and A.B; validation, S.K.A., G.K. and H.T.J.; formal analysis, M.S.M. and S.K.A.; investigation, S.K.A., H.T.J., R.A.L., A.G.S., T.M., G.K. and A.B.; resources, S.K.A.; data curation, S.K.A., H.T.J., G.K. and M.S.M.; writing—original draft preparation, S.K.A.; writing—review and editing, S.K.A., H.T.J., M.S.M., T.M., G.K., A.B. and K.U.M.; visualization, S.K.A., M.S.M., H.T.J. and K.U.M.; supervision, R.A.L., A.G.S. and T.M.; project administration, R.A.L., S.K.A., G.K. and H.T.J.; funding acquisition, G.K. All authors have read and agreed to the published version of the manuscript.

**Funding:** This research received no external funding.

**Institutional Review Board Statement:** Not applicable.

**Informed Consent Statement:** Not applicable.

**Data Availability Statement:** The data used in this work are available on request from the corresponding author(s).

**Acknowledgments:** We are thankful to the Muhammad Sabir Khan, Shahab Pervez, Mirza Shahid Baig, Muhammad Basharat, Tayyab Riaz, and the Director of the Institute of Geology, University of Azad Jammu and Kashmir, Muzaffarabad, for their help and friendly attitudes during our research work, which was always a source of inspiration for us. Last but not the least, the authors are highly indebted to the five reviewers for their fruitful reviews, which helped us to improve the manuscript.

**Conflicts of Interest:** The authors declare no conflict of interest.

# References

1. Bilal, A.; Mughal, M.S.; Janjuhah, H.T.; Ali, J.; Niaz, A.; Kontakiotis, G.; Antonarakou, A.; Usman, M.; Hussain, S.A.; Yang, R. Petrography and provenance of the Sub-Himalayan Kuldana Formation: Implications for tectonic setting and Palaeoclimatic conditions. *Minerals* **2022**, *12*, 794. [CrossRef]
2. Mughal, M.S.; Zhang, C.; Du, D.; Zhang, L.; Mustafa, S.; Hameed, F.; Khan, M.R.; Zaheer, M.; Blaise, D. Petrography and provenance of the Early Miocene Murree Formation, Himalayan Foreland Basin, Muzaffarabad, Pakistan. *J. Asian Earth Sci.* **2018**, *162*, 25–40. [CrossRef]
3. Bossart, P.; Dietrich, D.; Greco, A.; Ottiger, R.; Ramsay, J. A new structural interpretation of the hazara-kashmir syntaxis, Southern Himalayas, Pakistan. *Kashmir J. Geol.* **1984**, *2*, 19–36.
4. Calkins, J.A.; Matin, A.A. *The Geology and Mineral Resources of the Garhi Habibullah Quadrangle and the Kakul Area, Hazara District, Pakistan*; US Geological Survey: Reston, VA, USA, 1973; pp. 1258–2331.
5. Baig, M.; Lawrence, R. Precambrian to early Paleozoic orogenesis in the Himalaya. *Kashmir J. Geol.* **1987**, *5*, 1–22.
6. Basharat, M.; Rohn, J.; Baig, M.S.; Khan, M.R.J.G. Spatial distribution analysis of mass movements triggered by the 2005 Kashmir earthquake in the Northeast Himalayas of Pakistan. *Geomorphology* **2014**, *206*, 203–214. [CrossRef]
7. Marks, P.; Ali, C.M. The geology of the Abbottabad area, with special reference to the Infra-Trias. *Geol. Bull. Punjab Univ.* **1961**, *2*, 47–56.
8. Greco, A.J.J.G. Stratigraphy metamorphism and tectonics of the Hazara-Kashmir syntaxis area, Kash. *J. Geol.* **1991**, *8*, 939–965.
9. Baig, M.S.; Iqbal Siddiqi, M. Paleocene-Eocene Biostratirgraphy of the Yadgar Area, Muzaffarabad, Azad Kashmir, Pakistan. *Pak. J. Hydrocarb. Res.* **2006**, *16*, 59–65.
10. Khan, S.; Ahmad, W.; Ahmad, S.; Khan, J.K. Dating and depositional environment of the Tredian Formation, western Salt Range, Pakistan. *J. Himal. Earth Sci.* **2016**, *49*, 14–25.
11. Aadil, N. Stratigraphy and structure of Sarda, Manil, Changpur and Naghal areas, district Kotli, Jammu & Kashmir. *J. Geol. Soc. India* **2013**, *82*, 639–648.
12. Ashraf, M.; Chaudhry, M.; Qureshi, K. Stratigraphy of Kotli area of Azad Kashmir and its correlation with standard type areas of Pakistan. *Kashmir J. Geol.* **1983**, *1*, 19–29.

13. Bilal, A.; Yang, R.; Fan, A.; Mughal, M.S.; Li, Y.; Basharat, M.; Farooq, M. Petrofacies and diagenesis of Thanetian Lockhart Limestone in the Upper Indus Basin (Pakistan): Implications for the Ceno-Tethys Ocean. *Carbonates Evaporites* **2022**, *37*, 1–19. [CrossRef]
14. Latif, M. A Cambrian age for the Abbottabad group of Hazara, Pakistan. *Geol. Bull. Panjab Univ.* **1974**, *10*, 1–20.
15. Hussain, A.; Yeats, R.S. Geological setting of the 8 October 2005 Kashmir earthquake. *J. Seismol.* **2009**, *13*, 315–325. [CrossRef]
16. Myrow, P.M.; Hughes, N.C.; Goodge, J.W.; Fanning, C.M.; Williams, I.S.; Peng, S.; Bhargava, O.N.; Parcha, S.K.; Pogue, K.R. Extraordinary transport and mixing of sediment across Himalayan central Gondwana during the Cambrian–Ordovician. *GSA Bull.* **2010**, *122*, 1660–1670. [CrossRef]
17. Mustafa, S.; Khan, M.A.; Khan, M.R.; Sousa, L.M.; Hameed, F.; Mughal, M.S.; Niaz, A. Building stone evaluation—A case study of the sub-Himalayas, Muzaffarabad region, Azad Kashmir, Pakistan. *Eng. Geol.* **2016**, *209*, 56–69. [CrossRef]
18. Iqbal, O.; Baig, M.S.; Pervez, S.; Siddiqi, M.I. Structure and stratigraphy of Rumbli and Panjar areas of Kashmir and Pakistan with the aid of GIS. *J. Geol. Soc. India* **2018**, *91*, 57–66. [CrossRef]
19. Medlicott, H.B. Note upon the Sub-Himalayan Series in the Jammu (Jammoo) Hills. *Rec. Geol. Surv. India* **1876**, *9*, 49–57.
20. Bossart, P.; Dietrich, D.; Greco, A.; Ottiger, R.; Ramsay, J.G. The tectonic structure of the Hazara-Kashmir syntaxis, southern Himalayas, Pakistan. *Tectonics* **1988**, *7*, 273–297. [CrossRef]
21. Baig, M.S.; Yeats, R.S.; Pervez, S.; Jadoon, I.; Khan, M.R.; Sidiqui, I.; Lisa, M.; Saleem, M.; Masood, B.; Sohail, A. Active tectonics, October 8, 2005 earthquake deformation, active uplift, scarp morphology and seismic geohazards microzonation, Hazara-Kashmir Syntaxis, Northwest Himalayas, Pakistan. *J. Himal. Earth Sci.* **2010**, *43*, 17–21.
22. Khan, S.; Shah, M.M. Multiphase dolomitization in the Jutana Formation (Cambrian), Salt Range (Pakistan): Evidences from field observations, microscopic studies and isotopic analysis. *Geol. Acta* **2019**, *17*, 1–18.
23. Shah, M.M.; Nader, F.; Garcia, D.; Swennen, R.; Ellam, R. Hydrothermal dolomites in the Early Albian (Cretaceous) platform carbonates (NW Spain): Nature and origin of dolomites and dolomitising fluids. *J. Oil Gas Sci. Technol. Rev. D'ifp Energ. Nouv.* **2012**, *67*, 97–122. [CrossRef]
24. Zhang, F.; Xu, H.; Konishi, H.; Shelobolina, E.S.; Roden, E.E. Polysaccharide-catalyzed nucleation and growth of disordered dolomite: A potential precursor of sedimentary dolomite. *Am. Mineral.* **2012**, *97*, 556–567. [CrossRef]
25. Bontognali, T.R.; Vasconcelos, C.; Warthmann, R.J.; Bernasconi, S.M.; Dupraz, C.; Strohmenger, C.J.; McKENZIE, J.A. Dolomite formation within microbial mats in the coastal sabkha of Abu Dhabi (United Arab Emirates). *Sedimentology* **2010**, *57*, 824–844. [CrossRef]
26. Machel, H.G.; Lonnee, J. Hydrothermal dolomite—A product of poor definition and imagination. *Sediment. Geol.* **2002**, *152*, 163–171. [CrossRef]
27. Warren, J. Dolomite: Occurrence, evolution and economically important associations. *Earth-Sci. Rev.* **2000**, *52*, 1–81. [CrossRef]
28. Ali, S.K.; Janjuhah, H.T.; Shahzad, S.M.; Kontakiotis, G.; Saleem, M.H.; Khan, U.; Zarkogiannis, S.D.; Makri, P.; Antonarakou, A. Depositional Sedimentary Facies, Stratigraphic Control, Paleoecological Constraints, and Paleogeographic Reconstruction of Late Permian Chhidru Formation (Western Salt Range, Pakistan). *J. Mar. Sci. Eng.* **2021**, *9*, 1372. [CrossRef]
29. Kazmi, A.H.; Jan, M.Q. *Geology and Tectonics of Pakistan*; Graphic Publishers: Santa Ana, CA, USA, 1997.
30. Kadri, I.B. *Petroleum Geology of Pakistan*; Pakistan Petroleum Limited: Karachi, Pakistan, 1995.
31. Baig, M.; Snee, L. Pre-Himalayan dynamothermal and plutonic activity preserved in the Himalayan collision zone, NW Pakistan: An thermochronologic evidence. *Proc. Geol. Soc. Am. Abstr. Programs* **1989**, *21*, A268.
32. Munir, H.; Baig, M.S.; Mirza, K. Upper Cretaceous of Hazara and Paleogene Biostratigraphy of Azad Kashmir, North-West Himalayas, Pakistan. *Geol. Bull. Punjab. Univ* **2006**, *40–41*, 69–87.
33. Dunham, R.J. *Classification of Carbonate Rocks According to Depositional Textures*; American Association of Petroleum Geosciences: Tulsa, OK, USA, 1962.
34. Sibley, D.F.; Gregg, J.M. Classification of dolomite rock textures. *J. Sediment. Res.* **1987**, *57*, 967–975.
35. Shah, S. Stratigraphy of Pakistan (memoirs of the geological survey of Pakistan). *Geol. Surv. Pak.* **2009**, *22*, 1–381.
36. Wadia, D.N. The geology of Poonch State (Kashmir) and adjacent portions of the Punjab. *Mem. Geol. Surv. India* **1928**, *51*, 185–370.
37. Flügel, E.; Munnecke, A. *Microfacies of Carbonate Rocks: Analysis, Interpretation and Application*; Springer: Berlin/Heidelberg, Germany, 2010; Volume 976.
38. Logan, B.W.; Rezak, R.; Ginsburg, R. Classification and environmental significance of algal stromatolites. *J. Geol.* **1964**, *72*, 68–83. [CrossRef]
39. Kalkowsky, E. Oolith und Stromatolith im norddeutschen Buntsandstein. *Z. Dtsch. Geol. Ges.* **1908**, *60*, 68–125.
40. Logan, B.W. *Pressure Responses (Deformation) in Carbonate Sediments and Rocks Analysis and Application, Canning Basin*; American Association of Petroleum Geosciences: Tulsa, OK, USA, 1984.
41. Gee, E.; Gee, D. Overview of the geology and structure of the Salt Range, with observations on related areas of northern Pakistan. *Geol. Soc. Am.* **1989**, *232*, 95–112.
42. Pogue, K.R.; Wardlaw, B.R.; Harris, A.G.; Hussain, A. Paleozoic and Mesozoic stratigraphy of the Peshawar basin, Pakistan: Correlations and implications. *Geol. Soc. Am. Bull.* **1992**, *104*, 915–927. [CrossRef]
43. Calkins, J.A.; Tw, O.; Skm, A. *Geology of the Southern Himalaya in Hazara, Pakistan and Adjacent Areas*; US Government Prointing State Office: Washington, DC, USA, 1975.

44. Qasim, M.; Khan, M.A.; Haneef, M. Stratigraphic characterization of the Early Cambrian Abbottabad Formation in the Sherwan area, Hazara region, N. Pakistan: Implications for Early Paleozoic stratigraphic correlation in NW Himalayas, Pakistan. *J. Himal. Earth Sci.* **2014**, *47*, 25–40.
45. Gregg, J.M.; Bish, D.L.; Kaczmarek, S.E.; Machel, H.G. Mineralogy, nucleation and growth of dolomite in the laboratory and sedimentary environment: A review. *Sedimentology* **2015**, *62*, 1749–1769. [CrossRef]
46. Manche, C.J.; Kaczmarek, S.E. A global study of dolomite stoichiometry and cation ordering through the Phanerozoic. *J. Sediment. Res.* **2021**, *91*, 520–546. [CrossRef]
47. Machel Hans, G. Concepts and models of dolomitization: A critical reappraisal. *Geol. Soc. Lond. Spec. Publ.* **2004**, *235*, 7–63. [CrossRef]
48. Latif, M. Lower Palaeozoic (? Cambrian) hyolithids from the Hazara Shale, Pakistan. *Nat. Phys. Sci.* **1972**, *240*, 92. [CrossRef]
49. Haldar, S.; Tišljar, J. Basic mineralogy. In *Introduction to Mineralogy and Petrology*; Elsevier: Oxford, UK, 2014; p. 338.
50. Rahim, H.-u.; Qamar, S.; Shah, M.M.; Corbella, M.; Martín-Martín, J.D.; Janjuhah, H.T.; Navarro-Ciurana, D.; Lianou, V.; Kontakiotis, G. Processes Associated with Multiphase Dolomitization and Other Related Diagenetic Events in the Jurassic Samana Suk Formation, Himalayan Foreland Basin, NW Pakistan. *Minerals* **2022**, *12*, 1320. [CrossRef]
51. Ahmad, I.; Shah, M.M.; Janjuhah, H.T.; Trave, A.; Antonarakou, A.; Kontakiotis, G. Multiphase Diagenetic Processes and Their Impact on Reservoir Character of the Late Triassic (Rhaetian) Kingriali Formation, Upper Indus Basin, Pakistan. *Minerals* **2022**, *12*, 1049. [CrossRef]
52. Fang, Y.; Xu, H. Dissolved silica-catalyzed disordered dolomite precipitation. *Am. Mineral.* **2022**, *107*, 443–452. [CrossRef]
53. Fang, Y.; Xu, H. Coupled dolomite and silica precipitation from continental weathering during deglaciation of the Marinoan Snowball Earth. *Precambrian Res.* **2022**, *380*, 106824. [CrossRef]
54. Yang, X.; Tang, H.; Wang, X.; Wang, Y.; Yang, Y. Dolomitization by penesaline sea water in Early Cambrian Longwangmiao Formation, central Sichuan Basin, China. *J. Earth Sci.* **2017**, *28*, 305–314. [CrossRef]
55. Qing, H.; Bosence, D.W.J.; Rose, E.P.F. Dolomitization by penesaline sea water in Early Jurassic peritidal platform carbonates, Gibraltar, western Mediterranean. *Sedimentology* **2001**, *48*, 153–163. [CrossRef]
56. Yang, W.; Lehrmann, D.J. Milankovitch climatic signals in lower Triassic (Olenekian) peritidal carbonate successions, Nanpanjiang Basin, South China. *Palaeogeogr. Palaeoclimatol. Palaeoecol.* **2003**, *201*, 283–306. [CrossRef]
57. Banerjee, S.; Khanolkar, S.; Saraswati, P.K. Facies and depositional settings of the Middle Eocene-Oligocene carbonates in Kutch. *Geodin. Acta* **2018**, *30*, 119–136. [CrossRef]
58. Gao, X.; Wang, P.; Li, D.; Peng, Q.; Wang, C.; Ma, H. Petrologic characteristics and genesis of dolostone from the Campanian of the SK-I Well Core in the Songliao Basin, China. *Geosci. Front.* **2012**, *3*, 669–680. [CrossRef]
59. Budd, D.A.; Hiatt, E.E. Mineralogical stabilization of high-magnesium calcite; geochemical evidence for intracrystal recrystallization within Holocene porcellaneous foraminifera. *J. Sediment. Res.* **1993**, *63*, 261–274.
60. James, N.; Choquette, P. Diagenesis 9. Limestones-the meteoric diagenetic environment. *Geosci. Can.* **1984**, *11*, 161–194.
61. Palmer, A.N. Origin and morphology of limestone caves. *Geol. Soc. Am. Bull.* **1991**, *103*, 1–21. [CrossRef]
62. Smith, M.; Soper, N.; Higgins, A.; Rasmussen, J.; Craig, L. Palaeokarst systems in the Neoproterozoic of eastern North Greenland in relation to extensional tectonics on the Laurentian margin. *J. Geol. Soc.* **1999**, *156*, 113–124. [CrossRef]
63. Compton, J.S. Porosity reduction and burial history of siliceous rocks from the Monterey and Sisquoc Formations, Point Pedernales area, California. *Geol. Soc. Am. Bull.* **1991**, *103*, 625–636. [CrossRef]
64. Hesse, R. Silica diagenesis: Origin of inorganic and replacement cherts. *Earth-Sci. Rev.* **1989**, *26*, 253–284. [CrossRef]
65. Berra, F.; Angiolini, L. The Evolution of the Tethys Region throughout the Phanerozoic: A Brief Tectonic Reconstruction. *Pet. Syst. Tethyan Reg.* **2014**, *106*, 1–7. [CrossRef]
66. Kaczmarek, S.E.; Sibley, D.F. On the evolution of dolomite stoichiometry and cation order during high-temperature synthesis experiments: An alternative model for the geochemical evolution of natural dolomites. *Sediment. Geol.* **2011**, *240*, 30–40. [CrossRef]
67. Manche, C.J.; Kaczmarek, S.E. Evaluating reflux dolomitization using a novel high-resolution record of dolomite stoichiometry: A case study from the Cretaceous of central Texas, USA. *Geology* **2019**, *47*, 586–590. [CrossRef]

**Disclaimer/Publisher's Note:** The statements, opinions and data contained in all publications are solely those of the individual author(s) and contributor(s) and not of MDPI and/or the editor(s). MDPI and/or the editor(s) disclaim responsibility for any injury to people or property resulting from any ideas, methods, instructions or products referred to in the content.

*Article*

# Sedimentary Facies, Architectural Elements, and Depositional Environments of the Maastrichtian Pab Formation in the Rakhi Gorge, Eastern Sulaiman Ranges, Pakistan

Mubashir Mehmood [1], Abbas Ali Naseem [1,*], Maryam Saleem [2], Junaid ur Rehman [3], George Kontakiotis [4,*], Hammad Tariq Janjuhah [5], Emad Ullah Khan [3], Assimina Antonarakou [4], Ihtisham Khan [3], Anees ur Rehman [3] and Syed Mamoon Siyar [6]

1. Department of Earth Science, Quaid-e-Azam University, Islamabad 45320, Pakistan; mubashirmehmood94@gmail.com
2. Department of Earth and Environmental Sciences, Bahria University, Islamabad 44000, Pakistan
3. Department of Geology, Abdul Wali Khan University, Mardan 23200, Pakistan
4. Department of Historical Geology-Paleontology, Faculty of Geology and Geoenvironment, School of Earth Sciences, National and Kapodistrian University of Athens, Panepistimiopolis, Zografou, 15784 Athens, Greece
5. Department of Geology, Shaheed Benazir Bhutto University, Sheringal 18050, Pakistan
6. Department of Geology, University of Malakand, Chakdarrah 18000, Pakistan
* Correspondence: abbasaliqau@gmail.com (A.A.N.); gkontak@geol.uoa.gr (G.K.)

**Abstract:** An integrated study of sediments was conducted to examine the facies architecture and depositional environment of the Cretaceous Pab Formation, Rakhi Gorge, and Suleiman Ranges, Pakistan. This research focused on analyzing architectural elements and facies, which are not commonly studied in sedimentary basins in Pakistan. To identify lithofacies, outcrop analysis and section measurement were performed. The identified lithofacies were then categorized based on their depositional characteristics and facies associations, with a total of nine types identified within a stratigraphic thickness of approximately 480 m. These facies were mainly indicative of high-energy environments, although the specifics varied by location. Sedimentary structures such as planar and trough crossbedding, lamination, nodularity, load-casts, and fossil traces were found within these facies, indicating high-energy environments with a few exceptions in calm environments. The identified facies were grouped into seven architectural elements according to their depositional environments: delta-dominated elements, including laminated shale sheet elements (LS), fine sandstone elements (SF), planar cross-bedded sandstone elements (SCp), trace sandstone elements (ST), and paleosol elements (Pa); and river-dominated elements, including trough cross-bedded sandstone elements (SCt), channel deposit elements (CH), and paleosol elements (Pa). These architectural elements, along with their vertical and lateral relationships, indicate a transitional fluvio-deltaic environment within the Pab Formation. In conclusion, by interpreting facies and architectural elements, it is possible to gain a better understanding of the depositional history of the formation and the distribution of reservoir units.

**Keywords:** sedimentary lithofacies; sequence stratigraphy; fluvial architecture; sedimentary processes; paleoenvironmental reconstruction

Citation: Mehmood, M.; Naseem, A.A.; Saleem, M.; Rehman, J.u.; Kontakiotis, G.; Janjuhah, H.T.; Khan, E.U.; Antonarakou, A.; Khan, I.; Rehman, A.u.; et al. Sedimentary Facies, Architectural Elements, and Depositional Environments of the Maastrichtian Pab Formation in the Rakhi Gorge, Eastern Sulaiman Ranges, Pakistan. *J. Mar. Sci. Eng.* **2023**, *11*, 726. https://doi.org/10.3390/jmse11040726

Academic Editors: Manuel Martín Martín and Dimitris Sakellariou

Received: 27 January 2023
Revised: 9 March 2023
Accepted: 24 March 2023
Published: 27 March 2023

**Copyright:** © 2023 by the authors. Licensee MDPI, Basel, Switzerland. This article is an open access article distributed under the terms and conditions of the Creative Commons Attribution (CC BY) license (https://creativecommons.org/licenses/by/4.0/).

## 1. Introduction

Architectural element analysis of the fluvial and fluvio-deltaic transitional sedimentary sequences has been studied by different researchers around the world during the last few years [1–4]. Field studies, outcrop analysis, and section measurement have been considered of prime importance in the study of architectural element analysis. Due to a series of cyclic depositional events, lateral and vertical facies migration, and the superposition of depositional units, the study of architectural element analysis has always been exceedingly difficult [4]. Facies description and sedimentary architecture elements are well studied throughout the world in different sedimentary basins. These studies are very important for the explanation of traditional scientific knowledge, but they are also important for economic purposes. These rocks can act as good reservoir rocks [5]. Research on architectural elements preserved in the fluvial and transitional marine deposits in Pakistan is not fully developed yet. However, clastic deposits are of great importance in terms of their hydrocarbon potential. The eastern Sulaiman ranges contain some excellent Cretaceous sedimentary succession exposures, including the Pab Formation. These rocks are very well exposed along the roadside cut. The lithologic units belonging to the study area are located in an active fold-belt region and are therefore very important in terms of their geology and reservoir properties [5] (Figure 1). Thick siliciclastic sequences of the Pab Formation are well exposed (480 m thickness) in the Rakhi gorge [6] and composed of thick sandstone beds with thinner interbedded horizons of shales, clays, and paleosols. Although the Pab Formation has been studied by various researchers in terms of its reservoir properties and geophysical studies, explanations regarding its fluvial-deltaic characters and architectural elements are not very well understood [7,8]. The sediments of this succession deliver one of the best prospects for sedimentological study in the eastern Sulaiman Ranges.

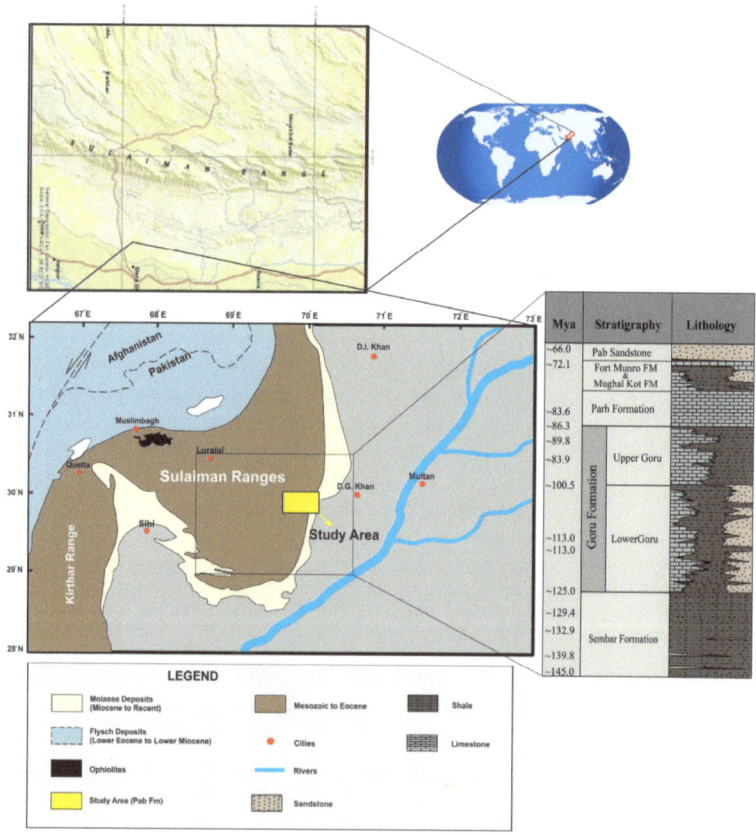

**Figure 1.** Map showing the outcrop belt of the Pab Sandstone in the Rakhi gorge, Eastern Sulaiman range of Pakistan, inset shows the generalized stratigraphic column.

This study is aimed at investigating the first detailed facies analysis within the Pab Formation in the Rakhi Gorge section. Understanding the facies will provide insights into the depositional environment in which the sandstone was formed. Moreover, this study will provide valuable information about the depositional processes and sedimentary environments of the past. The goal of the study also includes explaining the cyclic depositional events of the different facies, including architectural elements of the Pab Formation. The target of the architectural element analysis is to get a better understanding of the clastic input within the fluvio-deltaic succession as well as their lateral and vertical arrangement within the formation's units, which is also critical in assessing the reservoir quality of sandstone deposits. Porosity, permeability, and grain size distribution are key factors that determine the potential for hydrocarbon accumulation in sandstone reservoirs. Through the sedimentary facies identification and architectural element analysis, the present study uses the fluvial facies knowledge as a solid foundation for the future identification of multiple sedimentary facies in the field of lithofacies paleogeography. The optimized understanding of sandstone lithofacies and architectural elements is important for a range of geological applications, including hydrocarbon exploration, paleoenvironmental reconstruction, and stratigraphic correlations.

## 2. Geological Setting

Along the western edge of the Himalayan collisional belt, there are many faults and fold-thrust belts [9–11]. Because of these faults and fold and thrust belts, Pakistan is split into different tectonic belts that extend from the Salt Range in the north-east to the Kirthar and Sulaiman Ranges in the south-west. Vernant et al. [12] identified that the westernmost strike-slip fault has been extended to the fold and thrust sequences in the Makran subduction zone. This is where the Oceanic-Arabian plate passes under the Eurasian plate in Afghanistan. The Himalayas and Eurasia are drawing closer together in the north. This is assumed to be the result of the subduction. When the Indian and Eurasian plates collided about 30 Ma ago, they caused a fold and thrust belt to form in the Sulaiman Ranges. Because the Indian Plate moves counterclockwise and strikes the Eurasian Plate, the Sulaiman Ranges have some of the most complicated tectonic features in the world [13]. The Sulaiman lobe was formed by strike-slip movement to the left along the Chaman fault and southward thrusting along the western edge of the Indian subcontinent. The molasse layers kept changing because of prograde deformation, which moved the center of deposition to the south and east [14]. The study area and outcrop are located in an area called the "Rakhi Gorge", which is in the eastern Sulaiman Ranges of the Central Indus Basin (Figure 1). The Sulaiman Fold belt is a large tectonic structure close to the collision zone. Even though the Pab Formation has the same types of rocks in both the Central and Southern Indus Basins, it seems to have formed in different geographic locations in each basin. Numerous faults along with the fold-thrust belts are present along the western margin of the Himalayan collisional belt [13,14]. As a result of these faults and fold and thrust belts, Pakistan has been divided tectonically into different belts extending from the Salt Range in the north-east to the Kirthar and Sulaiman Ranges in the south-west. The westernmost strike-slip fault has been extended up to the fold and thrust sequences present in the Makran subduction zone, where the Oceanic Arabian plate is subducted under the Eurasian plate (Afghanistan) [12]. The ultimate consequence of the subduction has been linked with the Himalayan-Eurasian convergence in the north. The Sulaiman Ranges were formed as a fold and thrust belt due to the collision between the Indian and Eurasian plates about 30 Ma ago [12]. The Sulaiman Ranges have some of the most intricate tectonic features in the world as a result of the Indian Plate's counterclockwise rotation, resulting in the clash with the Eurasian Plate [13]. Due to left-lateral strike-slip motion along the Chaman fault and southward thrusting along the western edge of the Indian subcontinent, the Sulaiman lobe was formed through transgression. Prograde deformation kept changing the molasses layers, which moved the center of deposition to the south and east [14]. The study region (Pab Formation) and associated successions was deposited on the north-western margin of the Indian plate derived and originated from the Aravali and Deccan ranges [14]. These are now located in the Central Indus Basin's eastern Sulaiman Ranges (Figure 1). The Sulaiman Fold Belt is a significant tectonic structure close to the collision zone. Despite having identical lithological components in both the Central and Southern Indus Basins, the Pab Formation appears to have been deposited in different environments in both basins [14].

## 2.1. Stratigraphy of the Study Area

In the Central Indus Basin, the Cretaceous succession includes alternating carbonate and clastic intervals and includes the Mughal Kot, Ranikot, and Pab Formations with more than 1500 m thickness along the Rakhi Gorge anticline (Figures 1 and 2). The Pab Formation is generally devoid of fossil assemblages; however, Vredenburg [15] reported some Cretaceous Orbitoides minor species in the Pab Formation. No detailed biostratigraphic record is present or published yet. The Pab Formation (with interbeds of clay and shale) reaches a maximum thickness of 480 m along the eastern limb of the Mughalkot anticline in the eastern Sulaiman Ranges. The Paleocene and Cretaceous sections are divided by a locally preserved unconformity characterized by red lateritic nodules. The Ranikot Formation of lower Paleocene age was deposited in a variable setting and consists of strata that are sandy in the eastern part and shaly in the western part [5]. The Paleocene strata are represented by the Dunghan Formation, which is primarily limestone in the eastern and western portions with some shale units present in the central part.

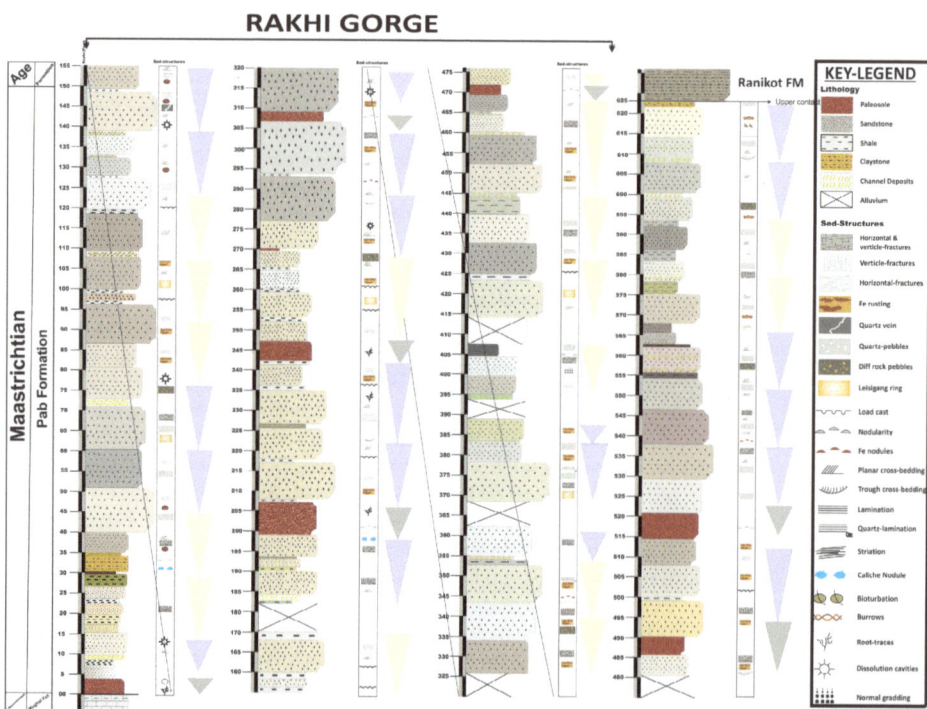

**Figure 2.** Showing the overall lithologic log of the Pab Formation in Eastern Sulaiman Ranges, the cyclic changes have been represented by the funnel shape with faded color representing fluvial and light color representing the deltaic sediments. The off-black color represents the development of a humid climate resulting in paleosol.

Pab Formation

The Eastern-Sulaiman Fold belt contains the double-plunging Fort Munro anticlinorium. The oldest rock exposed in Rakhi Gorge is the Late Cretaceous Mughal Kot Formation. The Late Cretaceous Mughal Kot (mudstone/marl, and sandstone), Fort Munro

Limestone and Pab Sandstone, and the latest Cretaceous Vitakri Formation (sandstone and red mudstone) of the Fort Munro Group are exposed in the Shadiani, Rakhi Gaj, and other gorges as core strata [15]. The Cretaceous Pab Formation consists of sandstone with minor shale units. The study area is present within the Rakhi Gorge section of the Eastern Sulaiman Ranges, where the lower contact of the Pab Formation is with the late Cretaceous Mughal Kot Formation and the upper contact is with the Paleocene Ranikot Formation. The first unit of the Pab Formation in the Rakhi Gaj area is the Paleosol unit. At the lower contact of the Pab Formation, the total thickness of the paleosol unit is 5.4 m. The paleosol varied in color from pale brown to rusty brown. The paleosol unit recognizes different layers at the base of color variation. The paleosol is recognized in the Pab Formation at different locations, and therefore, its thickness is varied at different locations in the field. The Pab Formation is dominantly sandstone. The sandstone unit is mostly medium grains to very coarse; some beds are fine grains, subangular to subrounded, and moderately sorted (Figures 2 and 3A,B). The Pab Formation has some thin shale units. The shale is a dusty yellowish green claystone that ranges from light olive gray to grayish olive gray. The lower part of the Pab Formation consists of intercalated claystone and shale. The claystone is sandy and very fine-grained locally.

## 3. Material and Methods

In Rakhi Gorge, Eastern Sulaiman Range, the Pab Formation of the Cretaceous age was examined in depth. A comprehensive field investigation was conducted to reconstruct the depositional environment of the Pab Formation by analyzing the facies and sedimentary architectural features. Sedimentary characteristics of the lithological assemblages and units were considered and studied in detail. The details from the sedimentological record of the formation and the characters were evaluated using the Miall classification schemes [16–18] for facies and architectural elements. The facies codes and classification scheme of Farrell et al. [19] were also used. Different depositional cycles and facies associations were identified. Using field data, a complete lithological record of the formation was compiled (Figure 2). On the log, the lithologic units were denoted and documented according to their representative properties. Based on their respective qualities, traditional sedimentology concepts have been used to distinguish the various units. By performing textural, mineralogical, depositional, and provenance analyses, sandstone samples were systematically collected to define the classification's key components. QFL diagrams were used to analyze the composition and origin of the examined samples [20]. Nevertheless, the exact depositional environment of the formation has been determined through a combined evaluation of petrography, architectural elements, and facies study. Individual architectural elements have been evaluated based on a detailed analysis of the corresponding facies set [18]. The cyclicity of the facies and architectural elements in this paper are presented based on a modified version of the classification system after Ghazi et al. [21]. Thirty thin-section investigation slides were prepared and studied under the petrographic microscope in the Petrographic Laboratory of the National Center of Excellence in Geology, University of Peshawar. The Gazzi and Dickinson point counting method was used, and a total of 400-point counts were used to study the thin section. Sandstone categorization models, which provide the percentages of three framework grains on a QFR ternary diagram [20–22], namely quartz, feldspar, and rock fragment, were used to analyze the detrital grain composition of the examined sandstone samples.

## 4. Results

### 4.1. Petrography

The late Cretaceous Pab Formation is dominated by fine to extreme coarse-grained and well-sorted sandstone. According to detrital mineral composition, Pab sandstone is mostly quartz arenite and partly sublitharenite. The average modal composition of the sandstone, based on the QFL diagram, was Q 93% F 0.4% L 3.7%, and the percentages derived based on the QmFLt diagram were Qm 91% F 0.33% L 2.3%. The petrographic study of Pab sandstone shows grain-to-grain contact that is concave to convex, point, sutured, and plane contacts. The Pab sandstone reservoir is dominantly a fine- to medium-grained, course-to very coarse-grained, moderately to well sorted, cross-bedded quartz arenite of possible fluvio-deltaic origin. Based on the QFR [22] and QmFLt diagrams of Dickinson and Suczek [20], the source area for Pab Formation deposits was Quartz recycled to the cratonic interior (Figure 3). The formation is present in an active fold belt, and the results from the microscopic study show that in some samples, the quartz exhibits concave–convex contact, with some showing plane contact (Figure 3A,B). The formation is composed of very coarse- to coarse-grained quartz (mono-crystalline); however, in some places fine-grained (poly-crystalline) sandstone is also found (Figure 3C,D). These results show that the sediments present in the Pab Formation are derived from an origin present at a greater distance (Figure 3E,F).

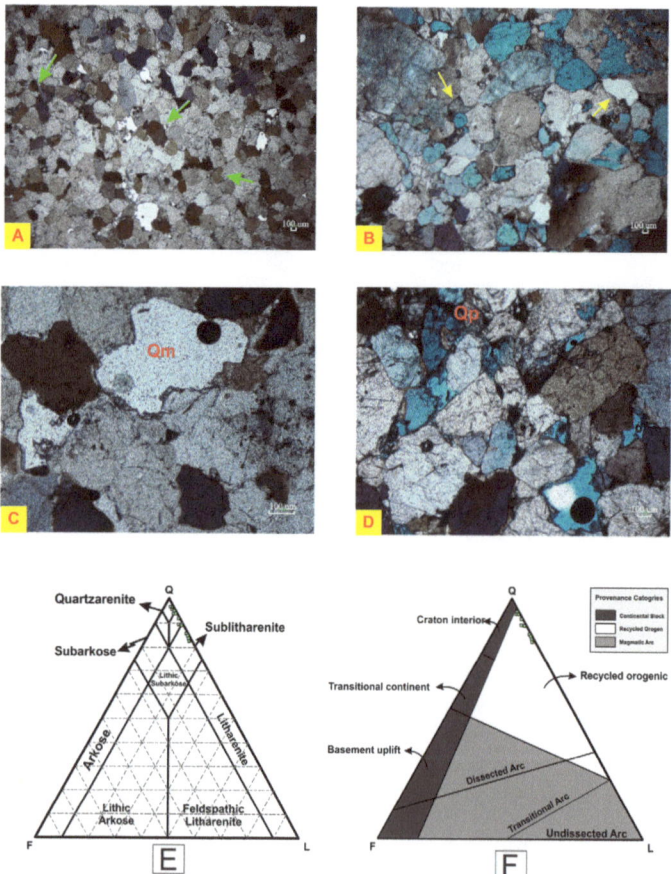

**Figure 3.** (**A**) Concave–convex contact highlighted by green arrows between medium- to coarse-grained sandstone; (**B**) plane contact highlighted by the yellow arrows; (**C**) (Qm) mono-crystalline quartz grains; (**D**) (Cp) poly-crystalline quartz grain; (**E**) classification of sandstone showing the quartz-arenite with a few samples belonging to sub-litharenite; and (**F**) classification showing the Cratonic interior origin of the sandstone [20–22].

### 4.2. Lithofacies

The Eastern Sulaiman Range's Pab Formation sediments are divided into different lithofacies that maintain a record of the depositional conditions. Using the Miall [23] classification technique, the Pab Formation can be divided into nine lithofacies. All these lithofacies are defined by their sedimentary structures, types of sediment, bed thickness, and sediment grain sizes (Table 1).

**Table 1.** Summary of sedimentary facies from the studied area (Pab Formation).

| | Grain Size | Bed-Thickness | Sed-Structures | Description | Interpret-Ation |
|---|---|---|---|---|---|
| 1. Paleosol Facies (Pf) | Fine soil | Thin to Thick bedded. From 0.30 m to 5.4 m. | Rootlets, Nodular Paleosol. | The Paleosol facies consists of fine-grained soil. The color of this facies is variable i.e., dark brown to rusty reddish, and in some places, the color is pale brown. Typically, the Paleosol are nodular and trace fossils of rootlets are found in this facies. | These lithofacies represents soil development in a humid climate. |
| 2. Sandstone with Channel Deposit Facies (Sch) | Coarse-grained sandstone, sand-gravel size in the channel deposits. | 0.08 m–0.16 m width and 8 m–10 m laterally extend along the beds | Channel deposits, Liesegang rings. | Channel deposits are in a layer and somewhere in the irregular form in the sandstone succession in the Pab Formation. It occurs at the base and also in the middle of sandstone beds. | Channel deposits indicate running water/Channel levee complex. |
| 3. Thin Bedded Clay stone Facies (Cf) | Very fine-grained clay. | Thin bedded, approximately 0.2 m to 0.4 m | None | The clay is almost sandy. The clay beds lie between the sandstone beds. The thickness of the clay beds is mostly thin-bedded. Color is varying, from med-brown-greyish to greenish-grey. | Suspension fall out/Flooding condition, overbank deposition. |
| 4. Thin-Medium Bedded Shale Facies (Sh) | Fine-grained sandy shale. | Thin to medium bedded. From 0.16 m up to 0.9 tm. | Load cast, Fe rusting. | Thin-medium bedded shale b/w the. The color of the shale is from greenish-grey to pale yellowish and in some places, the color is light brown. | Deep to semi-deep sedimentary settings/overbank deposits/fills of floodplain-drainage channels |
| 5. Thick bedded Sandstone Facies (St) | Coarse to very coarse | 2.1 m to 3 m. | Planar cross-bedding, Trough cross-bedding, Liesegang rings, post deformational fractures. | Weathered color is mostly med-brown, med-greyish to med-yellowish, also pinkish beds are found. Mostly post deformational vertical and horizontal fractures. Also, iron rusting is common in Sm facies. | Delta lobe |

Table 1. *Cont.*

| | Grain Size | Bed-Thickness | Sed-Structures | Description | Interpret-Ation |
|---|---|---|---|---|---|
| 6. Nodular sandstone Beds Facies (Sn) | Med-coarse grain and nodules diameter is typically 4-2-4 cm. | 0.25 m to 0.7 m. | Nodularity/Chaotic nature | Nodular Sandstone is present. The Fe and chert nodules are common in this facies. color is med-brown to dark greyish brown. | Slumping/sudden fall out along the margins and slope and rapid deposition. |
| 7. Coarse-grained Sandstone with Planar Cross-Bedding Facies (Sp) | Coarse to very coarse-grained size. | 5 m to 2.5 m. | Planar cross-bedding, Liesegang rings. Irregular post deformational fractures. | Consists of coarse-grained sandstone with planar crossbedding. The weathered color is varied, from dark greyish to dark brown, and at some places, the color is pinkish, while the fresh color is off-white, light greyish to pale yellow and brown. | Shelfal delta lobe |
| 8. Coarse-grained Sandstone with Trough Cross-Bedding Facies (Stc) | Medium-coarse-grained sandstone. | 0.30 m–1 m. | Trough cross-bedding, post deformational horizontal and vertical fractures. | Med-Coarse grained sandstone with crossbedding. The weathered color is a pale yellow to brownish while the fresh color is light grey to off-white. | Shoreface facies |
| 9. Sandstone with bioturbation and Trace Fossils Facies (Sb) | Medium to coarse-grained. | 0.30 m to 1.5 m. | Trace fossils, Bioturbation, Fe rusting. | The color of this facies is from light brown and greyish to dark brown and greenish-grey. The bed's thickness is about from 0.30 m to 1.5 m. The bioturbation occurs at the bottom of the sandstone beds. | Low-energy deltaic environment. |

4.2.1. Paleosol Facies (Pf)

Description: The Paleosol facies represents 10% of the total succession. The maximum thickness of the paleosol facies identified in the formation is around 5.4 m. However, the thickness is not uniform and varies at different places. The sediments in these facies are fragmented and devoid of sedimentary features. These facies consist of different features, including peds and cutans. Abundant visible cones and cone structures are found. These facies consist of fine-grained sediments, including clays and silt. The color of this facies is variable, i.e., from dark brown to rusty reddish and, in some places, pale brown (Figure 4A). The sedimentary characters of these facies are nodular and contain traces of ancient rootlets.

**Figure 4.** (**A**) Paleosol unit in the Pab Formation and their different colors variation; (**B**) field photograph showing the channelized sandstone unit in Pab Formation; (**C**) very fine-grained claystone unit; (**D**) shale unit interbedded in sandstone; (**E**) and (**F**) massive bedding sandstone unit; (**G**) nodules in sandstone; (**H**) and (**I**) planar cross-bedded sandstone; (**J**) large-scale trough cross-bedded sandstone; and (**K**) and (**L**) trace fossils and bioturbation in Pab Formation.

Interpretation: the characteristics and properties the paleosol consist of is indicative of the environment of formation. The properties of the paleosol exhibit the past climate, exposure of sediments and their time, the environmental influence including flooding conditions [24]. They also indicate the source rock characters and mineralogical characters of the parent material [25]. Root traces, soil structure, and soil horizons are the three characteristics that characterize paleosols. Paleosols are useful local marker beds, and as a result, they may provide essential data on basin stratigraphy. The presence of root traces is one of the most important diagnostic indicators. If there are no other indications of ancient soil formation, the presence of root remains on a rock indicates that it was formerly open to the atmosphere and inhabited by plants and is thus a soil by almost any definition [25]. This lithofacies is indicative of soil formation in a humid climate, as most of the authors suggest a humid climate for reddish paleosols with abundant bioturbation and root traces [26,27].

### 4.2.2. Sandstone with Channel Deposit Facies (Sch)

Description: The channel-deposit sandstone facies is common throughout the formation and occurs at different intervals. This facies accounts for approximately 11% of the total succession. The channel deposits range in thickness from 0.08 to 0.16 m and reach 8 to 10 m in some intervals. These facies are found in association with the fine-grained sandstone beds. The thickness of the channel deposit facies varies throughout the formation (Figure 4B). These facies consist of coarse-grained sediment particles. In certain instances, the sand-gravel size of the channel deposit facies is clearly different from the grain size of the associated upper and lower sandstone beds. This facies also contains quartz sediments with a 1-cm diameter. Cross-bedded sedimentary structures are common sedimentary structures found in channel deposit facies. However, these deposits are found in sandstone with coarse grains, and the channel grains are poorly sorted and have irregular shape and size. Fining upward sequences have been observed within the channel deposit facies at some locations.

Interpretation: The coarse-grained, channelized sandstone bodies are deposited as a result of small bodies of channels. The coarse grain particle usually represents the running water and energy conditions [28]. The existence of planar sedimentary structures represents the transitional environment, which is a kind of and intermediate zone between the deltaic channels and marine. The presence of large amounts of coarse pebbles represents high-energy conditions [29]. The high-energy rivers and channels are responsible for the transport of various loads, including suspended and bed loads [30]. The poor sorting of the sediments and the abundance of large-sized clasts suggest a high-energy environment and quick sedimentation [31]. The properties of this facies represent deposition from a high-velocity environment within a channel [32]. Based on the characteristics of these facies, it is assumed that they have been formed through a channel levee complex. It represents the deposition along the channel margins as a result of high-energy currents.

### 4.2.3. Thin Bedded Claystone Facies (Cf)

Description: This facies represents 2% of the whole succession. The thickness of the clay beds is mostly thin-bedded. The thickness is about 0.2 m to 0.4 m. These facies are composed of fine-grained claystone. The weathering color of this facies is greenish gray, while the fresh color is generally medium-brown to grey. The clay is almost sandy. The clay beds lie between the sandstone beds (Figure 4C). Sometimes they show gradational contact with the overlying sandstone beds.

Interpretation: Claystone is a lithified, non-fissile mud rock. To qualify as claystone, the rock must contain at least 50% clay particles [30,31]. The characteristics of the facies and lithologies indicate that they have been formed as a result of settling out from a suspension. The consistency of these facies suggest little or no interreference from strong currents [33]. These facies are interpreted to have been formed as suspension fall-out as the result of a flooding stage and associated subsequent deposition on the sides of the channels [34].

### 4.2.4. Thin-Medium Bedded Shale Facies (Sh)

Description: The Cretaceous Pab Formation consists of a few shale units, which in total account for 18% of the whole outcrop. The Sh facies is usually found as a thin- to medium-bedded unit; no thick-bedded units were preserved. In certain locations, the thickness of the bed varies between 0.1 m and 0.9 m. The Sh facies present in the formation consist of sand particles and are described as sandy shale (Figure 4D). The Sh facies is ranked from fine- to medium-grained based on the grain size analysis. In these facies, fissility is readily apparent. In some areas, the color of the shale is light brown; in general, the shale ranges in color from greenish gray to pale yellowish, and sometimes it is light brown. They occur in a sheetlike geometries. They represent a ball and socket structure at the contact within the overlying sandstone beds. In some places, cone and cone structures are present. Numerous caliche nodules are also present at some intervals.

Interpretation: Shale facies usually appear in a tranquil sedimentary setting. Shale layers in sand-braided stream deposits are frequently widespread as a result of flooding conditions and longstanding still water [35]. Typically, sediments originate in environments where muds, silts, and other sands were deposited and compacted by calm water conditions. The accumulation of fine sediments associated with low energy conditions results in the Sh type of facies [35]. Slow-moving currents that lack disturbance and are quite calm usually result in the formation of this kind of facies [36]. These facies consist of fine-grained uniform characters and a reddish brown to black color, indicating that they have been formed in a uniform depositional environment [37]. The lack of identified sedimentary structures was the result of a post-depositional event, for example cone and cone structures.

4.2.5. Thick Bedded Sandstone Facies (St)

Description: The Cretaceous Pab Formation consists of a large number of thick-bedded sandstone units. This St. facies contributes 32 percent of the overall succession. The beds are typically medium to thickly bedded. The thickness of these beds ranges from 2.1 to 5 m (Figure 4E,F). The weathered color is mostly medium-brown, grayish, to medium yellowish, with some pinkish beds. The grain size varies across these facies, being mostly medium- to coarse-grained and fine-grained in places. Coarse-grained pebbles with a size of more than 2 cm are also found. Vertical and horizontal deformational structures (fractures) are present. These facies are also comprised of abundant bioturbation in some places. The sandstone in these units is tabular, erosive, and has irregular basal contacts marked by balls and sockets. Some sandstone units within these facies are massive (St) and have no clear record of sedimentary structures. However, in some places, the sandstone consists of troughs and planar cross-beds.

Interpretation: The thick-bedded sandstone facies results from small channels caused by bank failure [38]. These types of facies are believed to have been formed as a result of the rapid accumulation of denser currents loaded with sediment. Arnot et al. [38] suggested that these types of facies developed as a result of the accumulation of sandy debris-loaded flow. Thick-bedded sandstone facies can also form as a result of a turbidite ramp system that is usually supplied with sediments from a deltaic source [35]. Based on the characteristics and interpretations of different authors, it is suggested that St. facies were formed as a result of direct feeding from a fluvial source in a deeper shelf setting [33–39]. The thick-bedded sandstone associated with thick burrowed units suggests they have been deposited as a result of continuous sediment aggradation [40]. Based on the observed characteristics and absence of slope facies, it is assumed that these facies formed within a shelfal delta lobe setting.

4.2.6. Nodular Sandstone Facies (Sn)

Description: The name "nodular sandstone facies" has been attributed to the amalgamated sandstone units that consist of nodular type sandstone. These sandstone beds are chaotic and irregular in their nature. The sandstone units of this facies are thin- to thick-bedded. Texturally, these sandstones are medium- to coarse-grained. These nodular sandstone units in the Pab formation comprise 4% of the whole sequence. They exhibit a light brown to dark grayish brown color. These sandstone units consist of small iron nodules. The diameter of the nodular sandstone in the section ranges from 2 to 4 cm (Figure 4G).

Interpretation: Nodules in sedimentary rocks are often understood as the result of post-depositional and secondary mineralization during diagenesis with an early diagenetic cementation history [37]. Sedimentary rock composed of scattered to loosely packed nodules in a matrix of similar or dissimilar composition is referred to as a nodule. The chaotic nature and nodularity developed in this sandstone facies are the result of different sedimentary processes [38]. In most cases, these types of facies are associated with the falling down of sediments deposited on slopes and the rapid deposition of these sediments. The coarse grain characteristics further suggest that these facies formed along the channel margin, where sediments frequently fall down the steep slope [39].

### 4.2.7. Coarse Grained Sandstone with Planar Cross Beds Facies (Sp)

Description: The Pab Formation has facies of coarse-grained sandstone with planar crossbedding that makes up 20% of the whole sequence. It consists of mostly coarse-grained sediments. Planar cross-bedded sedimentary structures are common in the entire facies' units. The length of the planar crossbedding is about 0.1 m on average of 0.1 m (Figure 4H,I). The weathered color of this facies is varied, from dark greyish to dark brown, and at some places, the color is pinkish, while the color on the fresh surface is off-white, light greyish, pale yellow, and brown. This facies consists of thick to massive bedding, ranging from 1.5 m to 2.5 m. Because of the coarse grain and high density, the lower portion of these beds is usually undulating. The grains are normal and graded in some places. The thickness of these facies changes with the change in sediment size.

Interpretation: High-energy conditions are thought to have formed sand with grain sizes ranging from coarse to extremely coarse, forming planar cross-bedded sedimentary structures. These high-energy conditions are possibly associated with the water flow linked to rivers and channels associated with delta [40]. Cross-bed sets are seen in granular sediments, notably sandstone, and they indicate that sediments were deposited in the form of flowing ripples or dunes because of water or air movement [41]. These facies are considered to form as shelfal delta lobe deposits, they are suggested to form as a result of uni-directional channel water flow laden with sediments toward the main body [42].

### 4.2.8. Coarse Grained Trough Cross Bedded Sandstone Facies (Stc)

Description: This sandstone facies is characterized by its ferrugenic character. The sandstone in these facies is medium grained. The grains are mostly sub-rounded to rounded, and sandstone incorporates shale in places. They represent poor sorting of sediments. The fresh color is off-white to light brown while the weathering color is dark brown to grey. The cross-bedded trough sandstone facies comprise 3% of the overall sequence. The thickness of the beds ranges between 0.3 m and 1 m. It is composed of sandstone with trough cross-beds having curved foresets. (Figure 4I). These facies are found within a sedimentary unit exhibiting a fining downward sequence. These facies consist of a large scale of trough cross-beds up to 1 m.

Interpretation: These facies were most likely deposited under high-energy conditions associated with rivers. Such currents are well-established on several contemporary wave-dominated shorefaces, where they provide laterally constrained, efficiently channelized paths across the shoreline region for sediment deposition offshore [43,44]. They are energized by the water currents derived from high-energy flow conditions. The stronger flow in the deeper areas of the channel produces subaqueous dunes in the sediment, and as the sand accumulates, trough or planar crossbedding forms. It is believed that high-energy unidirectional traction currents in the upper shore face formed the cross-bedded sandstone facies and resulted in the formation of migrating dunes.

### 4.2.9. Bioturbated, Trace Fossils Sandstone Facies (Sb)

Description: This facies comprises 10% of the overall succession. The color of this facies ranges from light brown and grayish to dark brown and greenish grey. The original color, a dark brown, has been changed to a greenish gray due to weathering. The bed's thickness is around 0.30 m to 1.5 m and is usually uniform. The sandstone in these facies is fine- to medium grained. Laminated sandstone beds have been identified. The bioturbation occurs near the bottom of the sandstone layers and is found at random places throughout the outcrop (Figure 4K,L). In the SB facies, trace fossils in sandstone are also observed in different areas. The intensity of bioturbation ranges between 61 and 90, with a grade of 4 [24].

Interpretation: The movement of the solutes and solids induced by the macrobenthos' movements and feeding resulted in the generation of bioturbated sandstone units. The organisms responsible for the generation of these traces include arthropods, annelids, and mollusks [45]. Very few primary sedimentary structures were preserved, since the reworking of sediments caused by the organism destroyed the primary characters [46].

These facies are believed to have formed in an inner shoreface setting under a normal wave condition with less disturbance resulting from turbulence [47].

*4.3. Architectural Element Analysis*

Architectural element analysis is a helpful tool that goes beyond facies analysis to identify genetic facies in addition to facies analysis [25,26] (Table 2). Individuals or groups of lithofacies are divided in architectural element analysis by the bounding surfaces of various hierarchies. Based on the Pab sandstone's sedimentary structures, geometry, paleocurrent indicators, and lateral and vertical arrangement of lithofacies, seven architectural elements were identified (Table 2).

**Table 2.** Summary and generalized table of architectural element analysis from the studied area (Pab Formation).

| Element (Code) | Geometry | Facies Assoc. | Description | Interpretation |
|---|---|---|---|---|
| Planar Cross-Bedded Sandstone Element. (SCp) | Tabular, Sheet like | Sp Sm Sch | The SCp consisits of medium- to coarse-grained sandstone with planar cross-beds. They are pebbly and lenticular in nature. The planar cross-bedding sandstone element is abundant laterally and thegrain size of sandstone becomes finer in the upward side. They have a gradational contact with the lower beds. | The intercalation of coarse-grained sediments with the lithofacies Sp may reflect a rapid change in flood regime or imply high-energy sheet floods into a lower energy environment. |
| Traces-Sandstone Element. (ST) | Lobate and sheet-like | Sb Sp Sch | Medium to coarse-grained, lobate geometry of ST element. They are laterally extended up to 10 m and their average thickness is 1 m to 2 m. Their upper contact is flat but erosional with Sch facies. The common sedimentary structures found in element ST is traces of fossils. | The lobate geometry and traces of different fossils indicate in low-energy deltaic environment. |
| Trough Cross-Bedding Sandstone Element. (SCt) | Lenticular geometry. | St Stc Sch | Extended to the lower and central portion of the Formation. The trough cross-bedding element is found in medium to coarse-grained sandstone. Abundant trough cross bedding structures are commom. Low angle planar cross-bedding, and some minor fractures. | SCt is interpreted as the product of three-dimensional dunes migrating in channels under lower flow regime conditions. |
| Fined-Sandstone Element. (SF) | Sheet like | Sch Sp Sm | The SF element is fined-grained sandstone. SF deposits have a sheet-like geometry, reflecting their origin by vertical aggradation. Trace fossils are found in the SF element. | Sheet-like geometry, together with the small-scale sedimentary structures and the fine-grained lithology suggests deposition as a bar-top or bar-flank sand sheet. |

Table 2. Cont.

| Element (Code) | Geometry | Facies Assoc. | Description | Interpretation |
|---|---|---|---|---|
| Channel Deposit Element. (CH) | Tabular | Sn Sch Sp | The CH element is up to 0.30 m thick and 10 m to 50 m wide. The element CH is present in between the Sp and St facies. The CH element comprises lithofacies Sch and Sn. The cavities and nodular sedimentary structures are found in the CH element. They have a coarse and erosive geometry. Their upper and lower contacts are not uniform. | Recognition of the CH element in a fluvial deposit depends on the ability to define the sloping channel margins. The presence of coarse-grained conglomerates may designate a sudden increase in the velocity of the depositional current. |
| Laminated Shale Sheet Element. (LS) | Tabular, lobate-like. | Sn Sh | The LS element is interbedded in the sandstone unit. Their lateral extension is up to 10 m and the average thickness is about 0.05 m up to 0.45 m. Having a deformational upper and lower contact. Abundant ball and socket structures at the upper contact. | Deep to semi-deep lake sedimentary settings are where shale element emerges. |
| Paleosol Element. (Pa) | Tabular | Sn Sb | The fine-grained paleosol element has lobate and tabular shape geometry. They are present in the upper and lower part of the Formation. Thay exhibit a variable geometry and thickness. Thicness is from 0.60 m to 5.4 m thick Sedimentary structures includes rootlets and cone and cone structures. | This element interpreted soil development in a humid climate. |

### 4.3.1. Planar Cross-Bedded Sandstone Element (Scp)

Description: each architectural element of Planar cross-bedded sandstone element is bounded at its base by an erosion surface that approximates paleohorizontal, as measured relative to an underlying shale bed and sandstone beds (Figure 5A). Their upper contact is irregular with coarse-grained sandstone beds. Such an element is common all over the formation and characterized by tabular geometry, up to 2 m thick and 150 m to 200 m wide, which can be traced laterally in a few instances for distances of 250 m (Figure 6A). The element SCp consists of different sandstone facies i.e., Sn and Sp, with sub-ordinate sets of facies St, all overlain by medium-coarse grained channel facies Sm and Sch. The most abundant sedimentary structure in SCp elements is planar cross-beds, which have set thicknesses up to 0.6 m thick. In the basal part of the SCp element, the planar crossbedding is found in coarse to very-coarse grained sandstone. As they extend upward, the grain size of the sandstone becomes finer.

**Figure 5.** (**A**) Planar cross-bedded sandstone element; (**B**) trace fossil sandstone element; (**C**) trough cross-bedded sandstone element; (**D**) fine sandstone element; (**E**) channel fill deposit element; (**F**) laminated shale sheet element; (**G**) Paleosol element.

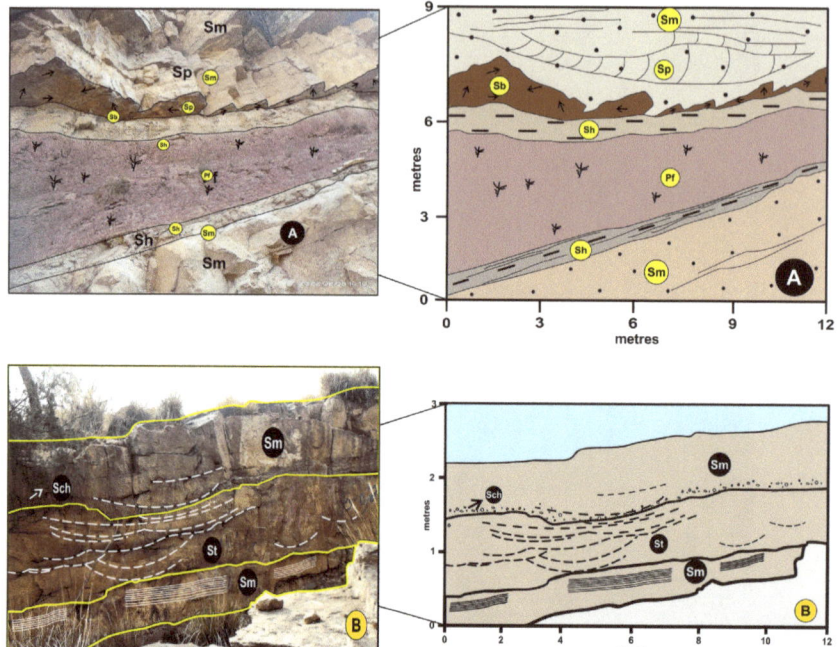

**Figure 6.** (**A**) Field sketch showing Planar cross-bedding element (SCp) and traces sandstone element (ST), asymmetrically filled with facies Sb, Pf, Sm, and Sh in Pab Formation. Arrows show the bioturbation in the Sb facies and heavy lines delineate the facies boundaries; and (**B**) photomosaic of trough cross-bedding element (SCt) in the middle part of the study area. This element is filled with facies Sm, St, and Sch in the Pab Formation. Heavy lines delineate the facies boundaries while the dashed lines show the trough crossbedding.

Interpretation: As a result of decreasing (possibly seasonal) flood events, coarse-grained, discontinuous materials accumulate as bar-top sand sheets and channel deposits [48,49]. Intercalation of coarse-grained sediments with the lithofacies Scp may reflect a rapid change in flood.

4.3.2. Trace Fossil-Sandstone Element (St)

Description: The element ST comprises facies Sb and Sp (Figure 5B). Their geometry is lobate or sheet-like. The traces are present in the medium- to coarse-grained sandstone. They are laterally extended up to 10 m and their average thickness is 1 to 2 m. The element ST have sharp erosional bases with SF overlain by Sp and Sch facies. Their upper contact is flat but erosional, with Sch facies.

The lithofacies Sb and Sch are recognized in this element. The common sedimentary structures found in element ST is traces of fossils.

Interpretation: In the case of trace fossil sandstone elements, the organism activity plays an important role in the development of this facies. Trace-sandstone elements such as Skolithos and Planolites are common in high-energy transitional marine environments [50–52]. Trace fossil diversity is low, in part due to high currents and sedimentation rates in tidal inlet settings [53].

4.3.3. Trough Cross-Bedded Sandstone Element (Sct)

Description: The trough crossbedding sandstone element is bounded at its base by alluvium with irregular contact and its upper contact with Sp and Sm facies in some places, while in the middle portion of the stratigraphic succession the upper contact marks the skyline (Figure 5C). It is characterized by a lenticular shape geometry, up to 3 m thick and 10 m to 20 m wide (Figure 6B). This element is mostly present in the middle of the stratigraphic succession, not present in the overall formation. The element SCt consists of different sandstone facies i.e., Sn, Sp, and St all overlain by medium coarse-grained channel facies Sch. The sedimentary structure found in the SCt element is trough crossbedding, low angle planar crossbedding, and some minor fractures. The trough crossbedding is found in medium- to coarse-grained sandstone.

Interpretation: It is believed that the element Sct results from the movement of three-dimensional dunes along channels during a low-flow regime. Their lenticular geometry, inclinations of their axes, moderate to poor sediment sorting, and scouring surfaces are indicative of the formation of troughs by migrating sinuous crested dunes [54]. While the smaller troughs were most likely formed by migrating dunes or mega ripples along with the lee sides of these bars, the larger foresets, with their progressively sloping dip and coarse-grained size, indicate deposition in low-angle bar fronts.

4.3.4. Fine Sandstone Element (Sf)

Description: The element Sf lies near the middle of the stratigraphic succession. The SF element deposits have a sheet-like geometry, reflecting their origin by vertical aggradation (Figure 5D). Their lower boundary is bounded by Sf facies with irregular-erosive contact (Figure 7A). The top surface is bounded by SCp element with flat contact and at some locations they are irregular. Characterized by sheet geometry, up to 3 feet thick and 50 m to 90 m in width. The SF element comprises lithofacies Sch, Sp, and Sm. Small-scale sedimentary structures of trace fossils are found in the SF element.

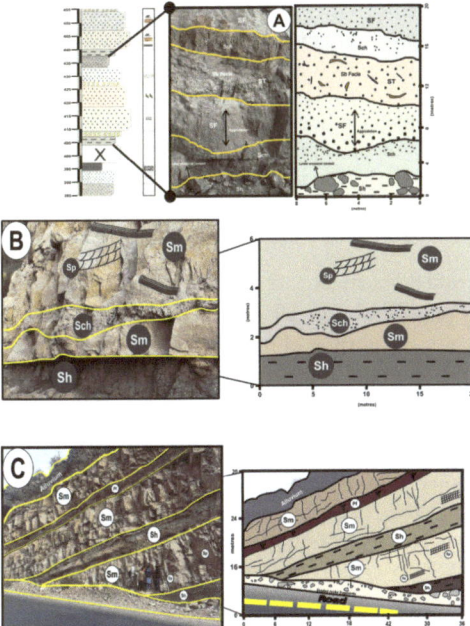

**Figure 7.** (**A**) Field sketch of fined-sandstone element (SF), filled with facies Sb, Sch, and Sh in the study area. The element SF is found in the lower and middle part of the Pab Formation, in the middle portion the aggradation is found in the Sf element; (**B**) photomosaic of channel deposit element (CH) interbedded with coarse-grained sandstone with Sm and Sp; and (**C**) field sketch of laminated shale element (LS) in the middle part of the Formation. This element presents in b/w the Sm facies all over the Pab Formation and filled with facies Sp, Pf, Sm, and Sh in Pab Formation.

Interpretation: These laminated sand-sheet sections were deposited by non-channelized rivers. Their thin, discontinuous, sheet-like shape, along with the small-scale sedimentary formations and fine-grained lithology, supports deposition as overbank flood sand sheet. When laminae accreted vertically during sheet flooding in an upper flow regime, they were deposited in the shallower sections of channels [55].

4.3.5. Channel Deposit Element (Ch)

Description: The channel deposit element is bounded at its basal contact by a sharp erosional boundary, and the top of the channel fill may be erosional or gradational with the Sn facies (Figure 5E). It is characterized by tabular geometry (Figure 7B). The CH element is up to 0.30 m thick and 10 m to 50 m wide. The element CH is present between the Sp and St facies. The CH element comprises the lithofacies Sch, Sn, and Sh. Cavities and nodular sedimentary structures are found in the CH element.

*Interpretation:* The presence of coarse-grained conglomerates may suggest a sudden increase in the velocity of the depositional current. This is often done by correlating closely spaced outcrop or subsurface sections, but since most deposits include a hierarchical network of channels of different sizes, such correlation may be difficult or impossible [55]. In the Pab Formation, the Ch element is filled by the lithofacies Sp, Sm, Sch, and Sh.

4.3.6. Laminated Shale Sheet Element (Ls)

Description: Laminated shale sheet element is found all over the study section (Pab Formation). These are the small-scale shale units interbedded in a sandstone (Figure 5F). Their lateral extension is up to 10 m and the average thickness is about 0.05 m up to 0.4 m The geometry of the LS element is tabular in between the Sp, St, and Sm facies (Figure 7C). The base contact is sharp with Sm facies, while at some places the lower contact is with gradational to paleosol Pf facies, and the top contact is also sharp with Sm and Sp facies. The Sn and Sm facies are associated with this element. The fissility sedimentary structure is well recognized in the LS element.

Interpretation: The fine laminated shale element is associated with the lateral extension of the bars [56]. These elements and associated facies are attributed to deposit in channels associated high energy environment [57]. However, these types of lithologies can also form in sheet floods [58]. In the channels and associated deposits these lithologies can occur as bar flank sheet deposits [59].

4.3.7. Paleosol Element (Pa)

Description: The paleosol element is found at the starting point of the Pab Formation and in the middle portion of the study area (Pab Formation) (Figure 5G). The thickness is varying from place to place, at the starting point, the thickness is more than that of the middle portion (Figure 8A). At the starting point, they are up to 2 m thick, while in the middle portion the thickness is reduced up to 0.6 m. The geometry of the Pa element is lobate, tabular-like. Their basal contact is the erosional boundary with the cretaceous Mughal Kot Formation at the starting point, while in the middle the contact is with the Sh and Sm facies. The top surface is bounded by Sm, Sh, and Sp facies. The rootlet's sedimentary structure is found in paleosol elements.

Interpretation: Paleosols are developed in floodplain units at the top of channel belts and stacked channel-belt complexes [60]. The preservation of shrink–swell features suggests that the paleosol developed under conditions of repeated wetting–drying cycles in a climate characterized by seasonal precipitation [61]. Abundant iron-oxide concentrations, iron-oxide coated grains and cross-cutting relationships indicating iron-oxide concentrations are secondary and likely formed during shallow burial. The dominance of clay content and the absence of large number carbonate nodules suggests that the paleosol formed on a well-drained floodplain without prolonged periods of aridity [62].

**Figure 8.** (**A**) Sedimentary cross bed representing fluvio-deltaic facies; (**B**) highly bioturbated bed red arrow associated with shelfal delta lobe; (**C**) shale unit highlighted by the meter stick of the delta front; (**D**) load cast belonging to channels; and (**E**) trough cross-bedded sandstone belonging to shore face environment.

## 5. Discussion

*5.1. Facies Association*

In the Rakhi Gorge section of the Pab Formation, five facies' relationships were observed. These facies are connected to one another based on their genetic and lithological similarities.

### 5.1.1. Fluvio-Deltaic Facies Associations

The Pab Formation contains facies that are associated with the fluvio-deltaic facies. The fluvio-deltaic facies association includes the Nodular sandstone facies (Sn). The fluvio-deltaic facies are characterized by the presence of large-scale sedimentary structures such as trough cross-bedded, cross-laminated, and hummocky sandstone, as well as bioturbated sandstone. These facies suggest that they were formed under conditions of strong tractional energy that varied from constant to periodic and are commonly found in settings ranging from deltaic to high-energy settings [63] (Figure 8A).

5.1.2. Shelfal Delta Lobe Facies Association

The predominant facies in this association are thick, massive sandstones, as well as bioturbated sandstones, hummocky sandstone, and mudstone. They include thick-bedded sandstone facies, coarse-grained sandstone with planar cross-bedding facies, thin bedded claystone facies. It has been proposed that these facies were deposited under a fair weather wave-base or storm wave base on the outer shelf, and most likely originated from a sand-rich delta [64]. These facies display characteristics of both shelfal delta lobe and flood-associated delta-front sandstones, as noted by Mutti et al. [65] (Figure 8B).

5.1.3. Delta Front Facies Association

The facies association includes thin-medium bedded shale facies (Sh), they are characterizing by thin- to medium-bedded shale. They also consist of some mudstone. The shales are bioturbated with no preserved sedimentary structures. These facies are usually sandwiched between the thick, massive sandstones consisting of hummocky sandstone beds [66]. These facies associations were deposited under fair weather wave base and quite environment, and the massive sandstones are the distal equivalent of the delta-front unit. It is likely that the source of these deposits was a suspension fall out from flood [67] (Figure 8C).

5.1.4. Channels Facies Association

These facies include the sandstone with channel deposit facies (Cf) and are composed of thick bedded, coarse-grained sandstone, with channelized sandstone deposits being a prominent feature. In some areas, the sandstone occurs in a lenticular shape. At the boundary between the sandstone and finer lithologies such as shale, sole marks such as grooves, flutes, and load casts are widespread. Poorly graded sandstone beds are common. Thick sandstone beds are common, with pinching and thinning occurring and being separated by intercalated fine mudstone. Mud clasts inclusions are abundant in these facies. The highly deformed sandstone beds in certain areas suggest the collapse of the channel margin. Based on the overall characteristics and features of these associated facies, it is believed that they were deposited in a channel by a high-density turbidity current [68] (Figure 8D).

5.1.5. Shore Face Facies Association

These facies association consist of the coarse sandstone with trough cross bed facies (Stc) and sandstone with bioturbation (Sb). These facies association of the shore face environment includes large-scale planar cross-bedded sandstones, trough cross-bedded sandstones, massive sandstones, bioturbated sandstones, and hummocky sandstones. The characteristics of these facies suggest that they were deposited in a high-energy fluvial deltaic environment. The presence of hummocky bedforms indicates the influence of storms [68] (Figure 8E).

*5.2. Depositional Model*

The Pab Formation was deposited on the western, northwest side of the Indian Plate. The early collision between the Indian plate and the Afghan block is when the Cretaceous and Eocene series emerged. Tertiary age deposits were created during the collision of the Indian plate with the Laurasian plate, and during that time transgressions and regressions were also seen [69,70]. The distribution and internal characteristics of the facies association and associated architectural elements (e.g., fluvio-deltaic facies, channel deposits facies) show that the Pab Formation belongs to a fluvio-deltaic dominated environment. The depositional environment, based on the observed lithofacies, was perceived as a fluvial deltaic. The depositional model for Pab Formation in the Rakhi Gorge section is shown in (Figure 9B). A total of nine architectural lithofacies were established in the Pab Formation. The high proportion of coarse sediments compares to the fluvial sedimentation, while the

Paleosol facies demonstrate the arid climate. The Pab sandstones are a remarkably lower flow regime than some facies (facies Sp, St). The interpretation of different sedimentary structures within facies (facies Sp, St, and Sb) are confirmed as a point bar sequence. The base lithological units have channeled deposits that interpret the inner channel bends. The fine sandstone facies (facies Sf) interpret the deep-sea environment. The associations of seven architectural elements also provide a good clue about the environment of high sinuosity fluvial to the deltaic environment. At the base of these, all lithofacies and architectural elements and various sedimentary structures demonstrate a fluvio-deltaic environment.

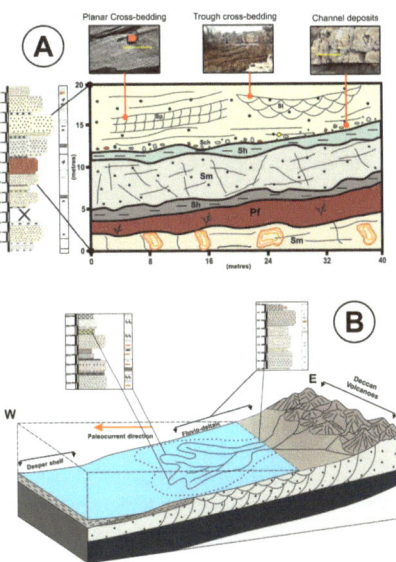

**Figure 9.** (**A**) Generalized section depicting the Pab architectural element b/w the Sh and Sm facies. This element consists of facies Sm, Sp, St, Sch, pf, and Sh. Showing the detailed lateral and vertical facies variations; and (**B**) depositional model of Pab sandstone in Eastern Sulaiman Ranges, Pakistan.

## 6. Conclusions

Based on a field experiment in the Maastrichtian Pab Formation, this study proposed a methodology for evaluating architectural elements and facies. The studied sandstone deposits at the base of the Miall classification system are classified by nine lithofacies (Pf, Sch, Cf, Sh, St, Sn, Sp, Stc and Sb). The lithofacies Sp, St, and Sn indicate that the succession was deposited during a lower flow regime, but the lithofacies Sm and Sb indicate a deltaic environment. The intercalation of a few thin strata of shale, denoted by facies Sh, indicates a deep to semi-deep lake sedimentary setting. The channel deposits (Sch facies) near the base of the Pab Formation suggest that these facies were deposited in a flowing-water and deltaic environment. The distinct types of seven architectural elements (SCp, ST, SCt, CH, LS, and Pa) were identified, with their respective geometries and vertical facies correlations. These multistory sandstone structures complement the fluvio-deltaic setting. The lateral study of channelized sandstone indicates the rapid migration and expansion of point bars and channels. The pattern of coarsening facies associations is one of the distinguishing features of a deltaic environment. The various sedimentary features, such as Fe rusting, nodules, and fractures, suggest subaerial exposure. The associated facies characteristics indicate that the formation's fluvio-deltaic origin can be inferred from examinations of its facies.

**Author Contributions:** M.M.; Conceptualization, Field work, writing—original draft preparation, A.A.N.; methodology, M.S.; validation, J.u.R.; Resources.; G.K. and H.T.J.; formal analysis and writing—review and editing, E.U.K., H.T.J., G.K. and A.A.; statistical analysis and data curation, I.K. and A.u.R.; Visualization, editing, S.M.S., administration. All authors have read and agreed to the published version of the manuscript.

**Funding:** This research received no external funding.

**Institutional Review Board Statement:** Not applicable.

**Informed Consent Statement:** Not Applicable.

**Data Availability Statement:** The data used in this work is available on request to the corresponding author(s).

**Acknowledgments:** The authors acknowledge Department of Earth Sciences, Quaid-e-Azam University for providing necessary support for this research.

**Conflicts of Interest:** The authors declare no conflict of interest (financial or non-financial).

# References

1. Ali, S.K.; Janjuhah, H.T.; Shahzad, S.M.; Kontakiotis, G.; Saleem, M.H.; Khan, U.; Zarkogiannis, S.D.; Makri, P.; Antonarakou, A. Depositional Sedimentary Facies, Stratigraphic Control, Paleoecological Constraints, and Paleogeographic Reconstruction of Late Permian Chhidru Formation (Western Salt Range, Pakistan). *J. Mar. Sci. Eng.* **2021**, *9*, 1372. [CrossRef]
2. Ghazi, S.; Mountney, N.P. Facies and architectural element analysis of a meandering fluvial succession: The Permian Warchha Sandstone, Salt Range, Pakistan. *Sediment. Geol.* **2009**, *221*, 99–126. [CrossRef]
3. Miall, A. *The Geology of Fluvial Sediments*; Springer: Berlin, Germany, 1996.
4. Zang, D.; Bao, Z.; Li, M.; Fu, P.; Li, M.; Niu, B.; Li, Z.; Zhang, L.; Wei, M.; Dou, L.; et al. Sandbody architecture analysis of braided river reservoirs and their significance for remaining oil distribution: A case study based on a new outcrop in the Songliao Basin, Northeast China. *Energy Explor. Exploit.* **2020**, *38*, 2231–2251. [CrossRef]
5. Khan, M.; Ghazi, S.; Mehmood, M.; Yazdi, A.; Naseem, A.A.; Serwar, U.; Zaheer, A.; Ullah, H. Sedimentological and provenance analysis of the Cretaceous Moro formation Rakhi Gorge, Eastern Sulaiman Range, Pakistan. *Iran. J. Earth Sci.* **2021**, *13*, 251–265.
6. Euzen, T.; Eschard, R.; Albouy, E.; Deschamps, R. *Reservoir architecture of a turbidite channel complex in the Pab Formation, Pakistan*; AAPG Studies in Geology 56: Atlas of Deep-Water Outcrops: Tulsa, OK, USA, 2007; Chapter 139. [CrossRef]
7. Mehmood, M.; Ghazi, S.; Naseem, A.A.; Yaseen, M.; Dar, Q.U.Z.; Khan, M.J.; Sarwar, U.; Zaheer, A. Petrofacies investigations of the Cretaceous Pab Formation Rakhi Gorge Eastern Sulaiman Range Pakistan—Implication for reservoir potential. *Bull. Geol. Soc. Malays.* **2021**, *72*, 37–46. [CrossRef]
8. Umar, M.; Friis, H.; Khan, A.S.; Kassi, A.M.; Kasi, A.K. The effects of diagenesis on the reservoir characters in sandstones of the Late Cretaceous Pab Formation, Kirthar Fold Belt, southern Pakistan. *J. Asian Earth Sci.* **2011**, *40*, 622–635. [CrossRef]
9. Reynolds, K.; Copley, A.; Hussain, E. Evolution and dynamics of a fold-thrust belt: The Sulaiman Range of Pakistan. *Geophys. J. Int.* **2015**, *201*, 683–710. [CrossRef]
10. Stein, S.; Sella, G.; Okal, E.A. *The January 26, 2001 Bhuj Earthquake and the Diffuse Western Boundary of the Indian Plate, in Plate Boundary Zones*; AGU: Washington, DC, USA, 2002; pp. 243–254.
11. Szeliga, W.; Bilham, R.; Kakar, D.M.; Lodi, S.H. Interseismic strain accumulation along the western boundary of the Indian subcontinent. *J. Geophys. Res. Atmos.* **2012**, *117*. [CrossRef]
12. Vernant, P.; Nilforoushan, F.; Hatzfeld, D.; Abbassi, M.R.; Vigny, C.; Masson, F.; Nankali, H.; Martinod, J.; Ashtiani, A.; Bayer, R.; et al. Present-day crustal deformation and plate kinematics in the Middle East constrained by GPS measurements in Iran and northern Oman. *Geophys. J. Int.* **2004**, *157*, 381–398. [CrossRef]
13. Kassi, A.M.; Kelling, G.; Kasi, A.K.; Umar, M.; Khan, A.S. Contrasting Late Cretaceous–Palaeocene lithostratigraphic successions across the Bibai Thrust, western Sulaiman Fold–Thrust Belt, Pakistan: Their significance in deciphering the early-collisional history of the NW Indian Plate margin. *J. Asian Earth Sci.* **2009**, *35*, 435–444. [CrossRef]
14. Ghazi, A.; Hafezi Moghadas, N.; Sadeghi, H.; Ghafoori, M.; Lashkaripour, G. The effect of geomorphology on engineering geology properties of alluvial deposits in Mashhad City. *Sci. Q. J. Geosci.* **2015**, *24*, 17–28.
15. Vredenburg, E.W. Report on the Geology of Sarawan, Jhalawan, Mekran and the State of Las Bela, Considered Principally from the Point of View of Economic Development. *Rec. Geol. Surv. India* **1909**, *3*, 189–215.
16. Miall, A.D. Facies architecture in clastic sedimentary basins. In *New Perspectives in Basin Analysis*; Springer: New York, NY, USA, 1988; pp. 67–81.
17. Miall, A.D. In defense of facies classifications and models. *J. Sediment. Res.* **1999**, *69*, 2–5. [CrossRef]
18. Miall, A.D. *Fluvial Depositional Systems*; Springer International Publishing: Cham, Switzerland, 2014; Volume 14, p. 316.
19. Farrell, K.M.; Harris, W.B.; Mallinson, D.J.; Culver, S.J.; Riggs, S.R.; Pierson, J.; Self-Trail, J.M.; Lautier, J.C. Standardizing Texture and Facies Codes for A Process-Based Classification of Clastic Sediment and Rock. *J. Sediment. Res.* **2012**, *82*, 364–378. [CrossRef]

20. Dickinson, W.R. Interpreting Provenance Relations from Detrital Modes of Sandstones. In *Provenance of Arenites*; Springer: Dordrecht, The Netherlands, 1985; pp. 333–361. [CrossRef]
21. Ghazi, S.; Mountney, N.P.; Sharif, S. Lower Permian fluvial cyclicity and stratigraphic evolution of the northern margin of Gondwanaland: Warchha Sandstone, Salt Range, Pakistan. *J. Asian Earth Sci.* **2015**, *105*, 1–17. [CrossRef]
22. McBride, E.F. A Classification of Common Sandstones. *J. Sediment. Res.* **1963**, *33*. [CrossRef]
23. Miall, A.D. Architectural-element analysis: A new method of facies analysis applied to fluvial deposits. *Earth-Sci. Rev.* **1985**, *22*, 261–308. [CrossRef]
24. Gustavson, T.C. *Arid Basin Depositional System and Paleosol: Fort Hancock and Camp Rice Formation (Pliocene-Pleistocene), Hueco Bolson, West Texas and adjacent Mexico*; Report of Investigation; Bureau of Economic Geology, Exploration Way: Austin, TX, USA, 1991; p. 198.
25. Retallack, G.J. Scoyenia burrows from Ordovician palaeosols of the Juniata Formation in Pennsylvania. *Palaeontology* **2001**, *44*, 209–235. [CrossRef]
26. Li, J.; Wen, X.; Huang, C. Lower Cretaceous paleosols and paleoclimate in Sichuan Basin, China. *Cretac. Res.* **2016**, *62*, 154–171. [CrossRef]
27. Tabor, N.J.; Myers, T.S.; Michel, L.A. Sedimentologist's guide for recognition, description, and classification of paleosols. In *Terrestrial Depositional Systems*; Elsevier: Amsterdam, The Netherlands, 2017; pp. 165–208.
28. Malaza, N.; Liu, K.; Zhao, B. Facies Analysis and Depositional Environments of the Late Palaeozoic Coal-Bearing Madzaringwe Formation in the Tshipise-Pafuri Basin, South Africa. *ISRN Geol.* **2013**, *2013*, 120380. [CrossRef]
29. Mayall, M.; Jones, E.; Casey, M. Turbidite channel reservoirs—Key elements in facies prediction and effective development. *Mar. Pet. Geol.* **2006**, *23*, 821–841. [CrossRef]
30. Schumm, S.A. *Evolution and Response of the Fluvial System, Sedimentologic Implications*; SEPM: Broken Arrow, OK, USA, 1981; pp. 19–29. [CrossRef]
31. Allen, J.R.L. Studies in fluviatile sedimentation: A comparison of fining upwards cyclothems, with particular reference to coarse member composition and interpretation. *J. Sediment. Petrol.* **1970**, *40*, 298–323.
32. Collinson, J.D. *Alluvial Sediments, In Sedimentary Environments and Facies*, 3rd ed.; Reading, H.G., Ed.; Blackwell Publishing: Oxford, UK, 1996; pp. 37–82.
33. Jackson II, R.G. Sedimentology of muddy fine-grained channel deposits in meandering streams of the American Middle West. *J. Sediment. Petrol.* **1981**, *51*, 1169–1192.
34. Hjellbakk, A. Facies and fluvial architecture of a high-energy braided river: The Upper Proterozoic Seglodden Member, Varanger Peninsula, northern Norway. *Sediment. Geol.* **1997**, *114*, 131–161. [CrossRef]
35. Ghazi, S.; Mountney, N.P. Subsurface lithofacies analysis of the fluvial early permian Warchha Sandstone, Potwar Basin, Pakistan. *J. Geol. Soc. India* **2010**, *76*, 505–517. [CrossRef]
36. Walker, R.G. Facies models and modern stratigraphic concepts. In *Facies Models: Response to Sea-level Change*; Walker, R.G., James, N.P., Eds.; Geological Association of Canada: St. John's, NL, Canada, 1992; pp. 1–14.
37. Kingsley, C.S. Stratigraphy and Sedimentology of the Ecca Group in the Eastern Cape Province, South Africa. Ph.D. Thesis, University of Port Elizabeth, Port Elizabeth, South Africa, 1977; p. 290.
38. Jamil, M.; Siddiqui, N.A.; Rahman, A.H.B.A.; Ibrahim, N.A.; Ismail, M.S.B.; Ahmed, N.; Usman, M.; Gul, Z.; Imran, Q.S. Facies Heterogeneity and Lobe Facies Multiscale Analysis of Deep-Marine Sand-Shale Complexity in the West Crocker Formation of Sabah Basin, NW Borneo. *Appl. Sci.* **2021**, *11*, 5513. [CrossRef]
39. Bhattacharya, J.P.; Miall, A.D.; Ferron, C.; Gabriel, J.; Randazzo, N.; Kynaston, D.; Jicha, B.R.; Singer, B.S. Time-stratigraphy in point sourced river deltas: Application to sediment budgets, shelf construction, and paleo-storm records. *Earth Sci. Rev.* **2019**, *199*, 102985. [CrossRef]
40. Arnot, M.J.; Browne, G.H.; King, P.R. *Thick-bedded Sandstone Facies in a Middle Basin-floor-fan Setting*; Mount Messenger Formation: Mohakatino Beach, New Zealand, 2007.
41. Jones, B.G.; Rust, B.R. Massive sandstone facies in the Hawkesbury Sandstone, a Triassic fluvial deposit near Sydney, Australia. *J. Sediment. Res.* **1983**, *53*, 1249–1259.
42. Fisher, W.L.; Galloway, W.E.; Steel, R.J.; Olariu, C.; Kerans, C.; Mohrig, D. Deep-water depositional systems supplied by shelf-incising submarine canyons: Recognition and significance in the geologic record. *Earth Sci. Rev.* **2021**, *214*, 103531. [CrossRef]
43. Stow, D.A.; Hernández-Molina, F.J.; Llave, E.; Sayago-Gil, M.; Del Río, V.D.; Branson, A. Bedform-velocity matrix: The estimation of bottom current velocity from bedform observations. *Geology* **2009**, *37*, 327–330. [CrossRef]
44. Selim, S.; El-Gwad, M.A.; Abu Khadrah, A. Sedimentology, petrography, hydraulic flow units, and reservoir quality of the bayhead delta reservoirs: Late Messinian Qawasim formation, Nile Delta, Egypt. *Mar. Pet. Geol.* **2021**, *130*, 105125. [CrossRef]
45. Boulesteix, K.; Poyatos-More, M.; Flint, S.S.; Hodgson, D.M.; Taylor, K.G.; Parry, G.R. Sedimentary facies and stratigraphic architecture of deep-water mudstones beyond the basin-floor fan sandstone pinchout. *J. Sediment. Res.* **2020**, *90*, 1678–1705. [CrossRef]
46. Aigbadon, G.O.; Akakuru, O.C.; Chinyem, F.I.; Akudo, E.O.; Musa, K.O.; Obasi, I.A.; Overare, B.; Ocheli, A.; Sanni, Z.J.; Bala, J.A., II. Facies analysis and sedimentology of the Campanian–Maastrichtian sediments, southern Bida Basin, Nigeria. *Carbonates Evaporites* **2023**, *38*, 27. [CrossRef]

47. Nichols, G. *Sedimentology and Stratigraphy*; John Wiley & Sons: Hoboken, NJ, USA, 2009.
48. Finthan, B.; Mamman, Y.D. The lithofacies and depositional paleoenvironment of the Bima Sandstone in Girei and Environs, Yola Arm, Upper Benue Trough, Northeastern Nigeria. *J. Afr. Earth Sci.* **2020**, *169*, 103863. [CrossRef]
49. Dar, Q.U.Z.; Renhai, P.; Ghazi, S.; Ahmed, S.; Ali, R.I.; Mehmood, M. Depositional facies and reservoir characteristics of the Early Cretaceous Lower Goru Formation, Lower Indus Basin Pakistan: Integration of petrographic and gamma-ray log analysis. *Petroleum* **2021**. [CrossRef]
50. Eschard, R.; Albouy, E.; Gaumet, F.; Ayub, A. Comparing basin floor fan versus slope fan depositional architecture in the Pab sandstone, Maastrichtian, Pakistan. *Geol. Soc. Lond. Spec. Publ.* **2002**, *222*, 159–185. [CrossRef]
51. Wang, C.-Z.; Wang, J.; Hu, B.; Lu, X.-H. Trace fossils and sedimentary environments of the upper cretaceous in the Xixia Basin, Southwestern Henan Province, China. *Geodin. Acta* **2016**, *28*, 53–70. [CrossRef]
52. Okoro, A.U.; Igwe, E.O.; Umo, I.A. Sedimentary facies, paleoenvironments and reservoir potential of the Afikpo Sandstone on Macgregor Hill area in the Afikpo Sub-basin, southeastern Nigeria. *SN Appl. Sci.* **2020**, *2*, 1–17. [CrossRef]
53. Ali, S.; Gingras, M.K.; Wilson, B.; Winter, R.; Gunness, T.; Wells, M. The influence of bioturbation on reservoir quality: Insights from the Columbus Basin, offshore Trinidad. *Mar. Pet. Geol.* **2023**, *147*, 105983. [CrossRef]
54. Stow, D.; Nicholson, U.; Kearsey, S.; Tatum, D.; Gardiner, A.; Ghabra, A.; Jaweesh, M. The Pliocene-Recent Euphrates river system: Sediment facies and architecture as an analogue for subsurface reservoirs. *Energy Geosci.* **2020**, *1*, 174–193. [CrossRef]
55. Coronel, M.D.; Isla, M.F.; Veiga, G.D.; Mountney, N.P.; Colombera, L. Anatomy and facies distribution of terminal lobes in ephemeral fluvial successions: Jurassic Tordillo Formation, Neuquén Basin, Argentina. *Sedimentology* **2020**, *67*, 2596–2624. [CrossRef]
56. Bjerstedt, T.W. Trace fossils from the early Mississippian price delta, southeast west Virginia. *J. Paleontol.* **1988**, *62*, 506–519.
57. Amireh, B.; Schneider, W.; Abed, A. Fluvial-shallow marine-glaciofluvial depositional environments of the Ordovician System in Jordan. *J. Asian Earth Sci.* **2001**, *19*, 45–60. [CrossRef]
58. Fielding, C.R. Upper flow regime sheets, lenses and scour fills: Extending the range of architectural elements for fluvial sediment bodies. *Sediment. Geol.* **2006**, *190*, 227–240. [CrossRef]
59. Maceachern, J.A.; Pemberton, S.G. Ichnological Aspects of Incised-Valley Fill Systems from the Viking Formation of the Western Canada Sedimentary Basin, Alberta, Canada. In *Incised-Valley Systems: Origin and Sedimentary Sequences*; SEPM Society for Sedimentary Geology: Broken Arrow, OK, USA, 1994. [CrossRef]
60. Boggs, S. *Principles of Sedimentology and Stratigraphy*; Pearson: London, UK, 2012.
61. Abdel-Fattah, Z.A. Fluvial architecture of the Upper Cretaceous Nubia Sandstones: An ancient example of sandy braided rivers in central Eastern Desert, Egypt. *Sediment. Geol.* **2021**, *420*, 105923. [CrossRef]
62. Capuzzo, N.; Wetzel, A. Facies and Basin architectural of the late Carboniferous Salvan-Dorenaz continental basin (western Alps, Switzerland/France). *Sedimentology* **2004**, *51*, 675–697. [CrossRef]
63. Olsen, H. The architecture of a sandy braided-meandering river system: An example from the Lower Triassic Solling Formation (M. Buntsandstein) in W Germany. *Geol. Rundsch.* **1988**, *77*, 797–814. [CrossRef]
64. Bordy, E.M.; Head, H.; Runds, M.J. Paleoenvironment and provenance in the early Cape Basin of southwest Gondwana: Sedimentology of the lower ORDOVICIAN Piekenierskloof Formation, Cape Supergroup, South Africa. *S. Afr. J. Geol.* **2016**, *119*, 399–414. [CrossRef]
65. Cant, D.J.; Walker, R.G. Fluvial processes and facies sequences in the sandy braided South Saskatchewan River, Canada. *Sedimentology* **1978**, *25*, 625–648. [CrossRef]
66. Kraus, M.J. Paleosols in clastic sedimentary rocks: Their geologic applications. *Earth-Sci. Rev.* **1999**, *47*, 41–70. [CrossRef]
67. Wilding, L.P.; Tessier, D. Genesis of Vertisols: Shrinkswell Phenomena. In *Vertisols: Their Distribution, Properties, Classification and Management*; Wilding, L.P., Puentes, R., Eds.; Texas A&M University Publishing Center: College Station, TX, USA, 1988; pp. 55–81.
68. Buol, S.W.; Southard, R.J.; Graham, R.C.; Mcdaniel, P.A. *Soil Genesis and Classification*, 5th ed.; Iowa State University Press: Ames, Lowa, 2003; 494p.
69. Shahzad, A.; Tan, J.; Ahsan, S.A.; Abbasi, I.A.; Shahzad, S.M. Identification of Potential Hydrocarbon Source Rocks Using Biological Markers in the Kohat-Potwar Plateaus, North Pakistan. In Proceedings of the 2022 Goldschmidt Conference, Honolulu, HI, USA, 11–15 July 2022.
70. Khan, S.; Nisar, U.B.; Ehsan, S.A.; Farid, A.; Shahzad, S.M.; Qazi, H.H.; Khan, M.J.; Ahmed, T. Aquifer vulnerability and groundwater quality around Brahma Bahtar lesser Himalayas Pakistan. *Environ. Earth Sci.* **2021**, *80*, 454. [CrossRef]

**Disclaimer/Publisher's Note:** The statements, opinions and data contained in all publications are solely those of the individual author(s) and contributor(s) and not of MDPI and/or the editor(s). MDPI and/or the editor(s) disclaim responsibility for any injury to people or property resulting from any ideas, methods, instructions or products referred to in the content.

Article

# Sedimentological Controls on the Reservoir Characteristics of the Mid-Triassic Tredian Formation in the Salt and Trans-Indus Surghar Ranges, Pakistan: Integration of Outcrop, Petrographic, and SEM Analyses

Kamil A. Qureshi [1,2], Mohamad Arif [3], Abdul Basit [4], Sajjad Ahmad [3], Hammad Tariq Janjuhah [5,*] and George Kontakiotis [6,*]

1. Department of Earth Sciences, COMSATS Islamabad, Abbottabad Campus, Abbottabad 22060, Pakistan; qureshika56@gmail.com
2. Department of Earth & Atmospheric Sciences, University of Houston, Houston, TX 77204, USA
3. Department of Geology, University of Peshawar, Peshawar 25120, Pakistan; arif_pkpk@yahoo.com (M.A.); dr.s_ahmed@uop.edu.pk (S.A.)
4. Geological Survey of Pakistan, Quetta 87300, Pakistan; a.basitgeo@gmail.com
5. Department of Geology, Shaheed Benazir Bhutto University, Sheringal, KP, 18000, Pakistan
6. Department of Historical Geology-Paleontology, Faculty of Geology and Geoenvironment, School of Earth Sciences, National and Kapodistrian University of Athens, Panepistimiopolis, Zografou, 15784 Athens, Greece
* Correspondence: hammad@sbbu.edu.pk (H.T.J.); gkontak@geol.uoa.gr (G.K.)

**Abstract:** The current study uses an integrated lithofacies, optical microscopy, and scanning electron microscopy (SEM) analysis to investigate the sedimentary processes, depositional architecture, and reservoir rock potential of the Tredian Formation's (Mid-Triassic) mixed siliciclastic and carbonate succession in the Salt and Trans-Indus Ranges. The formation has been divided litho-stratigraphically into two components: the lower Landa Member, which consists of fine-grained sandstone and shale, and the upper Khatkiara Member, which consists of coarse-grained sandstone. Based on sedimentary structures and lithology, four distinct types of lithofacies are identified. Two lithofacies representing sandstones interbedded with shale (LF1) and thick-bedded sandstone (LF2) lithofacies suggestive of fluvio-deltaic settings are among them. Another two lithofacies of thin-bedded sandstone (LF3) and dolomite (LF4) suggest a tidal flat depositional environment, correspondingly. The petrographic examination of the Tredian sandstones indicates a lithology ranging from sub-feldspathic arenite to feldspathic arenite with moderate packing. The presence of primary calcite cement, silica cement, and iron oxide/hydroxide cements were shown by the diagenetic investigation, which was supported by SEM studies. In addition, secondary cements include ferroan-dolomite, chlorite, and illite, which is linked with chemical alteration of unstable grains. The paragenetic sequence depicts the diagenetic evolution of the Tredian sandstone from early to late diagenetic phases. The reservoir quality of the LF1 and LF4 lithofacies has been destroyed by early-stage calcite cementation, but the lithofacies LF2 and LF3 have a strong reservoir potential owing to the scarcity of calcite cement, dissolution of unstable feldspar grains, and grain fracture.

**Keywords:** mixed siliciclastic and carbonate successions; reservoir heterogeneity; lithofacies; diagenetic evolution; cementation; regression; stratigraphic correlations; calcite and silica cement types; depositional environments

## 1. Introduction

Porosity and permeability are the key factors to assess the reservoir quality of siliciclastic rocks [1,2]. The interplay of geological factors such as tectonic history, provenance, and depositional evolution determines the mineralogical composition, grain size, shape, sorting of sandstone and control the reservoir rock's original porosity and permeability [3].

However, post-burial diagenetic processes such as compaction, cementation, dissolution, and authigenic mineralization may create or reduce the porosity of reservoir rocks [4,5].

The Upper Indus Basin (Kohat–Potwar) is one of the most prolific oil and gas-bearing basins in north Pakistan (Figure 1) [6]. Since the first commercial oil discovery in 1915 from the Miocene Siwaliks sandstone, numerous wells have been drilled in the Kohat–Potwar foreland basins, targeting multiple reservoirs. These reservoir rocks are composed of mixed siliciclastic and carbonate sequences of the Eocene, Paleocene, Cretaceous, and Jurassic ages [7–10]. However, underlying Triassic sequences were penetrated in only a few wells. The Triassic rocks consist of the Mianwali Formation (limestone), Tredian Formation (sandstone), and Kingriali Formation (dolomite), which are producing reservoirs in the Kohat–Potwar region. The Mid-Triassic Tredian Formation has been drilled in Isakhel-01, Chonai-01, and Makori-01 wells in the Upper Indus Basin and its thicknesses are 91 m, 52 m, and 22 m, respectively [11,12]. The formation tops of the Tredian sandstone were marked at a depth of 3400 m, 3795 m, and 4285 m in all these respective wells. Both Isakhel-01 and Chonai-01 are abandoned exploratory wells due to mechanical failure, while Makori-01 is the only well in the Kohat Basin in which Tredian Formation is the producing reservoir [11]. However, structural complexities involving multiple detachments and the presence or absence of salt lithofacies limit the drilling to shallow reservoirs in the fold-thrust belts of western Pakistan.

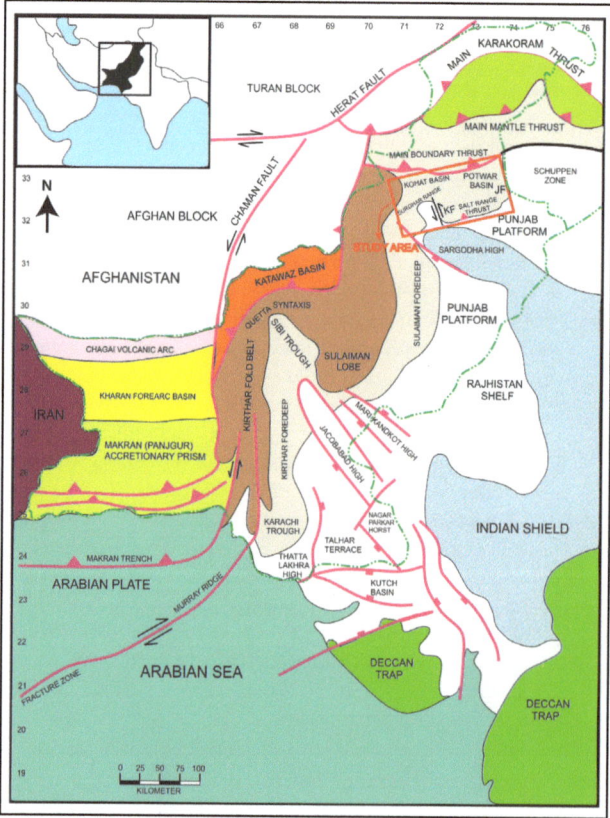

**Figure 1.** Three studied sections with their measured stratigraphic log and lateral correlation of various lithofacies of the Tredian Formation in the Salt and Trans-Indus Surghar Ranges. The measured stratigraphic sections include Landa Pasha (1), Gulakhel (2), and Nammal Nala (3), respectively. The stratigraphic thickness increases from east to west in the Landa Pasha section. Towards the west, lithofacies LF1 thickness increases and lithofacies LF2 decreases.

The Mid-Triassic Tredian Formation is well exposed in the western Salt Range and the Surghar Range. This gives scientists a unique chance to study the surface analog for predicting reservoir heterogeneity and facies variations [13,14]. Reservoir heterogeneity is caused by differences in depositional facies, diagenesis, and structural features (such as fractures or faults), and it happens on scales ranging from hundreds of meters to micrometers [14–16]. There are only a few studies conducted on the palynological and paleontological aspects of the Tredian Formation to determine its age and depositional environment [17–20]. However, details regarding vertical and lateral lithofacies variations and their diagenetic evolution are still lacking, which is important to understand the reservoir character, fluid flow, and recovery factor. The present study evaluated the reservoir quality of Tredian Formation sandstone in the three representative stratigraphic sections in the Salt and Trans-Indus Surghar Ranges (Figure 2).

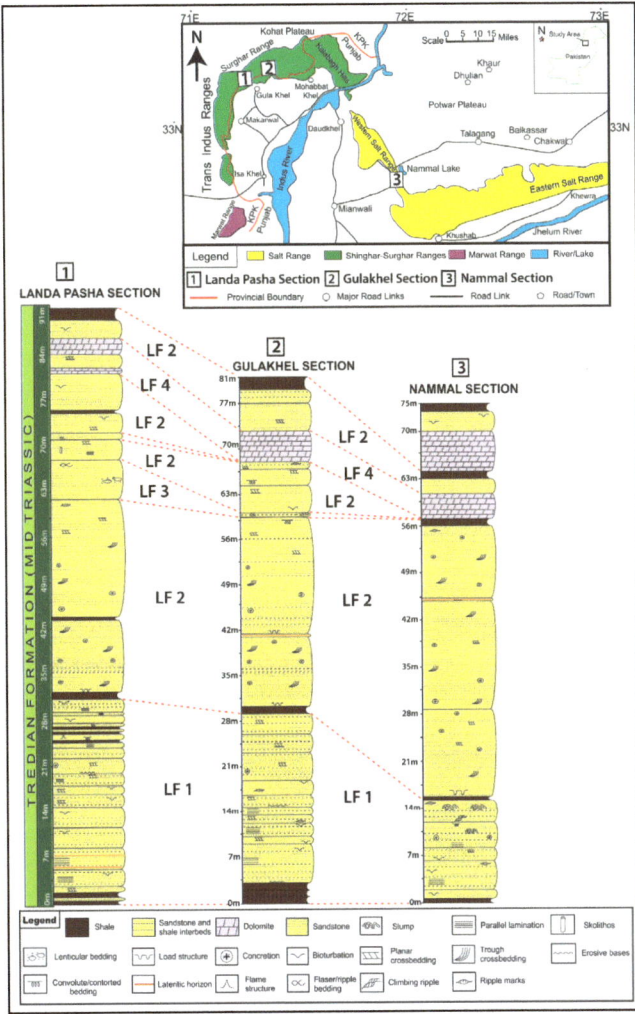

**Figure 2.** Tectono-stratigraphic framework of the western Himalayas in Pakistan. The north–south-oriented strike-slip Chaman Fault separates the western margin of the Indian plate from the Eurasian plate. The study area highlighted as red rectangle marks the foreland region as Salt and Trans-Indus Surghar Ranges and known as Main Frontal Thrust (MFT) (modified after Treloar and Izatt [21]). The numbers represent longitude and latitude, correspondingly. JF: Jhelum Fault, KF: Kalabagh Fault.

More explicitly, the current study aims to: (1) carry out lithofacies interpretation along with their vertical and lateral variations, (2) present a petrographic characterization of the Tredian Formation sandstone, (3) determine the extent and significance of diagenetic processes and its influence on reservoir quality using optical and scanning electron microscopy (SEM), and (4) provide an integrated depositional model.

## 2. Geological Framework

The Indo–Asia collision created the ~2500-kilometer-long seismically active Himalayan Mountain belt (55 Ma) [22]. The main frontal thrust (MFT) marks the southern boundary of this collisional zone and is represented by the Salt and Trans-Indus Ranges [23–25] (Figure 1). In comparison with the eastern and central Himalayas, MFT is a more than 100 km wide zone in the western Himalayas in Pakistan with a low degree of cross-sectional taper, resulting in the opening of wide basins, i.e., the Kohat and Potwar Basins [26,27]. The Salt Range has been subdivided into the eastern, central, and western sections, respectively [25,26]. The eastern termination of the Salt Range is the Jhelum Fault (left-lateral strike-slip), whereas the western termination is marked by the Kalabagh Fault (right-lateral strike-slip) [27] (Figure 1). The Salt Range Thrust (SRT) marks the leading edge of the Potwar Plateau. The sinuous shape of the Trans-Indus Ranges (TIR) demarcates the deformational front of the Kohat Basin in North Pakistan [25]. The TIR comprises the Surghar Range, the Shinghar Range, the Marwat–Khisor Ranges, the Manzai Range, and the Sheikh Badin Hills. The Surghar Range (SR) marks the eastern extremity of the Trans-Indus Ranges [28,29] (Figure 1).

Both the Salt and Trans-Indus Surghar Ranges have Triassic rocks cropping out roughly parallel to the Salt Range Thrust [30]. These successions are well exposed in different gorges, among which the Landa Pasha (Surghar Range), Gulakhel (Surghar Range), and Nammal Gorge sections (Western Salt Range) were focused during this work (Figure 2). The sedimentary successions of the Salt and Trans-Indus Surghar Ranges reflect the depositional environments of the Gondwana shelf. The depositional sites were probably located at 30° S [30,31]. The Tethyan Ocean was receiving siliciclastic sediments from the Indian subcontinent [7]. Exposed rocks in the Western Salt Range extend from the Precambrian (Salt Range Formation) to the Eocene (Sakessar Limestone). The Salt Range Thrust reveals Precambrian to Recent rocks. The Salt Range Formation represents the Precambrian age and is overlain by rocks of the Jhelum Group (Cambrian). The Permian-Eocene stratigraphic successions are well exposed in the study sections [32]. Exposed rocks in the Surghar Range show ages from the Late Permian (Wargal Formation) to Miocene (Chinji Formation). The generalized stratigraphy of the Western Salt Range (Nammal Gorge) and the Trans-Indus Surghar Range (Landa Pasha and Gulakhel sections) is shown in Figure 3.

The Triassic succession includes the Mianwali, Tredian, and Kingriali formations. The Triassic rocks have a disconformable lower contact (Permo-Triassic; P-T boundary) with rocks of the Zaluch Group, while upper contact with the Jurassic Datta Formation is conformable (Figure 3). The P-T boundary represents the great Permian extinction that occurred at about 252 Ma, during which 96% of marine species and 70% of terrestrial vertebrates became extinct [32–35]. The principal lithology of the Tredian Formation is sandstone with shale and some dolomite (Figure 2). It marks conformable contacts with the underlying Mianwali Formation and the overlying Kingriali Dolomite [7] (Figure 3). No fossils have been reported so far; however, some poorly preserved plant remains were observed by Balme [36–38].

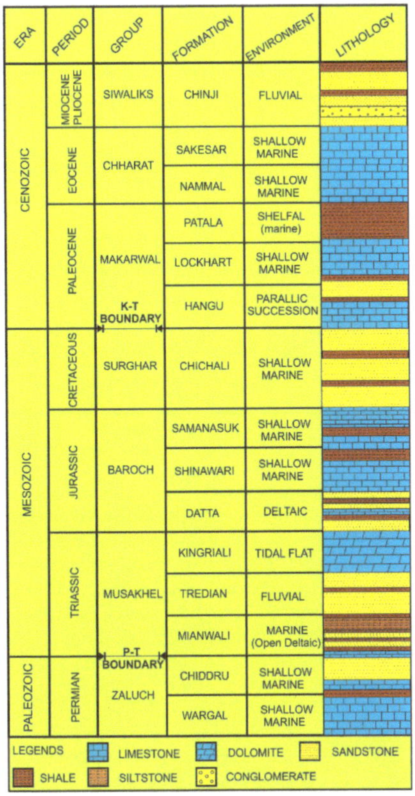

**Figure 3.** A generalized stratigraphic column in the Nammal Nala, Landa Pasha, and Gulakhel section in the Salt and Trans-Indus Surghar Ranges after Shah [7]. The rocks older than Permian Wargal Formation are missing in all these sections.

## 3. Materials and Methods

A detailed geological field trip was carried out to the Nammal Nala section (32°40′3.41″ N, 71°47′9.24″ E) in the western Salt Range and the Landa Pasha (32°58′0.00″ N, 71°12′0.00″ E) and Gulakhel sections in the Surghar Ranges, with the help of available geological maps [30], to record all pertinent field data (Figure 2). All stratigraphic sections were logged and measured with the help of a measuring tape. The measured thicknesses of the Tredian Formation at the Nammal Nala section are 75 m, the Gulakhel section is 81 m, and the Landa Pasha section is 91 m (Figure 2). A total of 46 samples were collected for petrographic analysis from the Landa Pasha section which represents a complete stratigraphic section. A 2 m sampling interval was chosen to cover all details of lithological variations. An outcrop-based lithofacies description was made using field observations, which included sedimentary structures, lithological variation, bedding, thickness, nature of contact, texture, color, fauna, and bioturbation.

After fieldwork, 35 representative rock samples were processed for thin section preparation in the rock cutting laboratory at the Hydrocarbon Development Institute of Pakistan (HDIP) for petrographic analysis. After cutting chips, casting resin and blue dye were used to impregnate samples, which determine the visual porosity percentage of the analyzed samples. Moreover, a chip was mounted on a slide made of thin glass with the help of petro-epoxy for the preparation of thin sections [39,40]. Detailed petrographic analysis was performed under the polarizing microscope available at the Petrography Laboratory of the Department of Earth Sciences, COMSATS Institute of Information Technology, Abbottabad campus, Pakistan. The SEM and EDX analyses of ten representative samples were performed to distinguish between primary and authigenic minerals. This investigation also helped in evaluating the effect of diagenesis on the Tredian sandstone reservoir potential. The SEM analyses were performed at the Centralized Resource Laboratory (CRL), Department of Physics, University of Peshawar, under the JEOL JSM-5910 Model SEM fitted with an Energy Dispersive X-ray Micro-analyzer (EDAX).

## 4. Results

### 4.1. Lithofacies Description

Four lithofacies have been identified in the Tredian Formation exposed in the studied sections by utilizing various sedimentological aspects. They are characterized as fluvio-deltaic and tidal flat lithofacies. A summary of all of them is provided in Table 1.

**Table 1.** Details of the lithofacies of the Tredian Formation along with their observed sedimentary structures, description, and interpretation in all the studied sections.

| Lithofacies | Sub-Lithofacies/Facies | Description | Interpretation |
| --- | --- | --- | --- |
| Lithofacies-1 (LF-1) Sandstone interbedded with shale lithofacies | Cross-bedded, parallel laminated, slumped, rippled, bioturbated sandstone facies | Medium-to-coarse grain, moderate-to-poor sorting, medium-to-thick-bedded sandstone, interbedded with carbonaceous black shales black. Commonly found sedimentary structures are cross bedding, parallel lamination, slump structures, bioturbation, symmetrical and asymmetrical ripples, flame structures, and channelized beds | The sandstone is deposited in distributary channels/channel margins of delta plain settings whereas the shale is deposited in low-energy settings of floodplain and interdistributary bay or marshes |
| Lithofacies-2 (LF-2) Thick bedded sandstone lithofacies | Planar cross-bedded, trough cross-bedded, rippled sandstone facies | Thick-bedded coarse-grain sandstone with planar cross bedding, trough cross bedding, ripple marks, basal erosional surface, and load marks. Pebbly bases of the beds were commonly observed. Well-developed channels are also common. Inter-bedded shale is dark greenish grey to black in color containing thin lamination of sand | The deposition of sandstone took place during high fluvial discharge in fluvial channel environment during active delta progradation, whereas the associated shale shows the deposition along the channel margins or levees |
| Lithofacies-3 (LF-3) Thin-bedded sandstone lithofacies | Parallel-laminated and ripple-laminated sandstone facies | Thin-bedded, greenish grey, medium-to-fine-grain sandstone. Commonly observed sedimentary structures are parallel and ripple lamination. Flaser and lenticular bedding are also common | Deposited in sand dominated delta front or tidal flats where fluctuation in sediment supply is common |
| Lithofacies-4 (LF-4) Dolomite lithofacies | | Medium-bedded yellowish brown dolomite. The dolomite is hard, compact, and brecciated and has laterally pinching channels. The lower part is sandy dolomite, whereas the upper part is pure dolomite. | Deposited during fluctuating depositional condition in shoreface |

#### 4.1.1. Fluvio-Deltaic Lithofacies

1. Sandstone Interbedded with shale lithofacies (LF1)

This lithofacies is present at the base and represents the Landa Member of the Tredian Formation, which is 31 m thick in the Landa Pasha Section, 29 m thick in the Gulakhel Section, and 15 m thick in the Nammal Section, respectively (Figure 2). It can be further subdivided into five sub-lithofacies based on diagnostic sedimentary characteristics (Table 1). It is predominantly composed of medium- to coarse-grained, moderately to poorly sorted, orange/yellow to maroon-colored, medium- to thickly bedded, hard, and compact sandstone interbedded with dark grey to black carbonaceous shale (Figure 4). Scattered pebbles can also be seen in some places. The ratio of sandstone to shale is 3:1. The pisolitic bed at the base of this lithofacies marks the conformable contact between the Mianwali and Tredian Formations. The most commonly observed sedimentary structures are parallel lamination, cross-bedding, slumps, ripple marks, convoluted or contorted bedding, flame structures, bioturbation, and iron concretions (Figure 4c–j). The large planar tabular cross-beds have a thickness varying from 8 to 14 cm. Slumps are observed only in the Nammal Nala section and were not encountered in the other two studied sections. Bioturbation is also very common (Figure 4k). In places, well-developed channels are present and show lateral pinching (Figure 4l).

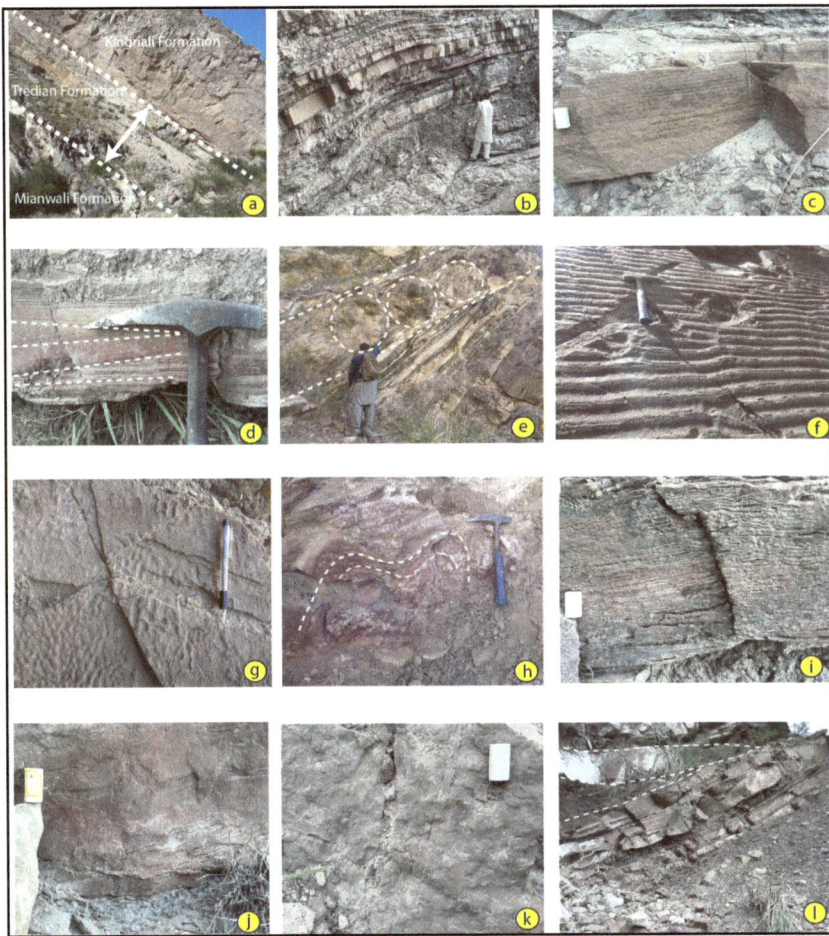

**Figure 4.** Field photographs showing various sedimentological features of Tredian Formation. (**a**) Outcrop exposure of Tredian Formation in Nammal Section with upper and lower conformable contacts. (**b**) Outcrop exposure of sandstone interbedded with shale lithofacies (LF-1). (**c**) Field photograph of parallel laminated medium bedded sandstone lithofacies. (**d**) Field photograph of cross-bedded sandstone lithofacies. (**e**) Field photograph showing the slumped facies of LF-1. (**f**) Symmetrical ripples. (**g**) Field photograph showing asymmetric ripples. (**h**) Convoluted bedding in the slumped facies of LF-1. (**i**) Contorted bedding. (**j**) Flame structures in LF-1. (**k**) Bioturbated sandstone facies of LF-1. (**l**) Well-developed channels and their lateral pinching.

2. Thick-Bedded Sandstone Lithofacies (LF2)

This lithofacies overlies the LF1 of the Tredian Formation and is predominantly comprised of light gray to whitish-colored, coarse-grain, thick-bedded sandstone, and carbonaceous shale (Figure 5a). The beds also have pebbly bases. The shale units are dark greenish gray to black, carbonaceous, and contain thin laminae of sand. Black carbonaceous shale marks the contact between LF1 and LF2 (Figure 5b). Lithofacies LF2 are approximately 31 m thick in the Landa Pasha and Gulakhel sections and 41 m thick in the Nammal Nala section (Figure 2). The sandstone-to-carbonaceous shale ratio is 9:1. The most commonly observed sedimentary structures are planar tabular cross-bedding, trough cross-bedding, erosional basal surfaces, and load structures. Ripple marks are commonly seen at the top of lithofacies, whereas trough and planar tabular cross bedding can be found throughout (Figure 5c–e). Syn-depositional features such as load structures are present at the base of this lithofacies. Well-developed channels are also present in lithofacies LF2 (Figure 5f).

**Figure 5.** Field photographs of various sedimentological features of Tredian Formation lithofacies. (**a**) Channelized sandstone of LF2 (the red rectangle highlight the Jacob Staff for scale). (**b**) Black carbonaceous shales mark the contact between LF1 and LF2. (**c**) Pebbly bases in LF2. (**d,e**) Planar and trough cross-bedded sandstone facies LF-2. (**f**) Thick-bedded sandstone lithofacies LF2 with channel pinching. (**g**) Thin-bedded sandstone with interbedded shale lithofacies (LF-3). (**h**) Rippled sandstone facies with flaser and lenticular bedding in LF3. (**i**) Thick-bedded dolomitic lithofacies LF4 on top of LF3. (**j**) Sandy dolomite lithofacies with laterally pinching channels. (**k**) Top unit of pure dolomitic lithofacies LF4.

4.1.2. Tidal Flat Lithofacies

1. Thin-Bedded Sandstone Lithofacies (LF3)

This lithofacies overlies LF2 in the Landa Pasha section and the Gulakhel section, whereas it is altogether missing in the Nammal Nala section (Figure 2). The measured thickness of LF3 in the Gulakhel and Landa Pasha sections is 1 m and 6 m, respectively. The ripple sandstone of LF2 grades upward into flaser bedded LF3 lithofacies. It comprises medium-to-fine-grained, greenish gray to whitish or brownish, moderately to well-sorted, and thin-bedded sandstone (Figure 5g). Sand lenses are 1 to 3 cm thick and show ripple laminations (Figure 5h). Further towards the top, the flaser and lenticular bedding are replaced by parallel lamination.

2. Dolomite Lithofacies (LF4)

This lithofacies overlies the thin-bedded sandstone lithofacies LF3 and contains sandy dolomite (Figure 5i). Dolomite is yellowish-brown to pinkish in color. The total thickness of this unit is 3 m in the Landa Pasha section, 5 m in the Gulakhel section, and 10 m in the Nammal section, respectively (Figure 2). The dolomite is hard, compact, brecciated, and has laterally pinching channels (Figure 5j). The lower part of the lithofacies is sandy dolomite, while the upper part is pure dolomite (Figure 5k).

*4.2. Petrographic Analysis*

The petrographic analysis of sandstone samples collected from the Tredian Formation revealed that it comprises fine-to-coarse framework grains and cement. Based on the model composition of Pettijohn [41], the sandstone is classified as sub-feldspathic arenite to feldspathic arenite. The predominant mineralogical constituents include quartz, feldspar, accessory minerals, mica, and lithic fragments. Most of the grains are sub-rounded and sub-angular to angular. The grains packing and sorting are moderate. Texturally, the sandstone is sub-mature to mature, whereas compositional maturity varies from sub-mature to immature. The most predominant constituent framework grain in the Tredian Formation is quartz, which has an average abundance of 49.4% (Table 2).

Grains shape is mostly sub-spherical with angular to sub-rounded outlines. Some well-rounded quartz grains are also present, which shows long distances of transport (Figure 6a). A clay rim was observed in a few quartz grains (Figure 6b). The grain contact can be pointed, long, concavo-convex, or suture-like (Figure 6b). Some fractured quartz grains are also present, but they are empty and not filled with cement. Some clean grains have quartz cement present as overgrowth (Figure 6a).

Medium-grained feldspar is the second most abundant mineral in the Tredian Formation, which exists in both plagioclase and K-feldspar forms with an average abundance of 14.3% (Table 2) (Figure 6c). Plagioclase, on the other hand, has the lowest concentration. The grain shape is sub-spherical to prismoidal with sub-angular to sub-rounded outlines. Some of the samples have feldspar grains that are partially or completely altered to clay minerals (Figure 6d,e). Lithic fragments include some minor fossils and siltstone lithics. The proportion of fossil fragments in the sandstone studied is less than 1% (Figure 6f). Calcite cement mostly replaces fossil fragments, while the margins of some fragments are altered by reacting with cementing fluids. Muscovite occurs in significant quantity in the studied thin sections. Its average abundance is 5.3%; however, in a few thin sections, it ranges up to 29.6%.

It mostly occurs as long individual flakes (Figure 6g). Biotite is present in trace amounts in the samples. They show clay alteration in some places. Its average abundance is 0.6%. The modal abundance of the accessory minerals in the Tredian sandstone is 0.85% (a trace amount), including the grains of tourmaline, rutile, and zircon (Figure 6h,i).

Table 2. Percentage modal proportions of framework elements in the Tredian Formation. Qt: Quartz total; Qm: Quartz monocrystalline; Qp: Quartz polycrystalline; Al.Feld: Alkali Feldspar; PlG: Plagioclase; Poros: Porosity; Musc: Muscovite; Biot: Biotite; Ceme: Cement. Accessory minerals include zircon, rutile, etc.

| SN | S# | Quartz | | | Feldspar | | Lithoc | Pores | Musc | Biot | Ceme | Ore | Accessort Minerals | Classification |
|---|---|---|---|---|---|---|---|---|---|---|---|---|---|---|
| | | Qt | Qm | Qp | ALFeld | PlG | | | | | | | | Pettijohn |
| 1 | T1 | 48 | 48 | 0 | 10 | 1 | 0 | 0 | 4 | 1 | 35 | 0 | 1 | Sub-Arkose |
| 2 | T2 | 46 | 46 | 0 | 8 | 0.5 | 0 | 0 | 3 | 0.5 | 42 | 0 | 0 | Sub-Arkose |
| 3 | T3 | 40 | 39.8 | 0.2 | 9 | 2.6 | 0 | 2 | 5.5 | 1 | 29.5 | 10 | 0.4 | Sub-Arkose |
| 4 | T4 | 65 | 65 | 0 | 9 | 3 | 0 | 2.5 | 2.5 | 1 | 10 | 5 | 2 | Sub-Arkose |
| 5 | T5 | 51 | 50.5 | 0.5 | 14 | 0.5 | 0.5 | 0 | 2 | 0 | 28 | 2.5 | 1.5 | Sub-Arkose |
| 6 | T6 | 33 | 33 | 0 | 18 | 1.5 | 0 | 2 | 4 | 0 | 39 | 2 | 0.5 | Arkose |
| 7 | T8 | 60 | 60 | 0 | 11 | 2 | 0.4 | 3 | 8 | 0.6 | 8 | 6 | 1 | Sub-Arkose |
| 8 | T9 | 58 | 58 | 0 | 7 | 1.2 | 0 | 1 | 4 | 2 | 22 | 4 | 0.8 | Sub-Arkose |
| 9 | T11 | 44 | 43 | 1 | 14.5 | 0.5 | 0 | 1.5 | 10.5 | 0.6 | 19 | 9 | 0.4 | Arkose |
| 10 | T12 | 60 | 60 | 0 | 17 | 3 | 0 | 3 | 1.6 | 1 | 1 | 12 | 1.4 | Arkose |
| 11 | T14 | 58 | 58 | 0 | 18 | 1 | 1 | 4 | 5 | 2 | 4 | 5 | 2 | Sub-Arkose |
| 12 | T16 | 56 | 55.5 | 0.5 | 15 | 3 | 0 | 10 | 1.5 | 0 | 6 | 8 | 0.5 | Sub-Arkose |
| 13 | T17 | 47 | 47 | 0 | 20 | 1.5 | 0.5 | 1 | 0.6 | 1.5 | 16 | 7 | 1 | Arkose |
| 14 | T18 | 52 | 52 | 0 | 22 | 2.5 | 0 | 6 | 0.5 | 0.6 | 8 | 4.5 | 2 | Arkose |
| 15 | T19 | 35 | 35 | 1.5 | 11 | 2 | 0 | 1 | 1 | 0.5 | 44 | 4 | 0.5 | Arkose |
| 16 | T20 | 84 | 84 | 0 | 7 | 0.3 | 0 | 4 | 0 | 1 | 2 | 0.7 | 1 | Sub-Arkose |
| 17 | T21 | 16 | 16 | 0 | 8 | 0.8 | 0 | 0 | 0 | 0 | 66 | 9 | 0.2 | Arkose |
| 18 | T22 | 14 | 14 | 0 | 9 | 1 | 0 | 6 | 0 | 0 | 64 | 6 | 0 | Arkose |
| 19 | T24 | 60 | 60 | 0 | 19 | 1.5 | 0 | 4 | 4.5 | 0 | 9 | 1 | 1 | Arkose |
| 20 | T25 | 38 | 36 | 2 | 14 | 2 | 0 | 0 | 3 | 1 | 36 | 5 | 1 | Arkose |
| 21 | NT5 | 55 | 55 | 0 | 24 | 2.4 | 0 | 3 | 8.5 | 0.5 | 2 | 3.6 | 1 | Arkose |
| 22 | NT6 | 66 | 65.2 | 0.8 | 12 | 0.5 | 0 | 2 | 12.5 | 0 | 6 | 1 | 0 | Sub-Arkose |
| 23 | NT7 | 36 | 36 | 0 | 10 | 2 | 0 | 0 | 20 | 0 | 27 | 5 | 0 | Sub-Arkose |
| 24 | NT9 | 32 | 31 | 1 | 13.5 | 1.9 | 0 | 0 | 13.4 | 1 | 29 | 7 | 2.2 | Arkose |
| 25 | NT10 | 30 | 30 | 0 | 11 | 1.5 | 0 | 0 | 2.8 | 0.2 | 45 | 8 | 1.5 | Arkose |
| 26 | NT12 | 78 | 78 | 0 | 7 | 1.8 | 0 | 13 | 0 | 0 | 0 | 0 | 0.2 | Sub-Arkose |
| 27 | NT13 | 71 | 70.5 | 0.5 | 7 | 0.5 | 0 | 5 | 0 | 0 | 15 | 1 | 0.5 | Sub-Arkose |
| 28 | NT14 | 80 | 80 | 0 | 5 | 1.2 | 0 | 8 | 0 | 0 | 5 | 0 | 0.8 | Sub-Arkose |
| 29 | NT17 | 19 | 19 | 0 | 20 | 1 | 0 | 0 | 29.6 | 2 | 28 | 0 | 0.4 | Arkose |

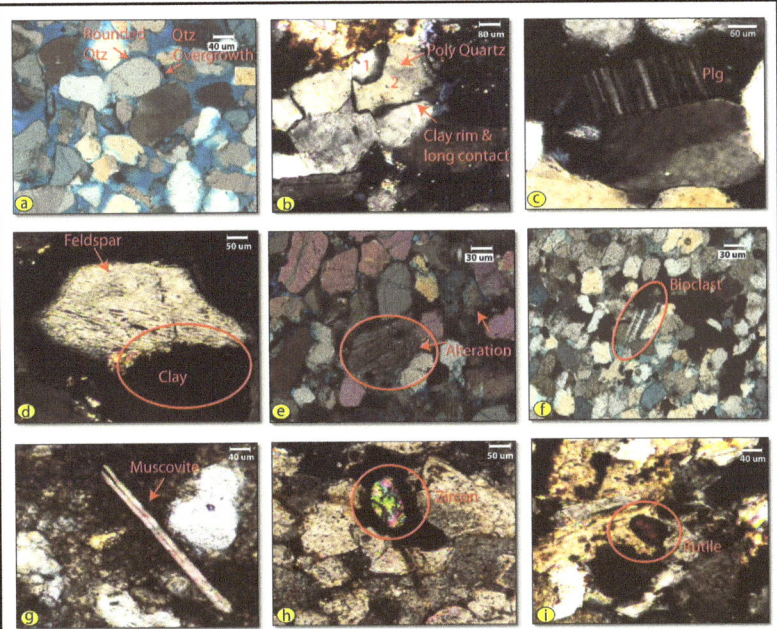

**Figure 6.** Petrographic analysis of framework grains. (**a**) Rounded quartz grains with quartz overgrowth in LF2. (**b**) Polycrystalline quartz with clay rim at grain boundaries and long grain contact in LF1. (**c**) Plagioclase grain with twinning in LF1. (**d**) Feldspar grain with clay alteration at the boundary and some replacement by cement in LF3. (**e**) Feldspar grain showing partial/complete alteration to clay in LF2. (**f**) Bioclast with some clay present at the boundary in LF2. (**g**) Long muscovite flake in LF1. (**h**) Accessory mineral zircon with clay rim at the boundary in LF1, and (**i**) rutile grain embedded in calcite cement in LF1.

*4.3. Diagenesis of the Tredian Sandstone*

4.3.1. Compaction

Compaction is considered the main factor affecting the reservoir as well as other physical properties of the Tredian Formation sandstone. The precipitation of early calcite cement marks the end of further compaction. The following types of grain packing explain the degree of compaction observed in Tredian Formation sandstone (Figure 7a–f). The depositional settings of various interpreted lithofacies are also described.

(a) Loosely packed grains: Loose packing of grains was observed at the basal part of the Tredian Formation, where precipitation of calcite cement ceased the further compaction of grains. The floating grains are observed with point and long-line contact, which also refers to the loose packing of grains (Figure 7a).

(b) Closely packed grains: This type of packing includes moderate-to-strong packing of grains. Line/long contact of grains is observed, where the packing is moderate, whereas concave-convex contact with pressure solutions refers to the strong packing of grains (Figure 7b).

(c) Directionally oriented packed grains: Flake mineral grains such as micas were mainly characterized by directional arrangement due to the pressure of overlying strata (Figure 7c).

(d) Plastic grains deformed by extrusion: The plastic grains such as micas and chlorites were observed to be bent or broken where the compaction was strong (Figure 7d).

(e) Rigid grains crushed by extrusion: Rigid grains such as feldspar and quartz are broken, forming micro-fractures, when the compaction further increases with increase in burial depth (Figure 7e).

(f) Chemically compacted fabric: Pressure-induced dissolution along the contacts of quartz grains provided the silica cement, which filled the pore spaces and resulted in the reduction of inter-granular porosity. This chemical compaction is observed mostly at the upper part of the Tredian Formation (Figure 7f).

**Figure 7.** Photomicrographs of Tredian sandstone showing various degrees of compaction and associated features. (**a**) Loosely packed grains with pointed and long contact. (**b**) Closely packed grains with long and concavo-convex contact between the grains showing moderate-to-strong degree of compaction, respectively. (**c**) Directionally oriented packed grains (red arrows) showing directional arrangement due to the pressure of overlying strata. (**d**) Bent grains of chlorite showing plastic grain deformation by extrusion. (**e**) Fractured grains in quartz and feldspar showing micro-fractures. (**f**) Chemically compacted fabric with pressure-induced dissolution (red arrows) along the contacts of quartz grains.

4.3.2. Cementation

After compaction, cementation is another important contributing factor that affects the reservoir quality of Tredian Formation sandstone by reducing the porosity and permeability. Precipitation of various types of cement (Figure 8a–d) reduced the pore spaces and the connectivity of pore throats, hence affecting reservoir quality. Petrographic studies and SEM analyses have revealed the occurrence of the following types of cement described below.

Calcite cement is the most dominant type of cement observed in Tredian Formation sandstone and occurs in both well-developed poikilotopic (Figure 8a) and fine-grained micritic forms (Figure 8b). The following two stages of calcite cementation are distinguishable based on distribution and texture: (a) Early diagenetic calcite cement developed between the floating framework grains in poikilotopic form at the basal part of the Tredian Formation (LF1); (b) Late diagenetic calcite cement was found in some samples from the Khatkiara Member. The late-stage fine-grain calcite cement has precipitated in the inter-granular pore spaces (LF2). Siliceous cementation in the form of quartz overgrowth was commonly observed in the samples taken from the Khatkiara Member of the Tredian Formation (LF2). The presence of quartz overgrowth indicates the availability of sufficient silica, a medium for its transport, and a clean surface of grains for its precipitation (Figure 8c). The early diagenetic calcite cementation in the Landa Member halted the silica precipitation in the form of quartz overgrowth; hence, quartz overgrowth was not observed in the Landa

Member of the Tredian Formation (Figure 8a). All the studied samples of the Tredian sandstone contain a significant amount of ferruginous cement (Figure 8d). It decreases the amount of secondary porosity by replacing the initial cement through dissolution. It is present both as pore-filling and grain-occluding phases.

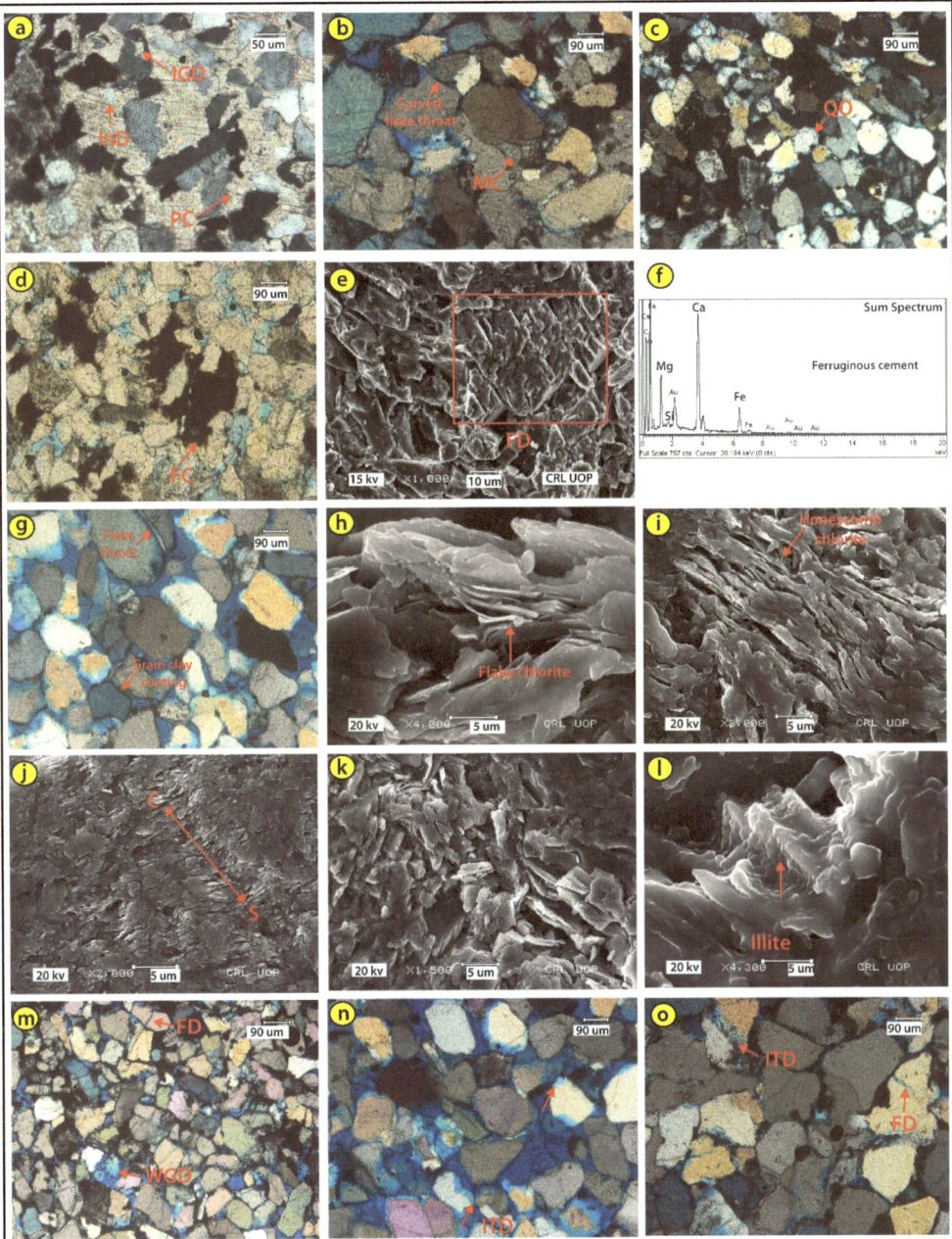

**Figure 8.** Photomicrograph showing various diagenetic features of Tredian sandstone. (**a**) Poikilotopic calcite cement (PC), interstitial dissolution (InD), and Intra-granular dissolution (IGD) in quartz grains.

(**b**) Connectivity between the pores through curved flake throat and micritic calcite cement (MC). (**c**) Strongly compacted grains with quartz overgrowth (QO). (**d**) Ferruginous cement (FC) as cementing material. (**e**) SEM image showing ferroan dolomite (FD). (**f**) EDX spectra showing ferruginous cement. (**g**) Pores connectivity through flake throat and clay coating around the grains. (**h**) SEM image showing flaky/rosettes chlorite. (**i**) SEM image showing honeycomb/leafy chlorite. (**j**) SEM image showing chloritization of smectite. (**k**) Small and thin platelets of mixed-layer illite-chlorite occurring as void filling. (**l**) SEM image showing early diagenetic illite. (**m**) Fracture dissolution (FD) and whole-grain dissolution (WGD). (**n**) Inter-granular dissolution (ITD) and connectivity of pores through neck throat. (**o**) Fracture dissolution and inter-granular dissolution.

Other types of cement include Ferroan dolomite which occurs in most of the samples as well-developed authigenic crystals, as identified petrographically and confirmed through SEM/EDX analysis (Figure 8e,f). The dolomite rhombs are fine grain in size and occur during the grain-replacing and pore-occluding phases. The ferroan dolomite in sandstone represents reducing and alkaline pore water conditions [42]. Chlorite is another commonly observed cement type that exists in both Rosettes (Figure 8h) and honeycomb (Figure 8i) as morphological forms. Chlorite occurs both as a grain coating and as pore filling (Figure 8g). Alteration of feldspars, direct precipitation of chlorite from pore fluids, and chloritization of other clay minerals are the most likely sources of chlorite formation [34,43]. Chloritization of smectite is observed through SEM analysis (Figure 8j). Chlorite with an irregular morphology is sometimes mixed with illite and forms a mixed layer of illite and chlorite. Small and thin platelets of illite-chlorite also occur as void filling (Figure 8k). Two types of chlorites have been recognized in studied samples: (a) Early diagenetic chlorite: In most of the samples, it lies perpendicular to the detrital grain surfaces suggesting their formation during early diagenesis [34,44]; (b) Late-stage burial diagenetic chlorite: The chlorite presents as pore-filling forms during burial diagenesis. Illite is present in the studied samples with a hair-like fibrous and ribbon crystal habit (Figure 8l). The typical illite is lath-like and represents kaolinite as its precursor. In the studied sandstone samples, illite occurs as pore-filling material. A markedly greater abundance of illite in the vicinity of altered detrital grains of feldspar strongly suggests the formation of illite by feldspar alteration. Illite is an exclusively burial diagenetic mineral, and its formation requires the availability of potassium and a temperature exceeding 70 °C [45].

4.3.3. Dissolution

Dissolution is a main contributing factor in enhancing the reservoir character of sandstone. Following are the different types of dissolution observed in the sandstone of the Tredian Formation: (a) Intra-granular dissolution is observed inside the grains, which have been dissolved by acidic fluids. The dissolved part of the grain either exists in the shape of spots or honeycomb shape (Figure 8a,f); (b) Whole-grain dissolution was common in some grains forming moldic pores (Figure 8f,m); (c) Dissolution along weak planes was commonly observed along the cleavage planes, cracks, and fractures of feldspar grains (Figure 8f); (d) Inter-granular dissolution is a kind of irregular dissolution that is observed along the edges of some soluble grains, such as feldspar and lithic grains (Figure 8n,o); (e) Interstitial dissolution is also observed, in which interstitial material, such as calcite cement, has been dissolved, forming interstitial pores (Figure 8a).

## 5. Discussion

*5.1. Lithofacies Depositional Environment*

5.1.1. LF-1 Interpretation

The presence of features such as ripple marks, planar cross-bedding, parallel lamination, and bioturbation indicates deposition of this lithofacies in distributary channels and inter-distributary bays, which are the principal components of a delta plain [46–48]. The presence of scattered pebbles also supports the deposition in delta plain settings [49]. The cross bedding is the indication of deposition under unidirectional flow, which is mostly associated with fluvial channels or distributary channels in a delta plain setting [50–52].

Parallel laminated sandstones represent deposition in upper-flow regime conditions [53,54]. This lithofacies has a close association with the shoreline/shoreface setting; however, the presence of slumps, which is uncommon in a shoreline environment, describes it as a delta plain facies [55]. The presence of wave ripples indicates that this lithofacies was also affected by shoreline/beach action; however, the presence of erosive bases, channel morphology, channel lag, cross-bedding, planar lamination, bioturbation, and slump structures supports deposition in distributary channels/channels or channel margins of delta plain settings, whereas the shale was deposited in low-energy settings of flood plains and inter-distributary bay/marshes [56].

5.1.2. LF-2 Interpretation

The well-developed channels with erosive and pebbly bases, load structures, large cross-bed sets (trough and planar), and coarse grain size in LF2 indicate deposition during high-energy tractional flows [55,57]. The migration of three-dimensional large bedforms results in trough cross bedding, which commonly occurs in the channel belt facies in upper and middle delta plain areas [52,58,59]. Hence, channel morphology, sedimentary structures, and coarse grain size help interpret its deposition in a delta plain or fluvial channel environment during active delta progradation. Deposition of the sandstone took place during high fluvial discharge. The associated carbonaceous shale/clay shows deposition along the channel levee/margin [60].

5.1.3. LF-3 Interpretation

This thin-bedded lithofacies has developed over the thick-bedded, channelized LF2 lithofacies. The presence of abundant ripples shows deposition in a coastal environment [61–63]. The lithofacies have flaser and lenticular bedding, reflecting fluctuating depositional conditions [64]. The rippled sands and mud drapes are associated with the current activity of sand-dominated tidal flats, and as the abundance of mud decreases, the bedding changes from lenticular to flaser type [65]. Moreover, flaser and lenticular bedding are common in tidal flats and in the delta front environment, where fluctuation in sediment supply is common [66]. It may also be interpreted to represent deposition in a tidal setting, i.e., tidal flats and tidal channels [67]. Its position above the fluvial lithofacies (LF2) and below the dolomitic lithofacies (LF4) shows the onset of marine conditions. Hence, it is concluded that this lithofacies was deposited in the tidal flat setting.

5.1.4. LF-4 Interpretation

The presence of dolomite marks the onset of marine conditions. Lithology grades from pure fluvial sand through sandy dolomite to pure dolomite and further up into the thick dolomite of the Kingriali Formation. This change in lithology marks the change in environmental conditions, i.e., the dominant siliciclastic system of the Mid-Triassic Tredian Formation was finally changed into the carbonate system of the Upper Triassic Kingriali Dolomite [68,69]. However, there is no obvious clue of any tectonic activity for this change of the system. The transgressive sequence of this lithofacies marks the delta abandonment, as also documented in the Rhone and Niger deltas [51,70], as well as the Mississippi Delta [71]. The process of transgression takes place due to delta switching after the abandonment of the delta, with components of reworking of shoreface retreat and submergence [72]. The presence of ripple marks and dolomite indicates fluctuating depositional conditions and suggests shallow beach or near-shore depositional settings due to reworked sands of retreating shorefaces [49,64].

*5.2. Depositional Settings*

The dominant characteristics like facies association, composition, and distribution of the Tredian sandstone suggest its deposition in a fluvio-deltaic environment (fluvial-dominated delta) (Figure 9). The siliciclastic platform was in regression during the Middle Triassic. The Lower Landa Member of the Tredian Formation shows deposition of sandstone in distributary channels or inter-distributary bays with associated shales in a low-energy setting, such as a floodplain or marshes (Figure 9). It is followed by the deposition of thick channelized sandstone in the Khatkiara Member with a dominant fluvial character. The thick sandstone of the Khatkiara Member is the result of high-density channel flow showing a fining upward sequence. This period is marked by maximum regression, during which the depositional environment changed from deltaic to fluvio-deltaic. Directional features, i.e., cross-beds, indicate channel flow from the southeast. The presence of carbonaceous clays indicates deposition in an overbank area (swamps and marshes). The top of the channelized sandstone shows a transition from thick channelized deposits to thin tide-dominated deposits, i.e., rhythmites (Figure 9). The top of the Khatkiara Member contains a few dolomite beds indicating tidal environments, followed up by the thick dolomite deposits of the Kingriali Formation. This shows that the deposition of Mid-Triassic rocks took place at a time marked by a fall in regional base level and more clastic input. Finally, during the Late Triassic, a major rise in the regional base level terminated the siliciclastic system and started depositing carbonates in the overlying Kingriali Formation.

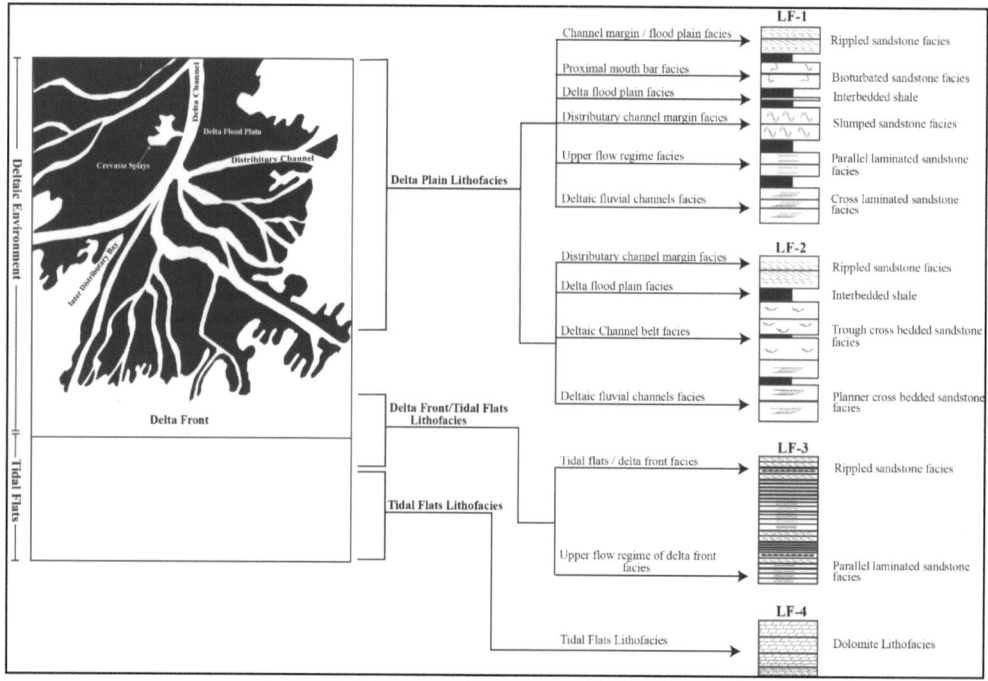

**Figure 9.** Depositional model of Tredian Formation indicative of a fluvio-deltaic environment showing the regional location with its dominant characteristics such as facies association, composition, and distribution of the studied sandstones.

### 5.3. Paragenetic Sequence

The paragenetic sequence of the Tredian Formation was determined based on petrographic and SEM observations. The paragenetic sequence inferred for the Tredian sandstone is shown in Figure 10. The Tredian sandstone has survived intense and complex diagenetic phases (early, late, and uplift). The earliest diagnostic events occurred soon after deposition, and the latest ones are those occurring currently.

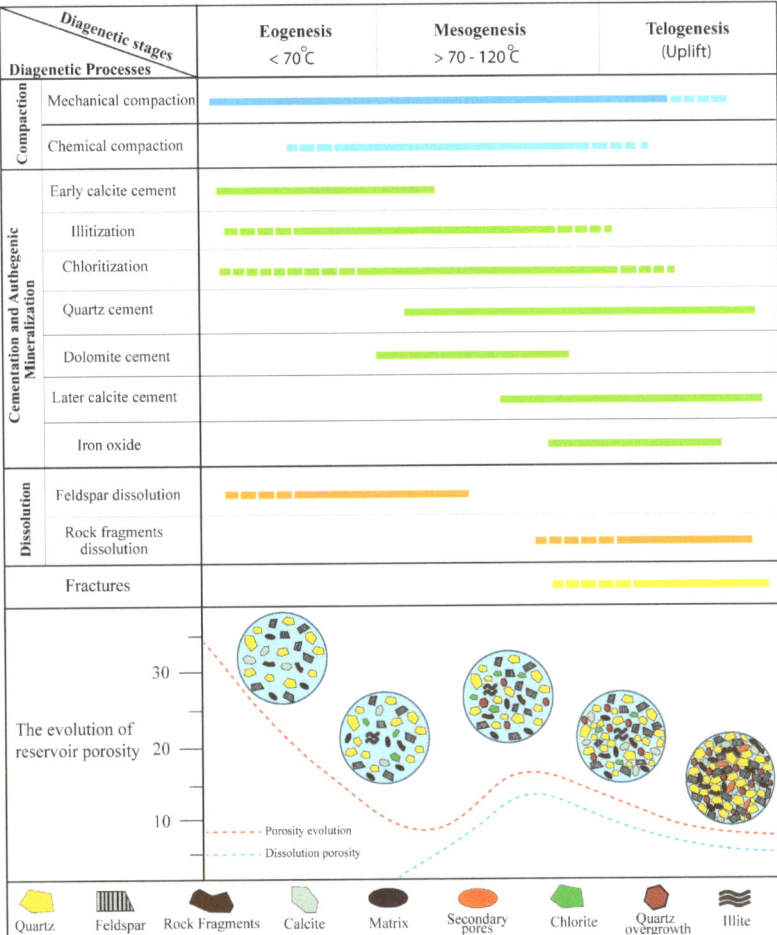

**Figure 10.** Diagenetic model showing the relationship of various diagenetic stages and evolution of porosity with time [73].

The sequence of inferred major diagenetic processes distinguished in the studied sandstone is as follows:

Early mechanical compaction is the first diagnostic event, which is evident from the tight grain packing of sandstone [74] (Figure 7b,f). Mechanical compaction typically occurs from the early-to-late diagenetic stages [75]. However, the early precipitation of massive poikilotopic calcite cement in the Landa Member prevented further compaction (Figure 8a). The rarity of calcite cement made possible the continuation of compaction through later diagenetic stages in the Khatkiara Member. Mechanical compaction is also evident from the breaking down of grains (Figure 7m). After compaction and calcite cementation, dissolution of unstable grains, such as feldspars or alteration of feldspar to clays, possibly due to the influx of meteoric water, is an important event (Figure 8o) [76,77].

Illitization and chloritization constitute the next diagenetic stage and are time- and temperature-dependent [34]. Illite may have been formed by decomposing the feldspar grains or by clay mineral transformation, i.e., smectite and kaolinite illitization. The absence of quartz overgrowth from samples containing chlorite as grain coating indicates the latter to be an earlier diagenetic product (Figure 8g). The relationship between illite and chlorite is not clear. The occurrence of chlorite on quartz grains without overgrowth shows chlorite obstruction in the formation of overgrowth. Quartz overgrowth is late-stage diagenetic cement and usually precipitates on clean surfaces (Figure 8c). The quartz overgrowth is impeded by the precipitation of dolomite. The possible sources of silica may include pressure solution, dissolution of framework grains, feldspar replacement, or transported silica-rich fluids. In addition to the early calcite cementation, another generation of pore-filling calcite cement (formed during burial diagenesis) took place after the quartz and dolomite cementation and led to corrosion of the dolomite cement. Following the precipitation of iron oxide or hydroxide and the dissolution of grains and cements, grain fracturing was the most recent diagenetic event. Grain fracturing postdates late-stage calcite cement as the fractures are not filled and may be related to uplifting (Figure 8o). During the process of uplifting (telogenesis) [6], ferruginous cement precipitated and engulfed other diagenetic products, especially calcite and dolomite cements. The patchy appearance and uneven distribution of iron-oxide cements indicate surface weathering-related infiltration during the telogenetic stage [78]. The secondary porosity has been created in the studied sandstone due to the dissolution of grains as well as authigenic cements.

## 5.4. Reservoir Characterization

Reservoir characterization of sandstone refers to the amount of porosity and permeability it retained after diagenetic processes and is a function of depositional and diagenetic control [79,80]. Burial diagenesis most likely affects the reservoir quality of sandstone. The diagenetic processes, which may either reduce porosity and permeability by physical compaction and authigenic cementation or enhance porosity by the unstable grains and cement dissolution, largely determine the reservoir quality of sandstone [81–83]. The real benefit of secondary porosity for reservoirs is rare as the dissolved material may precipitate as cement in the sandstone [84]. Thin sections stained with blue epoxy were used to estimate the visual porosity of sandstone by point counting along with some SEM images. Although the porosity of examined samples of sandstone is highly affected by compaction, cementation, and authigenic clay mineral formation, it is dominated by inter- and intra-granular pores, dissolution, and fractured pores (secondary porosity). The estimated visual point-count porosity values for the sandstone vary between 0.5 and 16%, averaging 6%. The processes of mechanical compaction and early calcite cementation in the Landa Member have appreciably reduced the primary porosity of the Tredian sandstone (Figure 8a). Due to the rarity of calcite cement, the Khatkiara Member of the Tredian sandstone continued to experience mechanical compaction to greater depths and thus attained closer grain packing (long to concavo-convex) (Figure 8b). The initial porosity reduction was due to mechanical compaction that continued to take place in the samples containing little calcite cement. The inter-granular volume (IGV) of sandstone is decreased by the processes of physical and chemical compaction, as evidenced by the tight grain packing.

The role of chlorite precipitation, either in pore spaces or on grain surfaces, may also contribute to either the reduction or preservation of porosity in the sandstone of the Tredian Formation. The porosity is preserved in the sandstone samples, where quartz overgrowth was prevented due to precipitation of chlorite on grain surfaces (Figure 8g). Chlorite precipitation in the pore spaces reduces permeability, but the rosette form of chlorite possesses inter-crystalline porosity (Figure 11).

The fibrous illite blocks the pore spaces and thus reduces the reservoir properties. Fractures produced due to physical compaction also opened due to dissolution and contributed significantly to the overall porosity of the sandstone [85]. An excessive number of unstable grains have undergone dissolution during the diagenesis of sandstone (Figure 11c). Dissolution of feldspar grains, volcanic fragments, and cements created secondary porosity during the early-to-late stages of diagenesis (Figure 11d). It is observed that the authigenic (calcite) cements partly to completely fill the secondary pores produced as a result of grain dissolution during early diagenesis. Additionally, the late-stage grain and cement dissolution has created and preserved good visual secondary porosity, as observed in the thin sections, and further confirmed through SEM images (Figure 11a–f).

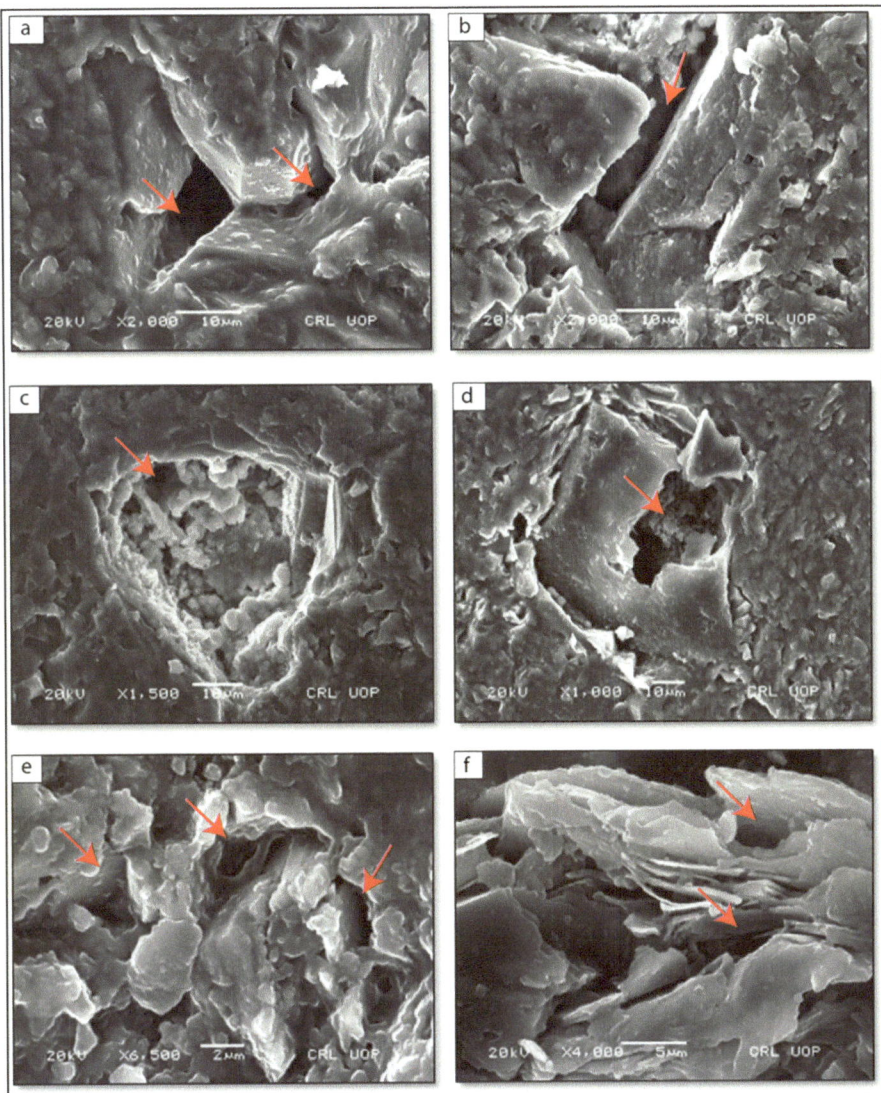

**Figure 11.** SEM images showing different types of porosity (Red arrow). (**a**) SEM image showing inter-granular porosity between the framework grains in tight packing (arrow indicate porosity). (**b**) SEM image showing porosity between the framework grains. (**c**) SEM image showing dissolution porosity filled with chlorite. (**d**) SEM image showing secondary porosity generated as a result of partial dissolution of carbonate grain. (**e**) SEM image showing secondary porosity between the framework grains (Red arrow). (**f**) SEM image showing inter-crystalline porosity between the rosette chlorite (Red arrow).

According to Khan et al. [86], the distribution of attributes such as the thickness of the sandstone body and its sand percentage along with the internal architecture, heterogeneity, sand body geometries, and their constituent facies may also influence the reservoir characteristics. Thicker sandstone strata usually show high porosity values; however, prevailing depositional environments and sediment influx determine the sand body thickness. Table 3 shows the reservoir quality of the Tredian Formation lithofacies.

Table 3. Reservoir quality characteristics of different lithofacies of Tredian Formation.

| Lithofacies | Type of Porosity | Degree of Compaction | Cementing Material and Authigenic Clays | Reservoir Quality |
| --- | --- | --- | --- | --- |
| Lithofacies 1 (LF-1) | No visual porosity because all the intergranular pores are filled with cement | Moderate to tight compaction of grains | Poikilotopic cement is filled in pore spaces, quartz overgrowth is not seen | Poor reservoir |
| Lithofacies 2 (LF-2) | Intergranular and dissolution porosity | Moderately packed grains | Pores are filled with late-stage ferruginous cement. Chlorite is seen as grain coating. Illite is also present in pore spaces | Good reservoir |
| Lithofacies 3 (LF-3) | Secondary fracture and dissolution porosity is observed | Loose–moderate packing of grains | Well-developed quartz overgrowth. Clay rims around few grains | Good reservoir |
| Lithofacies 4 (LF-4) | Fractures or any other dissolution cavities are not observed | Highly compacted | Calcite and dolomite cement is observed | Poor reservoir |

## 6. Conclusions

The studied Tredian Formation has been sub-divided into sandstone interbedded with shale lithofacies (LF1), thick-bedded sandstone lithofacies (LF2), thin-bedded sandstone lithofacies (LF3), and dolomite lithofacies (LF4). The distinguished facies associations were interpreted as fluvio-deltaic and tidal flat settings. The characteristic four lithofacies show deposition in delta plain, fluvial channels, and tidal flat environments. The Tredian Formation in the Salt and Trans-Indus Surghar Ranges was the result of overall regression (progradation) during the Mid-Triassic. The falling sea level has supplied the required abundant sandy input. The environment of deposition for the study of Tredian Formation is a fluvial dominated delta. The characteristics depositional processes are both fluvial and marine. The sandstone of the Tredian Formation has been classified as sub-feldspathic arenite to feldspathic arenite based on petrography. Petrographic observations show that the Tredian sandstone has experienced all phases of diagenesis, i.e., shallow to deep burial followed by uplifting. These diagenetic changes have affected the reservoir properties of the Tredian Formation. The major diagenetic signatures observed in the Tredian sandstone are chemical and mechanical compaction, authigenic mineralization, cementation (calcite, silica, dolomite, and ferruginous), replacement, grain fracturing and dissolution. Early cementation has destroyed all the visible porosity in the lithofacies LF1 by enclosing all the grains. This lithofacies entirely consist of the Landa Member. No further clues of later dissolution were observed in samples from this lithofacies. Lithofacies LF2 and LF3, part of the Khatkiara Member, display good visible porosity, which is also confirmed by SEM analysis. The presence of a chlorite rim around some grains served to preserve porosity by inhibiting quartz overgrowth; however, quartz cementation and overgrowth are present in some samples. Moreover, the processes of dissolution and grain fracturing (later in diagenesis) have created enough secondary porosity and made the sandstone a suitable hydrocarbon reservoir. Lithofacies LF4 is dolomitic facies containing some floating framework grains but it shows no visible porosity and suggests the start of marine conditions after regression. Finally, ferruginous cementation during the last stage of diagenesis has destroyed the reservoir properties of the Tredian sandstone in specific zones; however, dissolution of framework grains, fractures, and cements (calcite, silica, dolomite, and ferruginous) have created secondary porosity.

**Author Contributions:** Conceptualization, K.A.Q. and M.A.; methodology, K.A.Q., M.A. and A.B.; software, K.A.Q., M.A., H.T.J. and G.K.; validation, K.A.Q., M.A., H.T.J., G.K. and S.A.; formal analysis, K.A.Q., M.A. and S.A.; investigation, K.A.Q., M.A. and G.K.; resources, M.A. and S.A.; data curation, K.A.Q., M.A. and G.K.; writing—original draft preparation, K.A.Q. and M.A.; writing—review and editing, H.T.J. and G.K.; visualization, M.A.; supervision, M.A.; project administration, M.A.; funding acquisition, H.T.J. and G.K. All authors have read and agreed to the published version of the manuscript.

**Funding:** This research was funded by the Higher Education Commission (HEC) of Pakistan under Access to Scientific Instrumentation Programme (ASIP) (HEC Award Letter No: 20-2(3)/HDIP, Ibd/ASIP/R&D/HEC/2016/630).

**Institutional Review Board Statement:** Not applicable.

**Informed Consent Statement:** Not applicable.

**Data Availability Statement:** The authors confirm that the data supporting the findings of this study are available within the article.

**Acknowledgments:** K.A.Q. is extremely thankful to Nowrad Ali (University of Peshawar) for his help during fieldwork. We are thankful to the Department of Earth Sciences, COMSATS Abbottabad for providing access to Lab instrumentation.

**Conflicts of Interest:** The authors declare no conflict of interest.

# References

1. Miall, A.D. Stratigraphic sequences and their chronostratigraphic correlation. *J. Sediment. Res.* **1991**, *61*, 497–505.
2. Bjørlykke, K. Relationships between depositional environments, burial history, and rock properties. Some principal aspects of diagenetic process in sedimentary basins. *Sediment. Geol.* **2014**, *301*, 1–14. [CrossRef]
3. Dickinson, W.R. Interpreting provenance relations from detrital modes of sandstones. In *Provenance of Arenites*; D. Reidel Publishing Company: Dordrecht, The Netherlands, 1985; pp. 333–361.
4. Dixon, S.A.; Summers, D.M.; Surdam, R.C. Diagenesis and preservation of porosity in Norphlet Formation (Upper Jurassic), southern Alabama. *AAPG Bull.* **1989**, *73*, 707–728.
5. Taylor, T.R.; Giles, M.R.; Hathon, L.A.; Diggs, T.N.; Braunsdorf, N.R.; Birbiglia, G.V.; Kittridge, M.G.; Macaulay, C.I.; Espejo, I.S. Sandstone diagenesis and reservoir quality prediction: Models, myths, and reality. *AAPG Bull.* **2010**, *94*, 1093–1132. [CrossRef]
6. Khan, S.D.; Chen, L.; Ahmad, S.; Ahmad, I.; Ali, F. Lateral structural variation along the Kalabagh Fault Zone, NW Himalayan foreland fold-and-thrust belt, Pakistan. *J. Asian Earth Sci.* **2012**, *50*, 79–87. [CrossRef]
7. Shah, S. *Stratigraphy of Pakistan (Memoirs of the Geological Survey of Pakistan)*; Geological Survey of Pakistan: Quetta, Pakistan, 2009; Volume 22.
8. Bilal, A.; Yang, R.; Janjuhah, H.T.; Mughal, M.S.; Li, Y.; Kontakiotis, G.; Lenhardt, N. Microfacies analysis of the Palaeocene Lockhart limestone on the eastern margin of the Upper Indus Basin (Pakistan): Implications for the depositional environment and reservoir characteristics. *Depos. Rec.* **2023**, *9*, 152–173. [CrossRef]
9. Bilal, A.; Mughal, M.S.; Janjuhah, H.T.; Ali, J.; Niaz, A.; Kontakiotis, G.; Antonarakou, A.; Usman, M.; Hussain, S.A.; Yang, R. Petrography and Provenance of the Sub-Himalayan Kuldana Formation: Implications for Tectonic Setting and Palaeoclimatic Conditions. *Minerals* **2022**, *12*, 794. [CrossRef]
10. Qureshi, K.A.; Hussain, H.; Shah, A.A.; Meerani, I.A.; Fahad, S.; Basit, A. Hydrocarbon Source and Reservoir Rock Potential of the Paleocene Hangu Formation in the Himalayan Foreland Basin, North West Pakistan: Insight from Geochemical and Diagenetic Study: Geochemical and Diagenetic Study of Hangu Formation. *Pak. J. Sci. Ind. Res. Ser. A Phys. Sci.* **2019**, *62*, 157–166. [CrossRef]
11. Ishaq, K.; Wahid, S.; Yaseen, M.; Hanif, M.; Ali, S.; Ahmad, J.; Mehmood, M. Analysis of subsurface structural trend and stratigraphic architecture using 2D seismic data: A case study from Bannu Basin, Pakistan. *J. Pet. Explor. Prod.* **2021**, *11*, 1019–1036. [CrossRef]
12. Khan, M.Z.; Khan, M.R.; Raza, A. Reservoir Potential of Marwat and Khisor Trans Indus Ranges, Northwest Pakistan. In Proceedings of the AAPG International Conference & Exhibition, Istanbul, Turkey, 14–17 September 2014.
13. Shahtakhtinskiy, A.; Khan, S. Quantitative analysis of facies variation using ground-based lidar and hyperspectral imaging in Mississippian limestone outcrop near Jane, Missouri. *Interpretation* **2020**, *8*, T365–T378. [CrossRef]
14. Shahtakhtinskiy, A.; Khan, S. 3D stratigraphic mapping and reservoir architecture of the Balakhany Suite, Upper Productive Series, using UAV photogrammetry: Yasamal Valley, Azerbaijan. *Mar. Pet. Geol.* **2022**, *145*, 105911. [CrossRef]
15. Morad, S.; Al-Ramadan, K.; Ketzer, J.M.; De Ros, L. The impact of diagenesis on the heterogeneity of sandstone reservoirs: A review of the role of depositional facies and sequence stratigraphy. *AAPG Bull.* **2010**, *94*, 1267–1309. [CrossRef]
16. Salah, M.K.; Janjuhah, H.; Sanjuan, J.; Maalouf, E. Impact of diagenesis and pore aspects on the petrophysical and elastic properties of carbonate rocks from southern Lebanon. *Bull. Eng. Geol. Environ.* **2023**, *82*, 67. [CrossRef]
17. Sitholey, R. Plant remains from the Triassic of the Salt Range, Punjab. *Proc. Nat. Acad. Sci. India* **1943**, *13*, 300–327.
18. Sarjeant, W. Acritarchs and Tasmanitids from the Mianwali and Tredian Formations (Triassic) of the Salt and Surghar Ranges, West Pakistan. 1973. Available online: https://eurekamag.com/research/037/751/037751535.php (accessed on 1 May 2023).

19. Hermann, E.; Hochuli, P.A.; Bucher, H.; Roohi, G. Uppermost Permian to Middle Triassic palynology of the Salt Range and Surghar Range, Pakistan. *Rev. Palaeobot. Palynol.* **2012**, *169*, 61–95. [CrossRef]
20. Khan, S.; Ahmad, W.; Ahmad, S.; Khan, J.K. Dating and depositional environment of the Tredian Formation, western Salt Range, Pakistan. *J. Himal. Earth Sci.* **2016**, *49*, 14–25.
21. Treloar, P.J.; Izatt, C.N. Tectonics of the Himalayan collision between the Indian plate and the Afghan block: A synthesis. *Geol. Soc. Lond. Spec. Publ.* **1993**, *74*, 69–87. [CrossRef]
22. Yin, A. Cenozoic tectonic evolution of the Himalayan orogen as constrained by along-strike variation of structural geometry, exhumation history, and foreland sedimentation. *Earth-Sci. Rev.* **2006**, *76*, 1–131. [CrossRef]
23. Ali, S.K.; Lashari, R.A.; Sahito, A.G.; Kontakiotis, G.; Janjuhah, H.T.; Mughal, M.S.; Bilal, A.; Mehmood, T.; Majeed, K.U. Sedimentological and Petrographical Characterization of the Cambrian Abbottabad Formation in Kamsar Section, Muzaffarabad Area: Implications for Proto-Tethys Ocean Evolution. *J. Mar. Sci. Eng.* **2023**, *11*, 526. [CrossRef]
24. Islam, F.; Ahmad, M.N.; Janjuhah, H.T.; Ullah, M.; Islam, I.U.; Kontakiotis, G.; Skilodimou, H.D.; Bathrellos, G.D. Modelling and Mapping of Soil Erosion Susceptibility of Murree, Sub-Himalayas Using GIS and RS-Based Models. *Appl. Sci.* **2022**, *12*, 12211. [CrossRef]
25. Kazmi, A.; Qasimjan, M. *Geology and Tectonics of Pakistan*; Sl Graphic Publishes: Karachi, Pakistan, 1997.
26. Jaumé, S.C.; Lillie, R.J. Mechanics of the Salt Range-Potwar Plateau, Pakistan: A fold-and-thrust belt underlain by evaporites. *Tectonics* **1988**, *7*, 57–71. [CrossRef]
27. McDougall, J.W.; Khan, S.H. Strike-slip faulting in a foreland fold-thrust belt: The Kalabagh Fault and Western Salt Range, Pakistan. *Tectonics* **1990**, *9*, 1061–1075. [CrossRef]
28. Bilal, A.; Yang, R.; Mughal, M.S.; Janjuhah, H.T.; Zaheer, M.; Kontakiotis, G. Sedimentology and Diagenesis of the Early–Middle Eocene Carbonate Deposits of the Ceno-Tethys Ocean. *J. Mar. Sci. Eng.* **2022**, *10*, 1794. [CrossRef]
29. Zaheer, M.; Khan, M.R.; Mughal, M.S.; Janjuhah, H.T.; Makri, P.; Kontakiotis, G. Petrography and Lithofacies of the Siwalik Group in the Core of Hazara-Kashmir Syntaxis: Implications for Middle Stage Himalayan Orogeny and Paleoclimatic Conditions. *Minerals* **2022**, *12*, 1055. [CrossRef]
30. Gee, E.; Gee, D. Overview of the geology and structure of the Salt Range, with observations on related areas of northern Pakistan. *Geol. Soc. Am. Spec. Pap.* **1989**, *232*, 95–112.
31. Ahmad, S.; Ali, A.; Khan, M.I. Imprints of transtensional deformation along Kalabagh fault in the vicinity of Kalabagh Hills, Pakistan. *Pak. J. Hydrocarb. Res.* **2005**, *15*, 35–42.
32. Powell, C.M. *A Speculative Tectonic History of Pakistan and Surroundings*; Geodynamics of Pakistan: Quetta, Pakistan, 1979; pp. 5–24.
33. Ahmad, I.; Shah, M.M.; Janjuhah, H.T.; Trave, A.; Antonarakou, A.; Kontakiotis, G. Multiphase Diagenetic Processes and Their Impact on Reservoir Character of the Late Triassic (Rhaetian) Kingriali Formation, Upper Indus Basin, Pakistan. *Minerals* **2022**, *12*, 1049. [CrossRef]
34. Humphreys, B.; Smith, S.; Strong, G. Authigenic chlorite in late Triassic sandstones from the Central Graben, North Sea. *Clay Miner.* **1989**, *24*, 427–444. [CrossRef]
35. Golonka, J.; Ford, D. Pangean (late Carboniferous–Middle Jurassic) paleoenvironment and lithofacies. *Palaeogeogr. Palaeoclimatol. Palaeoecol.* **2000**, *161*, 1–34. [CrossRef]
36. Basit, A.; Umar, M.; Jamil, M.; Qasim, M. Facies analysis and depositional framework of Late Permian-Jurassic sedimentary successions, Western Salt Range, Pakistan: Implications for sequence stratigraphic trends and paleogeography of the Neo-Tethys Sea. *Kuwait J. Sci.* **2023**, *50*, 1–12. [CrossRef]
37. Ali, S.K.; Janjuhah, H.T.; Shahzad, S.M.; Kontakiotis, G.; Saleem, M.H.; Khan, U.; Zarkogiannis, S.D.; Makri, P.; Antonarakou, A. Depositional Sedimentary Facies, Stratigraphic Control, Paleoecological Constraints, and Paleogeographic Reconstruction of Late Permian Chhidru Formation (Western Salt Range, Pakistan). *J. Mar. Sci. Eng.* **2021**, *9*, 1372. [CrossRef]
38. Balme, B.E. *Palynology of Permian and Triassic Strata in the Salt Range and Surghar Range, West Pakistan*; University Press of Kansas: Lawrence, KS, USA, 1970; Volume 4.
39. Janjuhah, H.T.; Alansari, A.; Santha, P.R. Interrelationship between facies association, diagenetic alteration and reservoir properties evolution in the Middle Miocene carbonate build up, Central Luconia, Offshore Sarawak, Malaysia. *Arab. J. Sci. Eng.* **2019**, *44*, 341–356. [CrossRef]
40. Garzanti, E.; Vezzoli, G.; Andò, S.; Paparella, P.; Clift, P.D. Petrology of Indus River sands: A key to interpret erosion history of the Western Himalayan Syntaxis. *Earth Planet. Sci. Lett.* **2005**, *229*, 287–302. [CrossRef]
41. Pettijohn, F. *Sedimentary Rocks*, 3rd ed.; Herper & Row: New York, NY, USA, 1975; Volume 614.
42. Morad, S. Carbonate cementation in sandstones: Distribution patterns and geochemical evolution. In *Carbonate Cementation in Sandstones: Distribution Patterns and Geochemical Evolution*; Wiley Publisher: Hoboken, NJ, USA, 1998; pp. 1–26.
43. Burton, J.; Krinsley, D.; Pye, K. Authigenesis of kaolinite and chlorite in Texas Gulf Coast sediments. *Clays Clay Miner.* **1987**, *35*, 291–296. [CrossRef]
44. Hayes, J.B. Polytypism of chlorite in sedimentary rocks. *Clays Clay Miner.* **1970**, *18*, 285–306. [CrossRef]
45. Warren, E.; Curtis, C. The chemical composition of authigenic illite within two sandstone reservoirs as analysed by ATEM. *Clay Miner.* **1989**, *24*, 137–156. [CrossRef]

46. Coleman, J.M.; Gagliano, S.M. Cyclic Sedimentation in the Mississippi River Deltaic Plain. 1964. Available online: https://www.semanticscholar.org/paper/Cyclic-sedimentation-in-the-Mississippi-River-Gulf-Coleman-Gagliano/ce137f9553108982f02f6680db1e11c40a47c19c (accessed on 1 May 2023).
47. Olariu, C.; Bhattacharya, J.P. Terminal distributary channels and delta front architecture of river-dominated delta systems. *J. Sediment. Res.* **2006**, *76*, 212–233. [CrossRef]
48. Kästner, K.; Hoitink, A.; Vermeulen, B.; Geertsema, T.; Ningsih, N. Distributary channels in the fluvial to tidal transition zone. *J. Geophys. Res. Earth Surf.* **2017**, *122*, 696–710. [CrossRef]
49. Reading, H.; Levell, B. Controls on the sedimentary rock record. *Sediment. Environ. Process. Facies Stratigr.* **1996**, *3*, 5–36.
50. Oomkens, E. *Depositional Sequences and Sand Distribution in the Postglacial Rhone Delta Complex*; SEPM Sociert of Sedmentart Geology: Broken Arrow, OK, USA, 1970; Volume 15.
51. Oomkens, E. Lithofacies relations in the Late Quaternary Niger delta complex. *Sedimentology* **1974**, *21*, 195–222. [CrossRef]
52. Coleman, J. *Deltas: Processes and Models of Deposition for Exploration*; Burgess CEPCO Division: Minneapolis, MN, USA; Springer: Heidelberg, The Neathearlands, 1981; Volume 124.
53. Swift, D.J.; Figueiredo, A.G.; Freeland, G.; Oertel, G. Hummocky cross-stratification and megaripples; a geological double standard? *J. Sediment. Res.* **1983**, *53*, 1295–1317.
54. Boggs, S., Jr.; Boggs, S. *Petrology of Sedimentary Rocks*; Cambridge University Press: Cambridge, UK, 2009.
55. Eriksson, P.G.; Condie, K.C.; Tirsgaard, H.; Mueller, W.; Altermann, W.; Miall, A.D.; Aspler, L.B.; Catuneanu, O.; Chiarenzelli, J.R. Precambrian clastic sedimentation systems. *Sediment. Geol.* **1998**, *120*, 5–53. [CrossRef]
56. Nichols, G. SELLEY, RC 2000. Applied Sedimentology, x+ 523 pp. San Diego, San Francisco, New York, Boston, London, Sydney, Tokyo: Academic Press. Price US $82.50 (hard covers). ISBN 0 12 636375 7. *Geol. Mag.* **2001**, *138*, 619–630. [CrossRef]
57. Mehmood, M.; Naseem, A.A.; Saleem, M.; Rehman, J.; Kontakiotis, G.; Janjuhah, H.T.; Khan, E.; Antonarakou, A.; Khan, I.; Rehman, A.; et al. Sedimentary Facies, Architectural Elements and Depositional Environments of the Maastrichtian Pab Formation in the Rakhi Gorge, Eastern Sulaiman Ranges, Pakistan. *J. Mar. Sci. Eng.* **2023**, *11*, 726. [CrossRef]
58. Allison, M.A.; Khan, S.; Goodbred, S.L., Jr.; Kuehl, S.A. Stratigraphic evolution of the late Holocene Ganges–Brahmaputra lower delta plain. *Sediment. Geol.* **2003**, *155*, 317–342. [CrossRef]
59. Scasso, R.; Aberhan, M.; Ruiz, L.; Weidemeyer, S.; Medina, F.; Kiessling, W. Integrated bio-and lithofacies analysis of coarse-grained, tide-dominated deltaic environments across the Cretaceous/Paleogene boundary in Patagonia, Argentina. *Cretac. Res.* **2012**, *36*, 37–57. [CrossRef]
60. Aitken, J.F.; Flint, S.S. *High-Frequency Sequences and the Nature of Incised-Valley Fills in Fluvial Systems of the Breathitt Group (Pennsylvanian), Appalachian Foreland Basin, Eastern Kentucky*; Special Publication No. 51; Society of Economic Paleontologists and Mineralogists: Tulsa, OK, USA, 1994; pp. 353–368.
61. Reineck, H.; Singh, I. Der Golf von Gaeta (Tyrrhenisches Meer); III, Die Gefuege von Vorstrand-und Shelfsedimenten. *Senckenberg. Marit.* **1971**, *3*, 185–194.
62. Wunderlich, F. Georgia Coastal Region, Sapelo Island, USA: Sedimentology and Biology. III. Beach Dynamics and Beach Development. 1972. Available online: https://pascal-francis.inist.fr/vibad/index.php?action=getRecordDetail&idt=PASCALGEODEBRGM732262573 (accessed on 1 May 2023).
63. Desjardins, P.R.; Buatois, L.A.; Limarino, C.O.; Cisterna, G.A. Latest Carboniferous–earliest Permian transgressive deposits in the Paganzo Basin of western Argentina: Lithofacies and sequence stratigraphy of a coastal-plain to bay succession. *J. S. Am. Earth Sci.* **2009**, *28*, 40–53. [CrossRef]
64. Boggs, S., Jr. *Principal of Sedimentology and Stratigraphy 4th Edition, Hal 550–553*; Pearson Prentice Hall: Upper Saddle River, NJ, USA, 2006.
65. Reineck, H.E.; Wunderlich, F. Classification and origin of flaser and lenticular bedding. *Sedimentology* **1968**, *11*, 99–104. [CrossRef]
66. Reineck, H.-E.; Singh, I.B. *Depositional Sedimentary Environments: With Reference to Terrigenous Clastics*; Springer Science & Business Media: Berlin/Heidelberg, Germany, 2012.
67. Terwindt, J. *Lithofacies of Inshore Estuarine and Tidal Inlet Deposits: Geologie & Mijnbouw*; Publicatie uit het Geografisch Instituut der Rijksuniversiteit te Utrecht: Utrecht, The Netherlands, 1971; Volume 50.
68. Hermann, E.; Hochuli, P.A.; Bucher, H.; Vigran, J.O.; Weissert, H.; Bernasconi, S.M. A close-up view of the Permian–Triassic boundary based on expanded organic carbon isotope records from Norway (Trøndelag and Finnmark Platform). *Glob. Planet. Chang.* **2010**, *74*, 156–167. [CrossRef]
69. Rahim, H.-U.; Qamar, S.; Shah, M.M.; Corbella, M.; Martín-Martín, J.D.; Janjuhah, H.T.; Navarro-Ciurana, D.; Lianou, V.; Kontakiotis, G. Processes Associated with Multiphase Dolomitization and Other Related Diagenetic Events in the Jurassic Samana Suk Formation, Himalayan Foreland Basin, NW Pakistan. *Minerals* **2022**, *12*, 1320. [CrossRef]
70. Lagaaij, R.; Kopstein, F. Typical features of a fluviomarine offlap sequence. In *Developments in Sedimentology*; Elsevier: Amsterdam, The Netherlands, 1964; Volume 1, pp. 216–226.
71. Penland, S.; Boyd, R.; Suter, J.R. Transgressive depositional systems of the Mississippi Delta plain; a model for barrier shoreline and shelf sand development. *J. Sediment. Res.* **1988**, *58*, 932–949.
72. Carrión-Torrente, Á.; Lobo, F.J.; Puga-Bernabéu, Á.; Mendes, I.; Lebreiro, S.; García, M.; van Rooij, D.; Luján, M.; Reguera, M.I.; Antón, L. Episodic postglacial deltaic pulses in the Gulf of Cadiz: Implications for the development of a transgressive shelf and driving environmental conditions. *J. Sediment. Res.* **2022**, *92*, 1116–1140. [CrossRef]

73. Dowey, P.J.; Taylor, K.G. Diagenetic mineral development within the Upper Jurassic Haynesville-Bossier Shale, USA. *Sedimentology* **2020**, *67*, 47–77. [CrossRef]
74. Barshep, D.V.; Worden, R.H. Reservoir Quality of Upper Jurassic Corallian Sandstones, Weald Basin, UK. *Geosciences* **2021**, *11*, 446. [CrossRef]
75. Salah, M.K.; Janjuhah, H.T.; Sanjuan, J. Analysis and Characterization of Pore System and Grain Sizes of Carbonate Rocks from Southern Lebanon. *J. Earth Sci.* **2023**, *34*, 101–121. [CrossRef]
76. Wang, Y.; Cheng, H.; Hu, Q.; Liu, L.; Hao, L. Diagenesis and pore evolution for various lithofacies of the Wufeng-Longmaxi shale, southern Sichuan Basin, China. *Mar. Pet. Geol.* **2021**, *133*, 105251. [CrossRef]
77. Fathy, D.; El-Balkiemy, A.F.; Makled, W.A.; Hosny, A.M. Organic geochemical signals of Paleozoic rocks in the southern Tethys, Siwa basin, Egypt: Implications for source rock characterization and petroleum system. *Phys. Chem. Earth Parts A/B/C* **2023**, *130*, 103393. [CrossRef]
78. Fathy, D.; Abart, R.; Wagreich, M.; Gier, S.; Ahmed, M.S.; Sami, M. Late campanian climatic-continental weathering assessment and its influence on source rocks deposition in southern Tethys, Egypt. *Minerals* **2023**, *13*, 160. [CrossRef]
79. Yang, Y. Reservoir characteristics and controlling factors of the Lower paleogene sandstones in the southeast part of Jiyang Sag, BohaiBay Basin, China. *Alex. Eng. J.* **2022**, *61*, 10277–10282. [CrossRef]
80. Jafarzadeh, N.; Kadkhodaie, A.; Bahrehvar, M.; Wood, D.A.; Janahmad, B. Reservoir characterization of fluvio-deltaic sandstone packages in the framework of depositional environment and diagenesis, the south Caspian Sea basin. *J. Asian Earth Sci.* **2022**, *224*, 105028. [CrossRef]
81. Chen, J.; Yao, J.; Mao, Z.; Li, Q.; Luo, A.; Deng, X.; Shao, X. Sedimentary and diagenetic controls on reservoir quality of low-porosity and low-permeability sandstone reservoirs in Chang101, upper Triassic Yanchang Formation in the Shanbei area, Ordos Basin, China. *Mar. Pet. Geol.* **2019**, *105*, 204–221. [CrossRef]
82. Burley, S.; Kantorowicz, J. Thin section and SEM textural criteria for the recognition of cement-dissolution porosity in sandstones. *Sedimentology* **1986**, *33*, 587–604. [CrossRef]
83. Janjuhah, H.T.; Kontakiotis, G.; Wahid, A.; Khan, D.M.; Zarkogiannis, S.D.; Antonarakou, A. Integrated Porosity Classification and Quantification Scheme for Enhanced Carbonate Reservoir Quality: Implications from the Miocene Malaysian Carbonates. *J. Mar. Sci. Eng.* **2021**, *9*, 1410. [CrossRef]
84. Gordon, J.B.; Sanei, H.; Pedersen, P.K. Secondary porosity development in incised valley sandstones from two wells from the Flemish Pass area, offshore Newfoundland. *Mar. Pet. Geol.* **2022**, *140*, 105644. [CrossRef]
85. Marghani, M.M.; Zairi, M.; Radwan, A.E. Facies analysis, diagenesis, and petrophysical controls on the reservoir quality of the low porosity fluvial sandstone of the Nubian formation, east Sirt Basin, Libya: Insights into the role of fractures in fluid migration, fluid flow, and enhancing the permeability of low porous reservoirs. *Mar. Pet. Geol.* **2023**, *147*, 105986.
86. Khan, A.; Kelling, G.; Umar, M.; Kassi, A. Depositional environments and reservoir assessment of Late Cretaceous sandstones in the south central Kirthar foldbelt, Pakistan. *J. Pet. Geol.* **2002**, *25*, 373–406. [CrossRef]

**Disclaimer/Publisher's Note:** The statements, opinions and data contained in all publications are solely those of the individual author(s) and contributor(s) and not of MDPI and/or the editor(s). MDPI and/or the editor(s) disclaim responsibility for any injury to people or property resulting from any ideas, methods, instructions or products referred to in the content.

*Article*

# Pb-210 Dating of Ice Scour in the Kara Sea

Osip Kokin [1,2,*], Irina Usyagina [3], Nikita Meshcheriakov [1,3], Roman Ananiev [4], Vasiliy Arkhipov [1,2], Aino Kirillova [1], Stepan Maznev [1,2], Sergey Nikiforov [4] and Nikolay Sorokhtin [4]

1. Geological Institute, Russian Academy of Sciences (GIN RAS), 7 Pyzhevsky per., 119017 Moscow, Russia
2. Faculty of Geography, Lomonosov Moscow State University (MSU), GSP-1, 1 Leninskie Gory, 119991 Moscow, Russia
3. Murmansk Marine Biological Institute, Russian Academy of Sciences (MMBI RAS), 17 Vladimirskaya St., 183010 Murmansk, Russia
4. Shirshov Institute of Oceanology, Russian Academy of Sciences (IO RAS), 36 Nakhimovsky Prospect, 117997 Moscow, Russia
* Correspondence: osip_kokin@mail.ru

**Citation:** Kokin, O.; Usyagina, I.; Meshcheriakov, N.; Ananiev, R.; Arkhipov, V.; Kirillova, A.; Maznev, S.; Nikiforov, S.; Sorokhtin, N. Pb-210 Dating of Ice Scour in the Kara Sea. *J. Mar. Sci. Eng.* **2023**, *11*, 1404. https://doi.org/10.3390/jmse11071404

Academic Editors: George Kontakiotis, Angelos G. Maravelis, Avraam Zelilidis and Antoni Calafat

Received: 24 March 2023
Revised: 30 June 2023
Accepted: 3 July 2023
Published: 12 July 2023

**Copyright:** © 2023 by the authors. Licensee MDPI, Basel, Switzerland. This article is an open access article distributed under the terms and conditions of the Creative Commons Attribution (CC BY) license (https:// creativecommons.org/licenses/by/ 4.0/).

**Abstract:** Ice scours are formed when the keels of floating icebergs or sea ice hummocks penetrate unlithified seabed sediments. Until now, ice scours have been divided into "relict" and "modern" according to the water depth that corresponds with the possible maximum vertical dimensions of the keels of modern floating icebergs. However, this approach does not consider climatic changes at the present sea level, which affect the maximum depth of ice keels. We present an application of $^{210}$Pb dating of the largest ice scour in the Baydaratskaya Bay area (Kara Sea), located at depths of about 28–32 m. Two sediment cores were studied; these were taken on 2 November 2021 from the R/V *Akademik Nikolay Strakhov* directly in the ice scour and on the "background" seabed surface, not processed via ice scouring. According to the results of $^{210}$Pb dating, the studied ice scour was formed no later than the end of the Little Ice Age. Based on the extrapolation of possible sedimentation rates prior to 1917 (0.22–0.38 cm/year), the age of the ice scour is estimated to be 1810 ± 30 AD. The mean rate of ice scour filling with 70 cm thick sediments from the moment of its formation is around 0.33 cm/year.

**Keywords:** ice gouge; iceberg ploughmark; sediment core; bathymetry; seismic profiles; sediment accumulation rate (SAR); specific activity; radionuclides; Baydaratskaya Bay

## 1. Introduction

Ice scours (ice gouges, ice ploughmarks) are widespread on the bottom of high- and mid-latitude seas and big freezing lakes at different water depths up to 500 m [1–12] and occasionally up to 1200 m [13,14]. They are formed when the keels of floating icebergs or sea ice hummocks (pressure ridges) collide with unlithified seabed sediments. The study of the ice scouring process has become especially important in recent times due to climate change. Additionally, these kinds of processes are a natural risk for the development of Arctic shelf oil and gas fields and the Northern Sea Route, adversely affecting the construction and operation of structures and the movement of ships in the Barents–Kara region.

For the purposes of marine engineering, it is important to understand in which areas ice scouring can still occur, and in which areas ice scours on the seabed are relicts. During the construction of submarine structures such as pipelines, it is also important to understand under what conditions (sea level, climate) the depth of penetration of ice keels into the bottom sediments is observed. To answer these questions, the dating of ice scours, which to our knowledge has not yet been achieved, could help. Only the dating of completely buried iceberg scour relicts (with the age around 31 kyr cal BP) in the subtropical North Atlantic is known [15]. Their age has been estimated based on radiocarbon dating of sediment samples above and below the iceberg scour surface.

To estimate the age of ice scours, the water depth at which they are observed is now the most widely used [3,7,8,12,13]. According to this, ice scours are divided into "relict" (formed at another relative sea level other than the present) and "modern" (formed at the present sea level). Ice scour relicts are located at a water depth significantly exceeding the possible maximum vertical dimensions of the keels of modern floating icebergs. The threshold between modern iceberg scours and iceberg scour relicts varies widely depending on the area and in the range of 40 to 550 m [3,4,7,8]. However, this approach does not take into account climatic changes at the present sea level which affect the maximum depth of ice keels and, accordingly, the depth of the seabed at which the ice scours were formed and located in the corresponding climatic period.

Another attempt to date ice scours is based on the identification of newly formed ice gouges because of a comparison of multitemporal bottom surveys carried out during monitoring [16,17]. This approach has its disadvantages: it requires many years of observations, does not consider the extreme sizes of ice keels and is limited in space.

The method of marine sediment dating using non-equilibrium lead ($^{210}Pb_{ex}$), which is widely used to determine sedimentation rates over the past 100–120 years, could expand the possibilities for estimating the age of ice scours [18]. However, we are not aware of any attempts to date ice scours based on measurements of the specific activity of the natural radionuclide $^{210}Pb$. In this regard, the main objective of the present study is to show the possibility of the well-established $^{210}Pb$ geochronology approach to ice scour dating.

## 2. Study Area

The study area is located in front of the entrance to Baydaratskaya Bay in the southwestern semi-enclosed part of the Kara Sea. The ice scours of the bay are the most studied in the entire Kara Sea [19]. They began to be investigated in 1988 as part of the design of a submarine gas pipeline crossing the bay. The largest ice scour known in this area was chosen for dating using the non-equilibrium lead method.

The studied ice scour is located at depths of about 28–32 m (Figure 1), which are close to the estimates of the maximum keel depths of sea ice features (stamukhi and hummocks) in the Kara Sea—no more than 28 m [20]. The maximum keel depth of modern icebergs in the Kara Sea is estimated to be no more than 180–200 m and the average is about 50 m [21]. There are known facts of iceberg penetration into Baydaratskaya Bay with depths of less than 20 m during periods of minimal ice cover in the Kara Sea in 1932 and 2007 [22].

From the point of view of the geomorphological structure of the bottom, the study area is located on a gently sloping coastal plain which extends to depths of 50–60 m and is bounded from the open sea side by a tectonic scarp [23]. By the beginning of the postglacial transgression, this plain was in subaerial conditions. During transgression, its surface from the scarp crest to the modern coastline was reworked by wave processes. It is assumed that about 10 kyr BP, there was a period of relative sea level stabilization at the modern sea depth of 32–34 m [23].

The Baydaratskaya Bay area is characterized by low average annual temperatures (from −7 to −10 °C) and a relatively short ice-free period (1–4 months a year) [19]. Over the past 50 years, there has been a tendency for the ice-free period to increase [24]. The sea ice of Baydaratskaya Bay consists of fast ice along the coast and drift ice offshore [25]. The maximum fast ice rim position generally varies in the range of sea depths of 10–20 m [26].

The wind and currents are the driving forces of ice drift. During the winter season, southerly and southwesterly winds prevail in the Baydaratskaya Bay area [25]. The currents here are associated with semi-diurnal tides and they are practically reversive and aligned along the axis of the bay. The maximum speed of the tidal current during the tidal cycle is 0.5 m/s, while the measured maximum current speed is 1 m/s [25]. Sea level fluctuations are associated with tides (up to 1.1 m) and storm surges (up to 2 m).

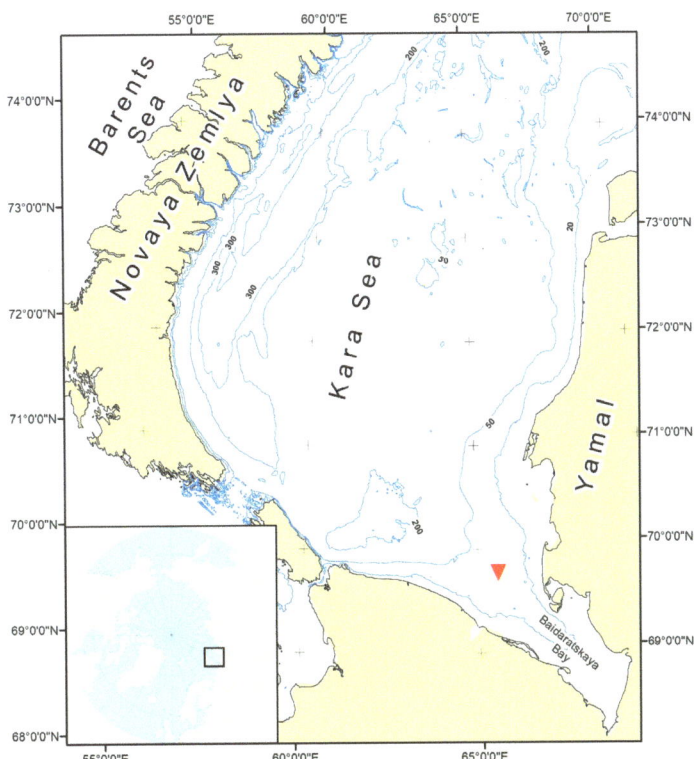

**Figure 1.** The overview map. The red triangle indicates the location of the studied area in the Kara Sea. The 20, 50, 200 and 300 m depths are indicated by the contour lines. The inset map shows the location of the Kara Sea.

Ice scours are widespread in Baydaratskaya Bay to depths of more than 12 m and can be up to 2 m deep, 50 m wide and several kilometers long [5]. In shallower areas, ice scouring processes are also active, but wind waves destroy their evidence during strong summer and autumn storms. The most intensive and deep ice scours occur at the range of depths from 16 to 19 m next to the Yamal coast fast ice rim, where ice hummocks continue during the whole of the cold season [5]. The predominant orientation of the ice scours corresponds to the direction of reverse tidal currents along the axis of the bay.

Sedimentation rates in the Kara Sea are presented mainly as average values for the thickness of the Holocene deposits [27]. More precisely, the sedimentation rate was determined using the method of accelerator mass spectrometry (AMS $^{14}$C) using carbonate shells of mollusks and foraminifers only for a few single horizons of the sedimentary section [28,29].

Sedimentation rates determined by $^{210}$Pb in the Kara Sea are presented in [30–32]. Fourteen cores of sediments up to 50 cm long were studied in key areas of the bottom deposit accumulation in the Kara Sea: the estuaries of the Ob and Yenisei, the East Novaya Zemlya Trench, the Voronin Trench and also in one of the northern bays of the Novaya Zemlya (Sedov Bay) [33]. However, all of them differ from Baydaratskaya Bay in terms of sedimentation conditions. The sedimentation rates were determined by measuring the specific activity of the natural $^{210}$Pb and technogenic $^{137}$Cs radionuclides. A close relationship was shown between the sedimentation rates and types of seabed sediments. The highest sedimentation rates are typical of terrigenous–estuary sediments, which are divided into traction sediments with sedimentation rates of 0.4–0.7 cm/year and sediments of "silt banks" with sedimentation rates of 0.7–1.0 cm/year [33].

## 3. Materials and Methods

The present study is based on the results of the seabed topography and sediment investigations of the Kara Sea by the Shirshov Institute of Oceanology RAS cruise 52 of the R/V *Akademik Nikolai Strakhov* (ANS) [34], which took place from 15 October to 10 November 2021. During the cruise, complex geological and geophysical, geomorphological and hydrophysical studies were carried out. One of the objectives of the cruise was a detailed study of the large ice scours in the Kara Sea.

### 3.1. Multibeam Echo Sounding

Multibeam echo sounding was carried out for the seabed topography mapping to obtain information about the morphology, morphometry and configuration of the ice scour. For this, a multibeam echo sounding system of the R/V *Akademik Nikolaj Strakhov* was used. It consists of a Reson 100 kHz SeaBat 8111, an Applanix POSMV integrating motion sensor and gyrocompass data. The multibeam transducers are installed in a hull-mounted gondola. The swath-bathymetric data were collected and processed using the PDS2000 software from Teledyne. The processed data were gridded at cell sizes of 4×4 m, and visualizations were carried out using PDS2000, ArcGIS and QGIS software. The survey area consisted of a series of single survey lines with a total length of about 35 km along and across the studied ice scour (Figure 2).

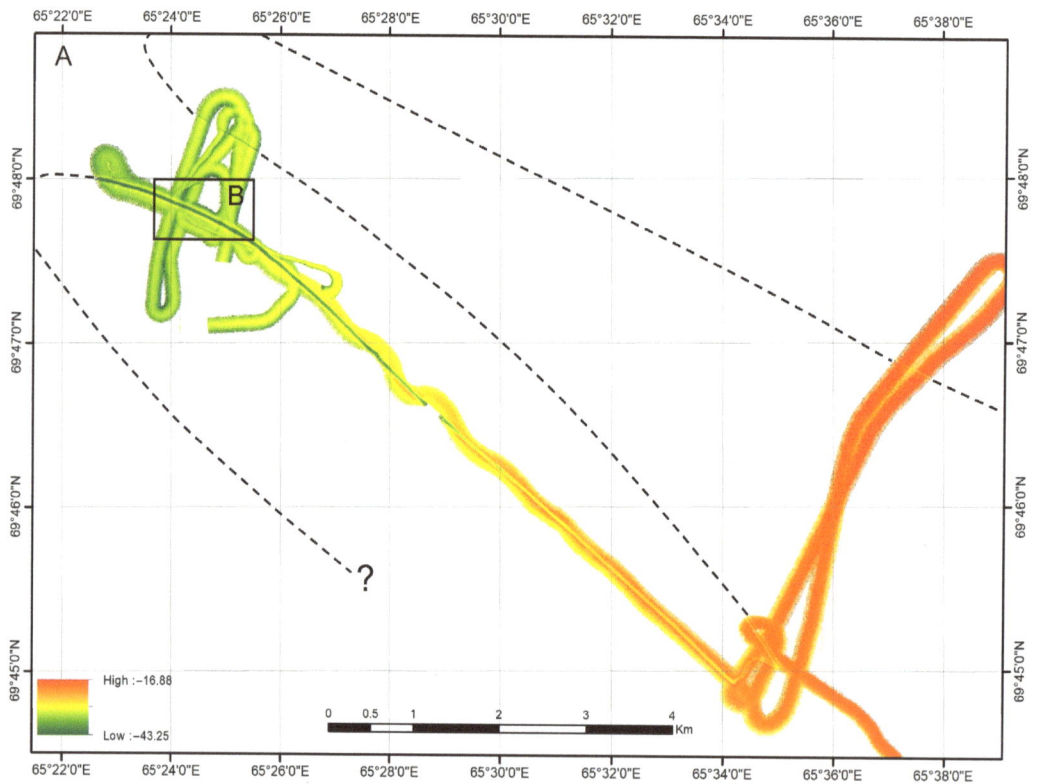

**Figure 2.** *Cont.*

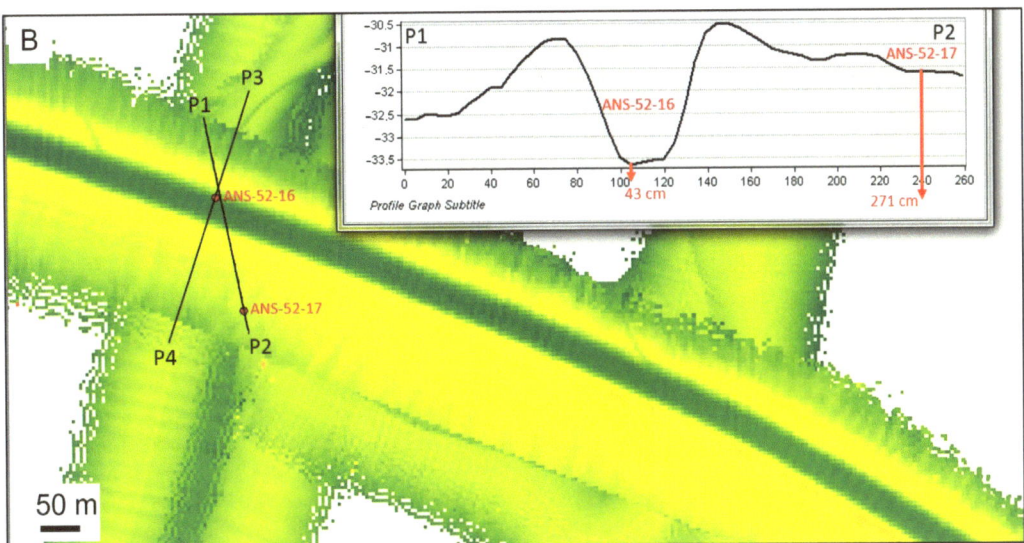

**Figure 2.** (**A**) Seabed topography in the area of the studied ice scour and estimated ice scour configuration outside of the survey area (dashed black line); (**B**) detailed seabed topography near the sampling site with a bathymetric profile P1–P2 drawn through the core sampling points ANS-52-16 ("in a gouge", sea depth 33.6 m) and ANS-52-17 ("background", sea depth 31.6 m). The black lines show the location of the bathymetric profile P1–P2 and the seismic profile P3–P4 (Figure 3). The red arrows on the bathymetric profile P1–P2 show the length of the sampled sediment cores.

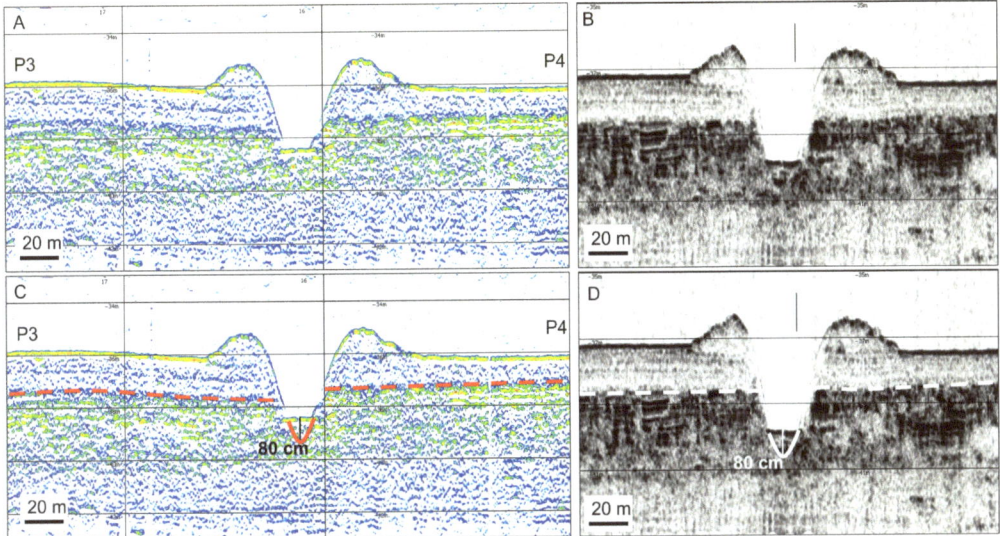

**Figure 3.** The seismic profiles crossing the ice scour in the core sampling area. (**A**) The seismic profile P3–P4 (the location is indicated in Figure 2B); (**B**) the seismic profile 400 m to SE from P3–P4; (**C**) the seismic profile P3–P4 with interpretation; (**D**) the seismic profile 400 m to SE from P3–P4 with interpretation. The solid line (red and white) highlights the deposits that filled the ice scour after the last one was formed, with indication of the measured thickness. The dotted line (red and white) shows the boundary between Late Holocene (above) and Late Pleistocene (below) deposits.

## 3.2. Seismic Profiling

Seismic profiling was carried out to assess the thickness of the deposits filling the ice scour and to obtain information about the structure of the sedimentary strata in which the ice scour was formed. For this, an EdgeTech 3300 high-resolution seismic profiler (sub-bottom chirp) and the seismoacoustic complex Geont-shelf with a sparker source were used. The sub-bottom transducers were installed in a hull-mounted gondola. In addition, the high-resolution parametric sub-bottom profiler SES-2000 Standard with frequencies of 8–10 kHz was installed on a pole on the starboard. The studies were carried out using a single-channel seismic streamer.

Seismic profiling was carried out along all multibeam echo sounding lines with a total length of about 35 km, but in this work, only seismic profiles located near the sampling site of the sediment core were used (Figure 2). In addition, profiles obtained near the sampling site during previous cruises were involved in the analysis.

## 3.3. Sediment Cores

In the present study, two sediment cores (ANS-52-16 and ANS-52-17) were studied (Table 1). They were taken on 2 November 2021 (2021.8 AD) using a 147 mm gravity corer from R/V *Akademik Nikolaj Strakhov* near the entrance of Baydaratskaya Bay in the Kara Sea (Figure 1). The first core (ANS-52-16, length 43 cm) was taken directly in the ice scour at a water depth of 33.6 m; the second one (ANS-52-17, length 271 cm) was taken at a depth of 31.6 m from the "background" seabed surface south of the ice scour, not reworked by ice ploughing (Figure 2). In the present study, the first core is called "ice scour", and the second is called "background". The distance between cores is about 140 m.

**Table 1.** Sediment cores studied in the present study.

| Sediment Core | Position | Sampler | Latitude (°N) | Longitude (°E) | Cruise | Water Depth, m | Length, cm |
|---|---|---|---|---|---|---|---|
| ANS-52-16 | Ice scour | 147 mm gravity corer | 69.797683 | 65.40115 | ANS-52 | 33.6 | 43 |
| ANS-52-17 | Background | 147 mm gravity corer | 69.796467 | 65.40185 | ANS-52 | 31.6 | 271 |

Since coring from a relatively narrow ice scour constitutes a complex task, the results of the core analysis from the "background" seabed surface that had not been reworked by ice ploughing should confirm or disprove that the first core was indeed taken from the ice scour. This approach is based on the fact that the sedimentation conditions within and outside of the ice scour should be different and therefore the results from the analyses should show different patterns in these cores.

The following technique was used to increase the probability of core sampling directly from the ice scour. The vessel approached the ice scour with a predetermined spatial configuration and a preselected sampling site (according to multibeam echo sounding and seismic profiling) at an angle that provided greater distance during the ice scour crossing. During the vessel drift, the gravity corer was hung out at a given horizon (approximately 20 m from the bottom); then, when the vessel was in position at the planned sampling point, on command from the Captain's Bridge, the gravity corer was released on the winch freewheel. At the moment when the gravity corer touched the seabed (determined by the sagging of the cable) the precise coordinates of the sampling point were recorded.

After the cores were recovered onboard the vessel, they were photographed and their lithology was visually described. For further laboratory analysis of the ice scour dating, the cores were packed into plastic cable channels 50 cm long, either completely (43 cm of "ice scour" core) or only the upper part of the core (50 cm of "background" core). Cores packed in cable channels were subjected to quick freezing ($-18\ °C$) onboard the vessel and sent frozen to the laboratory of the Murmansk Marine Biological Institute of the Russian Academy of Sciences.

The sampling for laboratory analysis of only the upper 50 cm of cores is explained by the fact that the expected age of deposits below 50 cm should be significantly more than 100–120 years, which exceeds the dating capabilities of the non-equilibrium lead method.

*3.4. Grain Size Analysis*

Grain size analysis was carried out in the laboratory of Murmansk Marine Biological Institute with wet samples using the wet sieve and decantation methods, which are considered to be the most accurate for fine sediments, as they are based on the hydraulic settling of particles according to the Stokes formula [35]. The accuracy of settling in each stage was controlled by a microscope. The classification of M.V. Klenova was used to interpret the types of bottom sediments based on the results of grain size analysis [36].

The sampled cores were divided into layers of 1–5 cm for detailed studies. A total of 44 samples were analyzed: 20 in the "ice scour" core (0–30 cm) and 22 in the "background" core (0–50 cm). Grain size is given in weight percent of the total sample. Moisture content was determined for each sample together with grain size analysis.

*3.5. Radiometric Measurements $^{210}Pb$, $^{226}Ra$, $^{137}Cs$ and $^{7}Be$*

The measurement of specific activity of $^{210}Pb$, $^{226}Ra$, $^{137}Cs$ and $^{7}Be$ radionuclides was carried out in the laboratory of the Murmansk Marine Biological Institute using a multichannel gamma-ray spectrometer for measuring X-ray and gamma radiation (Canberra Semiconductors NV, Olen, Belgium) with lead screen protection of the HPGe detector Ekran-2P manufactured by Aspect (Dubna, Russia). For the recording part, a BE5030 broadband detector made of ultrapure germanium planar type with a thin "carbon epoxy" entrance window (0.6 mm wide) and a crystal (diameter of 80 mm, an area of 5000 mm$^2$ and a thickness of 30.5 mm) was used. It records gamma quanta with energy from 3 keV to 3 MeV. The energy resolution along the 1332 keV $^{60}Co$ line is no less than 2.2 keV; along the $^{57}Co$ 122 keV isotope line, it is no less than 0.75 keV, and along the $^{55}Fe$ isotope line, it is no less than 0.5 keV. Spectral information was collected using a DSA-1000 pulse analyzer (Canberra Industries, Inc., Loches, France) with a resolution of 16K channels, corresponding to modern digital signal processing technology. Spectra processing took place and radionuclides were identified using Genie-2000 software (version 3.3).

The sampled cores were divided into layers of 1–2 cm for detailed studies. A total of 56 samples were analyzed: 36 in the "ice scour" core (0–43 cm) and 20 in the "background" core (0–25.5 cm). All sediment samples were dried before measurement, homogenized and left for 30 days in hermetically sealed vessels to ensure secular equilibrium between $^{226}Ra$ and $^{222}Rn$, as well as $^{214}Pb$ and $^{214}Bi$ [37,38]. The measurement of each layer was carried out in the same vessels after the onset of equilibrium between the indicated radionuclides. The measurement time was 85,000 s, which gave an error calculation of 5–15% in the upper layers of the cores. The activity of the supported $^{210}Pb$ determined from the main $^{226}Ra$ lines was subtracted to determine the excess $^{210}Pb$ activity ($^{210}Pb_{ex}$) from the total $^{210}Pb$ specific activity measured from its γ-line (46.5 keV). The quantification of $^{226}Ra$ was carried out by $^{214}Pb$ (295.2 keV and 351.9 keV) and $^{214}Bi$ (1120 keV). The $^{210}Pb_{ex}$ activity was corrected for the date of core collection (2021.8 AD) and the results are given on a dry weight basis, corrected for self-absorption and sample geometry [39]. The short-lived radionuclide $^{7}Be$ was measured in the uppermost layer to confirm that the surface layer was sampled correctly.

A quality check of the gamma spectrometer efficiency was carried out regularly with the help of two volumetric activity measures issued for special purposes by the D.I. Mendeleyev Institute for Metrology (VNIIM). The density of the the volumetric activity measure fillers was 1.01 g/cm$^3$. The activity of $^{210}Pb$ deposited on the filler was 1.64 kBq; the activity of $^{137}Cs$ was 1.4 kBq. The specific activity of $^{226}Ra$ was determined via the source of $^{152}Eu$, and the activity of $^{152}Eu$ in the source was 3.78 kBq.

## 3.6. Calculation of the Calendar Age and the Sediment Accumulation Rates Based on the Results of Radiometric Measurements

$^{210}$Pb dating of marine sediments is an accessible and widely used method used to estimate the rates of modern sedimentation in the Arctic seas [31–33,40–44]. The simultaneous determination of natural $^{210}$Pb and $^{226}$Ra and technogenic $^{137}$Cs in a sample became possible after the advent of modern gamma spectrometers for measuring X-ray and gamma radiation and detecting gamma rays with energies from 3 keV to 3 MeV. This significantly reduced the analysis time and made it possible to use the measured material for other studies.

The natural radionuclide $^{210}$Pb has a half-life ($T_{1/2}$) of 22.3 years and is a member of the $^{226}$Ra decay chain ($T_{1/2}$ = 1600 years) that is produced via the successive decay of the parent isotope $^{238}$U. $^{210}$Pb in marine sediments consists of supported $^{210}$Pb, which is continuously formed and, presumably, is in secular equilibrium with its initial radionuclide $^{226}$Ra and excess $^{210}$Pb. The last one enters the surface of water bodies and the surrounding drainage area, and then makes its way into the bottom sediments as a result of atmospheric deposition. The excess $^{210}$Pb is determined by subtracting the supported $^{226}$Ra activity from the total $^{210}$Pb activity and is used to determine the age of sediments in a sedimentation basin. However, compared to other isotopes used for geochronology, such as $^{14}$C ($T_{1/2}$ = 5730 years), excess $^{210}$Pb can only be used to date the age of recent deposits formed within the last 100–120 years [45,46].

Many studies have shown that the analysis of data and interpretation of the results obtained by dating via excess $^{210}$Pb are often associated with certain assumptions and limitations [47–50]. The age of the sediments is calculated using an exponential equation that describes the decrease in the total specific activity of $^{210}$Pb with depth, only if the flow of excess $^{210}$Pb to the sediment surface is constant and there are no processes leading to its mixing or redistribution (constant sedimentation) [49]. Often, these conditions are not met and significant fluctuations in the $^{210}$Pb content can be found in the vertical profiles of the deposits. To avoid questionable or inaccurate interpretations of $^{210}$Pb profiles, the chronology of $^{210}$Pb should always be confirmed using independent time markers (e.g., $^{137}$Cs) or any other available geochemical indicators [49,51].

The calendar age and the sediment accumulation rates of layers in the studied sediment cores were calculated using the constant flux (CF) model based on the data of the excess $^{210}$Pb activity [49]. The CF model makes it possible to consider the uneven supply of $^{210}$Pb with the flow of sediment mass to the seabed surface. Layers lying below the accepted equilibrium boundary are not considered during sediment age calculations. The formulae used for the calculation are presented in the Supplementary Materials.

The independent verification of chronology is essential to ensure a high level of confidence in the results. Therefore, the known independent key date method was used. Artificial radionuclides make it possible to determine several key dates in the last 70 years. The $^{137}$Cs chronostratigraphic marker is widely used to date sediment cores in the Arctic, as the radioactive fallout from atmospheric nuclear weapon testing was global. Thus, the presence of a concentration peak in most cases makes it possible to determine the depth of contamination with a known event date [52]. The calculated ages of studied sedimentary layers are verified using the results of measuring the specific activity of the technogenic radionuclide $^{137}$Cs, especially since the nuclear tests were carried out relatively close to the study area—on the Novaya Zemlya archipelago.

## 3.7. Estimation of Ice Scour Age

Ice scour age estimation is based on the fact that the sedimentation in it commences immediately after its formation as a result of the impact of drifting ice on the seabed. Therefore, if we determine the age of the lower sediment layers filling the ice scour, we can determine the age of the ice scour itself. The extrapolation of the sedimentation rate, determined from $^{210}$Pb in the upper layers, was used to estimate the age of the lower sediment layer of the ice scour in case it exceeded the capabilities of the $^{210}$Pb dating method (100–120 years).

## 4. Results

*4.1. Ice Scour Structure*

At depths of 28–32 m, a multibeam echo sounder mapped an almost rectilinear segment 10.8 km long of the ice scour (in front of the Baydaratskaya Bay entrance, Figure 2), oriented NW–SE at an angle to the isobaths and located almost on the continuation of the bay axis (Figure 1). The maximum depth of the ice scour cutting into the background surface was 3.2 m, and the maximum width did not exceed 35 m. Side berms up to 1.5 m high could be traced almost along the entire ice scour.

The ice scour made a U-turn of almost 180 degrees, at the SE end of the segment, and after which the survey was interrupted, making it impossible to trace the further configuration of the ice scour. However, on the tacks transverse to the main ice scour, the minor fragments of some ice scours were recorded, which presumably could be a continuation of the main ice scour. In this case, the ice scour had a serpentine-shaped plan configuration, changing its direction 2–3 times and reaching a length of at least 30–35 km (dashed black line in Figure 2A). Such a configuration indicates that the main driving force in the formation of the ice scour was reverse tidal currents. At the SE end of the segment where the U-turn occurs, the ice scour becomes double.

Core sampling was carried out in the NW part of the studied segment of the ice scour. Here, it had an asymmetric structure—the southern side was 0.5–1.0 m higher than the northern one (Figure 2B). At the core sampling site, the apparent depth of the ice scour cutting into the background seabed surface was about 2.4 m. According to seismic profiles, the thickness of sediments filling the ice scour was estimated at about 0.6–0.8 m (Figure 3). Thus, the initial depth of the ice scour cutting into the background seabed surface at this location could be up to 3.2 m.

Two sediment units are clearly visible on the seismic profiles crossing the ice scour in the core sampling area (Figure 3). The upper unit, 1.5–2 m thick, covers the underlying layers like a mantle and is displayed on the seismic profiles mainly as an acoustically transparent stratum, which is due to its water saturation and the relatively fine composition of sedimentary material. The almost-complete absence of acoustically pronounced boundaries in this unit may indicate that it was formed under the conditions of a stable water basin, i.e., without significant changes in sedimentation. This allowed us to assume their formation over the past 5–6 kyr BP, when the sea level became close to modern. Thus, we interpreted the age of the upper unit as Late Holocene.

The lower unit has a "chaotic" wavefield pattern, which is characterized by numerous short sub-horizontal and inclined axes of commonality (Figure 3). These deposits are characterized by a complex facies composition laterally and in depth. They are represented by lithological differences, from sands and sandy loams to clayey silts. We interpreted the age of the lower unit as Late Pleistocene. Thus, as can be seen in the seismic profiles (Figure 3), the ice scour completely cuts through Late Holocene deposits and cuts into Late Pleistocene deposits by 1.5–2 m (Figure 3).

*4.2. Sediment Grain Size Distribution*

Grain size analyses show that the sediments of both cores are composed exclusively of clayey silt. The pelite content (<0.01 mm) varies from 52.9 to 68.4% in the "ice scour" core, silt content (0.05–0.01 mm) ranges from 23.4 to 29.5% and heterogeneous sand (1–0.05 mm) is from 4.2 to 16.4% (Table 2). Some layers (1–2 and 15–27 cm) contain gravel grains. The highest content of fine gravel inclusions occurs in the upper part of the core (1–2 cm).

Table 2. Grain size composition and natural moisture content (W) of the "ice scour" core sediments (ANS-52-16).

| Layer, cm | Thickness, cm | >1 | 1–0.5 | 0.5–0.25 | 0.25–0.1 | 0.1–0.05 | 0.05–0.01 | <0.01 | W, % |
|---|---|---|---|---|---|---|---|---|---|
| 0–1 | 1 | 0 | 0 | 2.1 | 4.2 | 10.1 | 26.4 | 57.2 | 130 |
| 1–2 | 1 | 12.5 | 0.4 | 0.4 | 4.7 | 5.7 | 23.4 | 52.9 | 90 |
| 2–3 | 1 | 0 | 0 | 0.5 | 2.8 | 3.7 | 26.2 | 66.8 | 122 |
| 3–4 | 1 | 0 | 0 | 1.0 | 4.0 | 6.5 | 29.5 | 59.0 | 101 |
| 4–5 | 1 | 0 | 0 | 0 | 3.3 | 3.7 | 28.6 | 64.4 | 94 |
| 5–6 | 1 | 0 | 0 | 0.5 | 3.75 | 7.5 | 29.25 | 59.0 | 94 |
| 6–7 | 1 | 0 | 0 | 0.3 | 3.2 | 6.3 | 28.4 | 61.8 | 90 |
| 7–8 | 1 | 0 | 0 | 0 | 3.8 | 6.6 | 28.4 | 61.2 | 94 |
| 8–9 | 1 | 0 | 0 | 0 | 1.3 | 3.7 | 28.0 | 67.0 | 94 |
| 9–10 | 1 | 0 | 0 | 0 | 1.7 | 3.0 | 26.9 | 68.4 | 90 |
| 10–11 | 1 | 0 | 0 | 0 | 1.6 | 2.6 | 27.9 | 67.9 | 94 |
| 11–12 | 1 | 0 | 0 | 0 | 2.0 | 3.5 | 27.3 | 67.2 | 98 |
| 12–15 | 3 | 0 | 0 | 1.0 | 2.8 | 4.9 | 25.4 | 65.9 | 86 |
| 15–17 | 2 | 0.5 | 0.2 | 0.8 | 4.4 | 3.9 | 33.8 | 56.4 | 81 |
| 17–19 | 2 | 0.2 | 0.2 | 0.7 | 3.0 | 2.2 | 33.0 | 60.7 | 69 |
| 19–21 | 2 | 0 | 0 | 0.8 | 3.1 | 3.1 | 30.1 | 62.9 | 67 |
| 21–23 | 2 | 0.2 | 0.2 | 0.8 | 3.6 | 3.1 | 26.5 | 65.6 | 63 |
| 23–25 | 2 | 0 | 0 | 0.6 | 2.9 | 3.0 | 25.5 | 68.0 | 69 |
| 25–27 | 2 | 0.1 | 0.1 | 0.8 | 3.8 | 3.8 | 26.9 | 64.5 | 67 |
| 27–30 | 3 | 0 | 0 | 0.4 | 2.7 | 5.2 | 28.0 | 62.7 | 71 |

In the "background" core, the pelite content (<0.01 mm) varies from 56.6 to 72.7%, silt content (0.05–0.01 mm) ranges from 27.1 to 39.4% and the inclusion of heterogeneous sand and gravel is not significant (Table 3). Thus, the concentration of the coarse fraction in the sediments accumulated inside the ice scour is higher than in the sediments outside it, which is the main difference in the grain size composition of the studied cores. The natural moisture content of sediments (W) decreases from the upper layers to the lower ones in both cores. However, it is generally higher in the "ice scour" core (from 130 to 67%; Table 2) than in the "background" core (from 77 to 48%; Table 3).

Table 3. Grain size composition and natural moisture content (W) of the "background" core sediments (ANS-52-17).

| Layer, cm | Thickness, cm | >1 | 1–0.5 | 0.5–0.25 | 0.25–0.1 | 0.1–0.05 | 0.05–0.01 | <0.01 | W, % |
|---|---|---|---|---|---|---|---|---|---|
| 0–2 | 2 | 0 | 0 | 0 | 0 | 1.2 | 36.3 | 62.5 | 77 |
| 2–5 | 3 | 0 | 0 | 0 | 0 | 1.9 | 34.5 | 63.5 | 68 |
| 5–8 | 3 | 0 | 0 | 0.3 | 2.3 | 3.5 | 30.5 | 63.4 | 71 |
| 8–11 | 3 | 0.4 | 0.1 | 0.4 | 2.5 | 4.0 | 36.0 | 56.6 | 69 |
| 11–15 | 4 | 0 | 0 | 0 | 0.8 | 1.7 | 32.7 | 64.8 | 70 |
| 15–16 | 1 | 3.8 | 0 | 0 | 0 | 1.3 | 27.0 | 67.9 | 72 |
| 16–17 | 1 | 0 | 0 | 0 | 1.0 | 1.5 | 30.6 | 66.9 | 72 |
| 17–18 | 1 | 0 | 0 | 0 | 0 | 1.1 | 27.1 | 71.8 | 65 |
| 18–19 | 1 | 0 | 0 | 0 | 1.0 | 2.1 | 28.1 | 68.9 | 69 |
| 19–20 | 1 | 0 | 0 | 0 | 0.9 | 2.1 | 30.0 | 67.0 | 67 |
| 20–21 | 1 | 0 | 0 | 0 | 0 | 1.9 | 31.8 | 64.3 | 65 |
| 21–22 | 1 | 0 | 0 | 0 | 0.8 | 1.7 | 33.3 | 64.2 | 66 |
| 22–23 | 1 | 0 | 0 | 0 | 0 | 1.5 | 31.0 | 67.5 | 72 |
| 23–24 | 1 | 0 | 0 | 0 | 0 | 0.7 | 27.6 | 71.7 | 67 |
| 24–25 | 1 | 0 | 0 | 0 | 0 | 1.4 | 26.2 | 72.4 | 71 |
| 25–26 | 1 | 0 | 0 | 0 | 0.5 | 1.5 | 23.5 | 74.5 | 68 |
| 26–27 | 1 | 0 | 0 | 0 | 0.9 | 1.9 | 24.5 | 72.7 | 72 |
| 27–30 | 3 | 0 | 0 | 0 | 1.3 | 4.0 | 38.7 | 55.8 | 69 |
| 30–35 | 5 | 0 | 0 | 0.1 | 1.2 | 2.2 | 37.5 | 59.0 | 48 |
| 35–40 | 5 | 0 | 0 | 0.3 | 2.0 | 2.3 | 39.4 | 56.0 | 55 |
| 40–45 | 5 | 0 | 0 | 0.3 | 1.5 | 2.3 | 36.9 | 59.0 | 55 |
| 45–50 | 5 | 0.6 | 0.2 | 0.2 | 1.7 | 2.4 | 36.7 | 58.2 | 63 |

4.3. Distribution of $^{210}$Pb and $^{137}$Cs in Sediments of the Cores

The specific activity of $^{210}$Pb reaches a maximum value of 108 Bq/kg in the upper part of the "ice scour" core sediments (Table 4). It decreases to 48–55.9 Bq/kg in the lower layers (38–43 cm) without any significant outliers. The $^{226}$Ra content varies from 14.7 to 45.7 Bq/kg throughout the sediment core, with an average value of 28.6 Bq/kg. Equilibrium with the initial radionuclide $^{226}$Ra at the layer of 42–43 cm was not revealed in the "ice scour" core, but there was a tendency to approach it. The technogenic radionuclide $^{137}$Cs was determined in most layers. The values ranged from less than the minimum detectable activity (<MDA) to 47.3 Bq/kg. The MDA for $^{137}$Cs was 0.2 Bq/kg. The radionuclide was not found in sediment layers below 36 cm (Table 4). In the uppermost layer of the "ice scour" core (0–1 cm), short-lived radionuclide $^7$Be was found with specific activity of 18.4 ± 8.3 Bq/kg, which confirms that the surface layer was sampled correctly.

**Table 4.** The results of measurements of the specific activity of short-lived radionuclides and the calculation of excess $^{210}$Pb ($C_i$) in the "ice scour" core (ANS-52-16), Bq/kg.

| Layer, cm | Thickness, cm | $^{137}$Cs | U$^1$ ($^{137}$Cs) | $^{210}$Pb | U$^1$ ($^{210}$Pb) | $^{226}$Ra | U$^1$ ($^{226}$Ra) | $C_i$ | U$^1$ ($C_i$) |
|---|---|---|---|---|---|---|---|---|---|
| 0–1 | 1 | 6.0 | 2.4 | 93.7 | 9.2 | 22.1 | 5.6 | 71.6 | 10.8 |
| 1–2 | 1 | 9.3 | 0.4 | 74.8 | 3.0 | 24.2 | 2.7 | 50.6 | 4.0 |
| 2–3 | 1 | 10.3 | 0.6 | 107.0 | 4.9 | 24.1 | 5.4 | 82.9 | 7.3 |
| 3–4 | 1 | 11.3 | 0.8 | 90.7 | 5.4 | 31.3 | 3.8 | 59.4 | 6.6 |
| 4–5 | 1 | 10.9 | 0.8 | 73.1 | 5.4 | 28.0 | 3.7 | 45.1 | 6.5 |
| 5–6 | 1 | 11.6 | 0.7 | 78.9 | 5.1 | 30.0 | 3.6 | 48.9 | 6.2 |
| 6–7 | 1 | 9.1 | 0.8 | 73.0 | 5.6 | 17.7 | 3.8 | 55.3 | 6.8 |
| 7–8 | 1 | 10.6 | 0.8 | 87.7 | 6.0 | 35.4 | 4.3 | 52.3 | 7.4 |
| 8–9 | 1 | 12.6 | 0.8 | 92.7 | 5.4 | 34.9 | 4.0 | 57.8 | 6.7 |
| 9–10 | 1 | 13.7 | 0.9 | 108.0 | 6.3 | 43.2 | 4.6 | 64.8 | 7.8 |
| 10–11 | 1 | 14.9 | 0.9 | 102.0 | 5.9 | 39.5 | 4.6 | 62.5 | 7.5 |
| 11–12 | 1 | 12.1 | 0.5 | 82.4 | 4.0 | 45.7 | 4.3 | 36.7 | 5.9 |
| 12–13 | 1 | 11.8 | 0.9 | 83.9 | 6.2 | 41.9 | 4.6 | 42.0 | 7.7 |
| 13–14 | 1 | 9.4 | 1.4 | 83.8 | 10.3 | 25.5 | 5.0 | 58.3 | 11.4 |
| 14–15 | 1 | 13.6 | 0.9 | 72.5 | 6.4 | 31.0 | 4.3 | 41.5 | 7.7 |
| 15–16 | 1 | 15.3 | 2.2 | 83.0 | 14.7 | 20.8 | 5.9 | 62.2 | 15.8 |
| 16–17 | 1 | 17 | 1.1 | 92.3 | 8.2 | 23.2 | 4.1 | 69.1 | 9.2 |
| 17–18 | 1 | 17.4 | 1.8 | 89.4 | 12.1 | 26.1 | 5.3 | 63.3 | 13.2 |
| 18–19 | 1 | 13.4 | 0.9 | 84.9 | 6.7 | 27.9 | 3.6 | 57.0 | 7.6 |
| 19–20 | 1 | 11.9 | 1.8 | 66.1 | 12.7 | 22.7 | 4.6 | 43.4 | 13.5 |
| 20–21 | 1 | 2.9 | 0.6 | 71.5 | 5.5 | 24.4 | 3.5 | 47.1 | 6.5 |
| 21–22 | 1 | <MDA | - | 66.7 | 4.7 | 25.8 | 3.9 | 40.9 | 6.1 |
| 22–23 | 1 | <MDA | - | 57.3 | 5.9 | 22.4 | 4.8 | 34.9 | 7.6 |
| 23–24 | 1 | <MDA | - | 47.3 | 4.9 | 17.4 | 4.5 | 29.9 | 6.7 |
| 24–25 | 1 | <MDA | - | 43.7 | 4.9 | 14.4 | 3.5 | 29.3 | 6.0 |
| 25–26 | 1 | <MDA | - | 40.0 | 4.8 | 17.9 | 3.3 | 22.1 | 5.8 |
| 26–27 | 1 | 8.1 | 0.4 | 45.5 | 2.9 | 21.0 | 3.4 | 24.5 | 4.5 |
| 27–28 | 1 | <MDA | - | 56.8 | 5.9 | 26.4 | 4.2 | 30.4 | 7.2 |
| 28–30 | 2 | <MDA | - | 52.1 | 4.9 | 26.6 | 4.1 | 25.5 | 6.4 |
| 30–32 | 2 | 47.3 | 0.9 | 56.1 | 3.6 | 22.5 | 2.4 | 33.6 | 4.3 |
| 32–34 | 2 | 2.4 | 0.6 | 59.3 | 4.1 | 24.2 | 2.7 | 35.1 | 4.9 |
| 34–36 | 2 | 2.3 | 0.4 | 41.7 | 3.5 | 19.3 | 2.7 | 22.4 | 4.4 |
| 36–38 | 2 | <MDA | - | 43.9 | 6.3 | 23.7 | 3.3 | 20.2 | 7.1 |
| 38–40 | 2 | <MDA | - | 48.0 | 4.8 | 20.0 | 2.1 | 28.0 | 5.2 |
| 40–42 | 2 | <MDA | - | 55.9 | 4.8 | 25.0 | 3.9 | 30.9 | 6.2 |
| 42–43 | 1 | <MDA | - | 52.5 | 4.0 | 24.5 | 4.4 | 28.0 | 5.9 |

$^1$ U—standard uncertainty.

In the "background" core sediments, the highest specific activity of $^{210}$Pb is in the upper part, reaching 60 Bq/kg (Table 5). It decreases to 26.7–33.4 Bq/kg in the lower layers (17.5–25.5 cm). The $^{226}$Ra content varies from 20.5 to 30.4 Bq/kg throughout the sediment core, with an average value of 23.5 Bq/g. The equilibrium of $^{210}$Pb (32.9 Bq/kg) with the initial radionuclide $^{226}$Ra (31.1 Bq/kg) is observed in the layer of 19.5–21.5 cm. The underlying two layers (21.5–25.5 cm) show $C_i$ values (Bq/kg) of less than 1.0, considering the standard uncertainty. This allows us to take 19.5 cm as the equilibrium boundary for $^{210}$Pb and $^{226}$Ra. The layers below the accepted equilibrium boundary were not considered during sediment age calculations [49]. The $^{137}$Cs specific activity in all layers was below the MDA, except for layers of 1–2 cm and 10–11 cm (Table 5). The distribution of $^{210}$Pb excess in both sediment cores demonstrates the same pattern: an exponential decrease in depth with some fluctuations, whereby the peaks of which are correlated between the cores (Figure 4).

**Table 5.** Results of measurements of the specific activity of short-lived radionuclides and calculation of excess $^{210}$Pb ($C_i$) in the "background" core (ANS-52-17), Bq/kg.

| Layer, cm | Thickness, cm | $^{137}$Cs | U [1] ($^{137}$Cs) | $^{210}$Pb | U [1] ($^{210}$Pb) | $^{226}$Ra | U [1] ($^{226}$Ra) | $C_i$ | U [1] ($C_i$) |
|---|---|---|---|---|---|---|---|---|---|
| 0–1 | 1 | <MDA | - | 60.0 | 0.8 | 26.2 | 6.0 | 33.8 | 6.1 |
| 1–2 | 1 | 2.2 | 0.3 | 50.1 | 1.5 | 24.2 | 3.4 | 25.9 | 3.7 |
| 2–3 | 1 | <MDA | - | 36.0 | 5.0 | 25.3 | 4.9 | 10.7 | 7.0 |
| 3–4 | 1 | <MDA | - | 55.4 | 6.0 | 20.7 | 4.9 | 34.7 | 7.7 |
| 4–5 | 1 | <MDA | - | 35.5 | 3.6 | 16.1 | 4.3 | 19.4 | 5.6 |
| 5–6 | 1 | <MDA | - | 49.9 | 6.6 | 28.0 | 3.3 | 21.9 | 7.4 |
| 6–7 | 1 | <MDA | - | 49.3 | 4.6 | 28.0 | 5.2 | 21.3 | 6.9 |
| 7–8 | 1 | <MDA | - | 27.6 | 6.1 | 22.3 | 4.2 | 5.3 | 7.4 |
| 8–9 | 1 | <MDA | - | 45.3 | 3.6 | 21.5 | 5.0 | 23.8 | 6.2 |
| 9–10 | 1 | <MDA | - | 54.6 | 6.6 | 23.7 | 4.0 | 30.9 | 7.7 |
| 10–11 | 1 | 1.1 | 0.3 | 53.9 | 3.3 | 21.9 | 5.1 | 32.0 | 6.1 |
| 11–12 | 1 | <MDA | - | 47.0 | 7.3 | 30.4 | 4.4 | 16.6 | 8.5 |
| 12–13 | 1 | <MDA | - | 32.2 | 5.8 | 25.9 | 5.6 | 6.3 | 8.1 |
| 13–14 | 1 | <MDA | - | 30.1 | 4.7 | 25.9 | 5.5 | 4.2 | 7.2 |
| 14–15.5 | 1.5 | <MDA | - | 28.0 | 3.5 | 21.2 | 3.9 | 6.8 | 5.2 |
| 15.5–17.5 | 2 | <MDA | - | 34.6 | 2.5 | 25.5 | 3.9 | 9.1 | 4.6 |
| 17.5–19.5 | 2 | <MDA | - | 26.7 | 2.0 | 21.7 | 2.0 | 5.0 | 2.8 |
| 19.5–21.5 | 2 | <MDA | - | 32.9 | 3.2 | 31.1 | 3.0 | 1.8 | 4.4 |
| 21.5–23.5 | 2 | <MDA | - | 33.4 | 3.9 | 28.8 | 2.3 | 4.6 | 4.5 |
| 23.5–25.5 | 2 | <MDA | - | 28.6 | 3.0 | 23.7 | 2.7 | 4.9 | 4.0 |

[1] U—standard uncertainty.

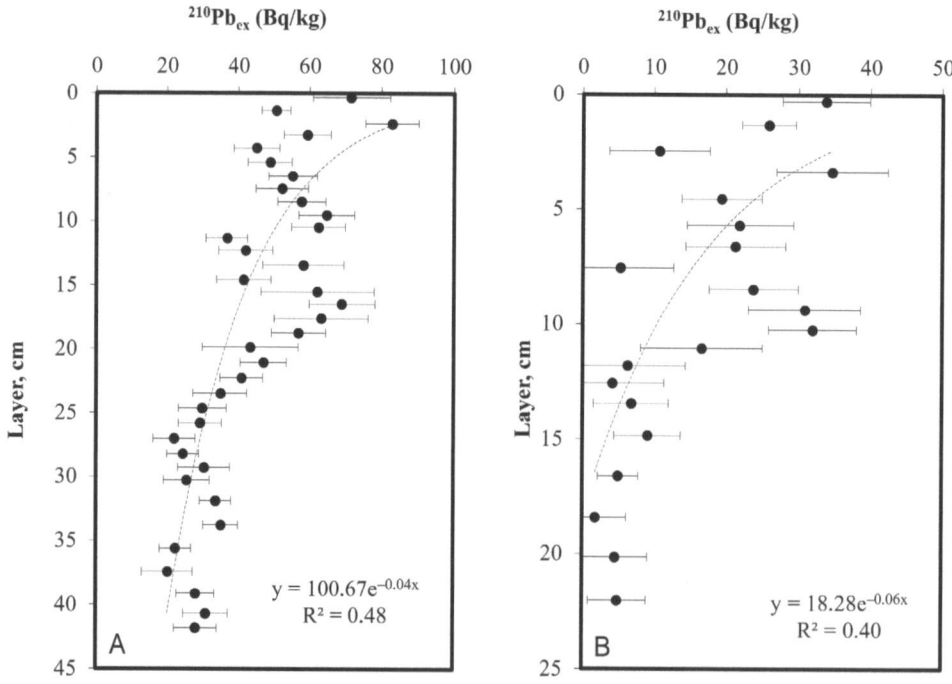

**Figure 4.** The distribution of excess $^{210}$Pb in the sediment cores: (**A**) "ice scour" (ANS-52-16); (**B**) "background" (ANS-52-17).

## 4.4. The Calendar Age and the Sediment Accumulation Rates

The calendar age and the sediment accumulation rate (SAR) of each sedimentary layer top in the studied sediment cores were calculated using the CF model [49], based on the measurements of the excess $^{210}$Pb activity (Figures 5 and 6, Tables S1 and S2 in Supplementary Materials). In the "ice scour" core, the calculations were based on the assumption that the equilibrium between $^{210}$Pb and $^{226}$Ra must be achieved in the underlying layers not penetrated by the corer, i.e., the supposed equilibrium border is at 43 cm (curve "Pb-210 age, ice scour, CF/43" in Figure 5). In the "background" core, the calculations were carried out taking into account the accepted equilibrium border at 19.5 cm (curve "Pb-210 age, background, CF/19.5" in Figure 5). The obtained calendar ages have the form of decimal fractions since they are calculated mathematically, relative to the date of the core selections—2 November 2021 (2021.8 AD).

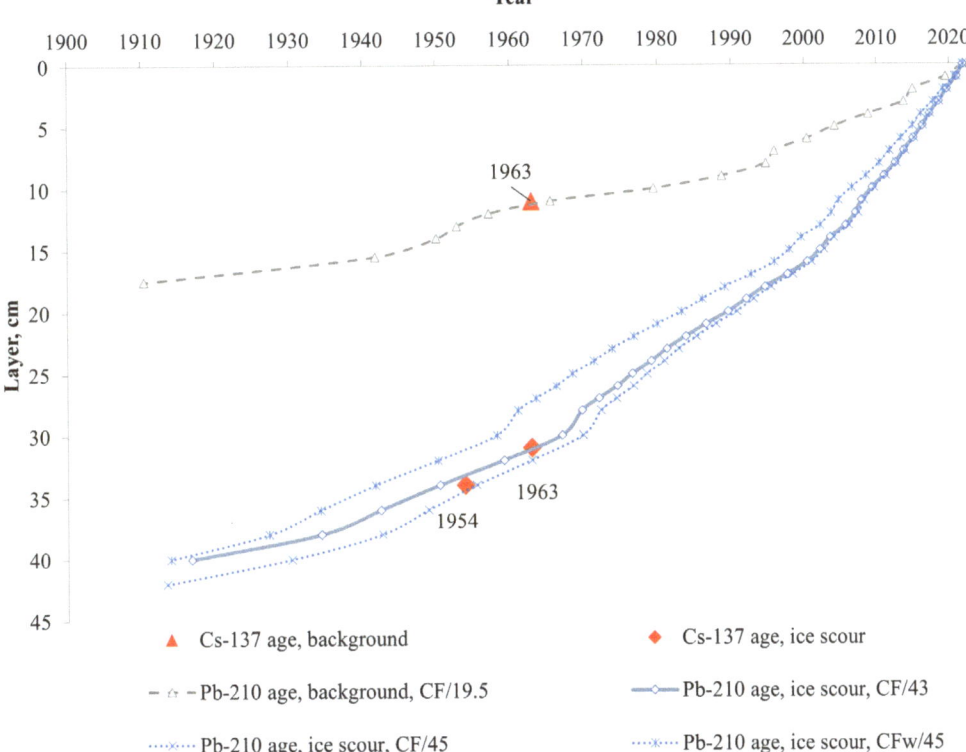

**Figure 5.** Distribution of the calendar ages of the layers by the depth of the cores ("ice scour" and "background"). Calendar ages are given according to $^{210}$Pb and $^{137}$Cs data. The designation "CF/43" contains the numerator information about the formulae used in the calculation (CF—constant flux model, CFw—modified formulae of the CF model, taking into account the change in the sedimentary layers' widths in the ice scour with depth), and the denominator contains information about the position of the equilibrium boundary, as used in the calculation (19.5; 43 and 45 cm).

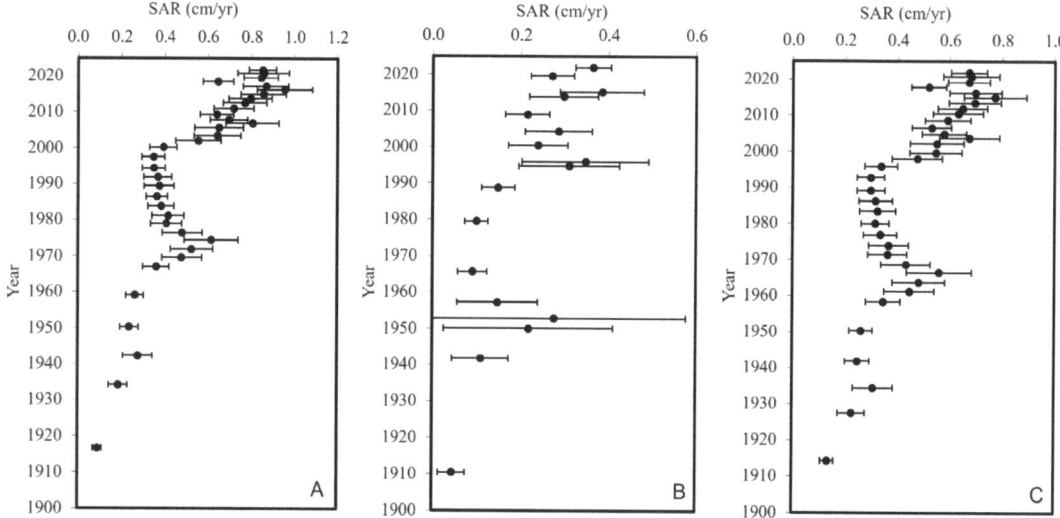

**Figure 6.** Changes in the sedimentation rate: (**A**) in the "ice scour" core (ANS-52-16) according to the CF model with the supposed equilibrium border at 43 cm (CF/43); (**B**) in the "background" core (ANS-52-17) according to the CF model with the accepted equilibrium border at 19.5 cm (CF/19.5); (**C**) in the "ice scour" core (ANS-52-16) according to the modified formulae of the CF model, taking into account the change in the sedimentary layers' widths in the scour, with the estimated boundary at 45 cm (CFw/45).

The calendar age of the lowest reliably dated levels is 1916.8 ± 4.9 AD (40 cm) in the "ice scour" core and 1910.4 ± 17.9 AD (17.5 cm) in the "background" core (Figure 5, Tables S1 and S2 in Supplementary Materials). The age value of 1884.1 ± 7.0 AD per 42 cm in the "ice scour" core is not considered, as it is beyond the limits of the method capabilities. Thus, the upper 40 cm in the "ice scour" core accumulated over the last 105 ± 4.9 years and the upper 17.5 cm in the "background" core accumulated over the last 111.4 ± 17.9 years.

The SARs in the "ice scour" core varied within the range of 0.09–0.95 cm/year in different years (Figure 6A, Table S1 in Supplementary Materials). The average SAR value for a thickness of 40 cm was 0.38 cm/year. The SARs in the "background" core were lower (0.04–0.38 cm/year), with an average value for a thickness of 17.5 cm of 0.16 cm/year (Figure 6B, Table S2 in Supplementary Materials). Comparing the distribution of the calendar ages and SAR values in both of the cores (Figures 5 and 6), one can notice the same trends in the change of rates but they occur at different times. The increase in the sedimentation rates in the ice scour occurs 10–20 years later compared to the background surface. For example, in the "ice scour" core, a sharp SAR increase occurred between 2000 ± 1 and 2002 ± 1, and in the "background" core it was between 1989 ± 4 and 1995 ± 3.

## 5. Discussion

### 5.1. Estimation of Missing Inventory of $^{210}Pb$ in the Ice Scour

Since there is no $^{210}Pb$ equilibrium with the initial radionuclide $^{226}Ra$ in the "ice scour" core, we estimated the missing inventory of $^{210}Pb$ (A(j), Bq/m$^2$) in the layers below 43 cm, not penetrated by the corer. To do this, we used two methods: the formula of the exponential dependence of the total $^{210}Pb$ on the mass ($m_j$, g/cm$^2$) $y = 159.44e^{-0.035x}$ and $^{137}Cs$ time marker estimation [49].

Using the formula of the first method, the total $^{210}Pb$ was reconstructed in the range of 43–51 cm. The mass of the layers was taken from the average value of the layers in the range of 32–40 cm (21.26 g). $^{226}Ra$ was calculated as the average value of the specific activity for all of the layers and amounted to 26.4 ± 7 Bq/kg. Subsequently, the calculation was carried out according to the CF model with discreteness of 2 cm. Equilibrium was reached on a layer of 50 cm, which dates back to 1806 AD, which cannot be taken as representative information, as it is beyond the limits of the method capabilities. However, the dates in the range of 0–40 cm did not change significantly.

As a time marker of the second method, we used the $^{137}$Cs peak (47.3 ± 0.9 Bq/kg), which was found in the 30–32 cm layer. According to the results of our measurements (Table S1 in Supplementary Materials), the inventory of non-equilibrium $^{210}$Pb ($\delta A_1$, Bq/m$^2$) that accumulated over 58 years from 1963 to 2021 (0–32 cm) was 15,341 Bq/m$^2$. The inventory of $^{210}$Pb ($A(j)_1$ = 2915 Bq/m$^2$) that accumulated before 1963 (layers from 32 cm to the boundary of equilibrium between $^{210}$Pb and $^{226}$Ra) was calculated using the formula $A(j)_1 = \delta A_1/(e^{\lambda t} - 1)$, where t = 58 yr and $\lambda$ = 0.03118 yr$^{-1}$ (disintegration constant). Based on the obtained values, the total inventory of non-equilibrium $^{210}$Pb from the seabed surface to the equilibrium boundary was 18,256 Bq/m$^2$ ($A(0) = \delta A_1 + A(j)_1$). According to the results of our measurements (Table S1 in Supplementary Materials), the inventory of non-equilibrium $^{210}$Pb ($\delta A_2$, Bq/m$^2$) accumulated in the layers (0–43 cm) was 17,879.5 Bq/m$^2$. Then, the missing inventory of $^{210}$Pb ($A(j)_2 = A(0) - \delta A_2$) is 376.5 Bq/m$^2$. Based on the activity of non-equilibrium $^{210}$Pb in the 42–43 cm layer, which is equal to 244 Bq/m$^2$, it can be assumed that the missing inventory of 377 Bq/m$^2$ accumulated in the 43–45 cm layer. In this case, the equilibrium boundary should be lowered by 3 cm to 46 cm.

The results of the layers' age calculations using the CF model, taking into account the estimation of the underlying missing inventory (curve "Pb-210 age, ice scour, CF/46"), are shown in the graph (Figure 5). This calculation makes the age a little younger: for layers in the range of 0–26 cm—no more than 2 years, in the range of 26–36 cm—no more than 5 years and in the range of 36–40 cm—no more than 8.5 years. The greatest age change is observed for the 40–42 cm layer, whose age has become 13.5 years younger.

*5.2. Application of the Constant Flux Model in the Ice Scour Conditions*

The CF model was designed for flat bottom sedimentation without any lateral limitations [49]. However, the accumulation of sediments in an ice scour occurs in a constrained environment. The sedimentation area at the bottom of the ice scour increases with time. In other words, as the ice scour is filled with sediments, the width of the ice scour bottom increases. This phenomenon is expressed in the cross section of the ice scour as an increase in the "width" of the layers upwards. Therefore, we calculated how the layer "width" changes with a step of 1 cm in an ice scour with a geometry such as ours, in order to assess the possible error in the results obtained from this model.

Calculations were made for an ice scour with $w_0/S$ = 15, where $w_0$—the ice scour width at the present bottom surface, and S—the thickness of the ice scour filling sediments (in our case, $w_0$ = 12 m and S = 0.8 m). This ratio shows the absolute value of the change in the ice scour width, which is constant and amounts to 15 cm per 1 cm. With this ratio, the angle of inclination of the ice scour sides is 7.6 degrees. The ice scour width at a given depth relative to the present bottom surface was calculated using the formula $w_i = 12 - 15 z_i$, where $w_i$ is the ice scour width at a given horizon, and $z_i$ is the depth of horizon. The 0–1 cm layer (the uppermost dated layer) will have a roof "width" of 12.0 m and a base "width" of 11.85 m. The change is 1.25%. In the 38–40 cm layer (the lowest dated layer), the "width" of the roof will be 6.3 m, and the "width" of the base will be 6.0 m. The change is 4.8%. There is a downward trend of increasing the magnitude of the layers' "width" relative changes which can reach 50–100% at the base of the filling sediments under the condition of an ideal cross section of the ice scour in the form of an isosceles triangle. However, we dated only the upper half of the filling sediments (40 cm out of 80 cm), where the "width" of 1 cm layers in the cross section changes by only 1.25–2.4% per 1 cm.

Such an insignificant change in the 1 cm layers' "width" in the cross section gives grounds in our case to neglect the influence of the ice scour geometry on sedimentation. Therefore, we consider it reasonable to use the CF model to calculate the age in both cores:

on the "background" surface, where the layers' "width" remains constant, and in 0–40 cm of the ice scour, where the layers' "width" changes no more than 2.4% per 1 cm. It is important to note that we realize that the results may have additional uncertainty, but it is not significant and does not exceed the standard uncertainty in the measurements themselves. The uncertainty increases with the increase in the thickness of the analyzed layers.

Another important fact is worth noting. As the ice scour fills with sediments, each successive layer becomes "wider". Therefore, at a constant sediment flux entering the ice scour, the sedimentation rate should decrease with time. However, we observe the opposite picture—the sedimentation rate increases with time. From this, it can be concluded that the sediment flux entering the ice scour has a greater effect on the sedimentation rate than the ice scour layers' "width", whose influence is insignificant.

An additional control of the reliability of the ice scour dating results can be undertaken by comparing them with the results of the "background" core dating. Due to the close location (only 140 m between the cores), the general tendencies of sedimentation in this bottom area should be traced in both of the cores, while the sedimentation rates can be determined according to local conditions (in our case, according to the features of the bottom topography). Indeed, in both cores, an exponential distribution of $^{210}$Pb was observed, whereby the fluctuations (peaks) of which are correlated between the cores (Figure 4). However, there is a 10–20 year lag of the sharp SAR changes in the ice scour relative to the background surface (Figure 6), which may be due to uncertainty due to the layers' width changes in the ice scour.

To eliminate this uncertainty, an attempt was made to modify the formulae of the CF model to take into account the change with depth in the sedimentary layers' width in the scour. The calculation using these modified formulae was designated as CFw (Table S3 in Supplementary Materials). Whereas the CF model calculation is based on the measured specific activity of radionuclides in the core, the CFw calculation is based on an estimation of the excess lead activity in all of the sediments filling the ice scour. To extrapolate the excess lead activity that had accumulated over the entire cross section of the ice scour, the excess lead activity in each layer was multiplied by the proportionality factor ($k_i$) of the ice scour width at a given depth to the ice scour width at the lowest horizon with the excess lead activity. Thus, at the lowest horizon (44 cm), $k_i$ is equal to 1, and at the highest horizon (the present surface of the bottom of the ice scour), $k_i$ reaches a value of 2.22. It was decided that $k_i$ be used, because it was noted that the results of the age and SAR calculations were affected by the rate of increase in values with each subsequent step in the horizon depth change, and not by the absolute values themselves, and by which, the excess lead activity in the core was multiplied. Thus, the calculation results were the same both when the excess lead activity was multiplied by the ice scour width at a given horizon depth, and when the excess lead activity was multiplied by the ratio of the ice scour width at a given horizon depth to the core diameter. The modified formulae used for the CFw calculation are presented in the Supplementary Materials.

The calendar age obtained from the CFw calculation turned out to be no more than 10 years earlier than the age obtained from the CF model (Figure 5, curve "Pb-210 age, ice scour, CFw/45"). Compared to the background surface, the lag of sharp changes in the SAR in the ice scour has decreased, but it still remains and has reached about 10 years (Figure 6C). This may be due to the lower detail and higher uncertainty in the "background" core compared to the "ice scour" core. The uncertainty in the calendar age calculation increases from the seabed surface down the core and reaches ±5 years in the ice scour and ±18 years on the background surface. Thus, the modified formulae (CFw) correct the age within the method uncertainty; so, this, together with what was noted above, gives us a reason to use the results obtained from the measured specific activities of radionuclides without taking into account the changing width of the ice scour (CF model) to estimate the ice scour age.

*5.3. Verification of Results with $^{137}Cs$*

The lowest layer from the "ice scour" core, where $^{137}Cs$ was found (34–36 cm), can be attributed to 1954 AD (1950.5 ± 3.2 AD at 34 cm according to the CF model, Figure 5). This is the year when the first high-performance thermonuclear weapon was detonated and atmospheric levels of $^{137}Cs$ in the Northern Hemisphere rose high enough to create a measurable signal that could be detected in soils and sediments in water bodies [53]. The $^{137}Cs$ specific activity increased approximately 20 times in the overlying layer (30–32 cm). Seventy-five atmospheric nuclear tests were carried out on the Novaya Zemlya archipelago between 1954 and 1962, including the largest thermonuclear charge ever detonated ("Tsar Bomba") [54]. The global number of atmospheric nuclear tests peaked at 110 explosions in 1962 [55]. The peak concentration of atmospheric radioactivity was recorded in 1963 as a result of atmospheric nuclear testing [56]; so, the calendar age of the 30–32 cm layer should be attributed to this year, which is comparable to the chronological data obtained using the CF model (1967.1 ± 2.2 AD at 30 cm and 1959.2 ± 2.6 AD at 32 cm, Figure 5).

The further distribution of $^{137}Cs$ in the "ice scour" core from a layer 26–27 cm to the surface layer did not make it possible to identify the increased specific activity of the radionuclide designated as a reference in the literature, for example, the transoceanic transport of radioactive discharges of nuclear fuel reprocessing plant Sellafield, UK, to the Irish Sea (1982–1983) and atmospheric fallout after the accident at the Chernobyl Atomic Electric Power Station (1986) [57]. Thus, according to the $^{137}Cs$ data, the upper 34 cm in the "ice scour" core accumulated after 1954 AD over the last 67.8 years, and the average SAR value was 0.5 cm/year for a thickness of 34 cm, which is higher, but for a lower thickness than according to $^{210}Pb$ (0.38 cm/year for 40 cm).

The "background" core is characterized by low $^{137}Cs$ specific activity (<MDA), except for two layers (1–2 and 10–11 cm). This is probably due to periodic seabed erosion, which prevents the accumulation of $^{137}Cs$ in seabed sediments. The appearance of $^{137}Cs$ in the 10–11 cm layer allows us to interpret the calendar age of this layer as 1963 AD, which is comparable with the age obtained from $^{210}Pb$ (1965.6 ± 8.0 AD at 11 cm). Thus, according to the $^{137}Cs$ data, the upper 10 cm in the "background" core accumulated after 1963 AD over the last 58.8 years, and the average SAR value was 0.17 cm/year for a thickness of 10 cm, which is comparable to $^{210}Pb$ results (0.16 cm/year for 17.5 cm) even for a lower thickness.

*5.4. Formation of Sediments and Dynamics of Their Accumulation*

The "background" core is more than six times longer than the "ice scour" core. We explain the small length of the "ice scour" core by the presence of a compacted horizon, which is associated with the pressure during gouging. Probably, the corer did not hit the ice scour axis, but the lower part of the ice scour wall, which is covered by only 43 cm (or so) of filling post-gouging sediments. The GPS coordinates of the sampling point that were correlated with the DEM of the bottom also show the displacement of the sampling site toward the ice scour wall relative to its axis (Figure 2B). Thus, we have reason to believe that 43 cm of the sampled sediments lie on pre-exaration deposits. Older post-exaration deposits that fill the ice scour are located in the axial part of the ice scour at a depth of 43–80 cm below the modern bottom of the ice scour.

According to the grain size analysis (Tables 2 and 3), sediment accumulation in the ice scour and at the "background" surface occurs mainly due to fine sediments (pelite and silt). In the ice scour, in contrast to the "background" surface, there is an inclusion of heterogeneous sand and fine gravel, which accumulated here as a result of slope subaqueous processes, saltation or ice rafting. The low content of sand and gravel, as well as the lower natural moisture content in the sediments of the "background" surface, may indicate higher hydrodynamics outside of the ice scour. Since the studied sediment cores are located quite close to each other (only 140 m), the quantity and quality (grain size composition) of sedimentary material entered in the ice scour and on the "background" surface are the same; however, active hydrodynamics outside of the ice scour lead to periodic erosion of the "background" surface and to the removal of sand and gravel. Under conditions of low hydrodynamics in the ice scour, coarse particles originating from ice rafting, slope processes and saltation are not able to move laterally and remain where they settled on the bottom. This fact explains the absence of the $^{137}$Cs marker in the sediments of the "background" core and the availability of this isotope in high concentrations in the ice scour (Tables 4 and 5). In other words, the sediments that have accumulated into the ice scour are stored there as in a sedimentological trap and the sediments of the "background" surface pass in transit, only being partially deposited. Thus, the obtained results additionally confirm that one of the sediment cores was indeed taken directly from the ice scour.

To quantify the sedimentological trap effect, the excess lead activity accumulated on the present seabed surface in the "background" core was multiplied by the proportionality factor of the ice scour width at the level of the present seabed surface to the ice scour width at the lowest horizon with excess lead activity ($k_0$ = 2.22). The obtained value was 6684 Bq/m$^2$ on the background surface, while in the ice scour it was 31,497 Bq/m$^2$. Thus, the studied ice scour accumulated 4.7 times more lead than the background surface.

The main sources of fine sediments accumulated in Baydaratskaya Bay are the erosion of its coasts and partly the runoff of the Ust-Kara River (Figure 1). The northern coast of the bay is mainly composed of homogeneous sandy deposits with relatively low ice content [58,59] and is actively eroded at an average rate of 0.3 to 1.2 m/year [60]. The southern coast of the bay is composed of medium ice-rich sandy loam deposits, underlain by medium or, in some places, strongly ice-rich boulder loams and clays [61], and is also actively eroded at an average rate of 0.5 to 2.4 m/year [60]. Taking into account the grain size composition of the studied sediment cores, it can be assumed that sedimentation in the ice scour and near it occurs due to the erosion of the southern coast of Baydaratskaya Bay.

Sediments of the last 100–120 years and older are composed of the same grain size. Consequently, the source of sedimentation material (erosion products of the southern coast of the bay) and lithodynamics (secondary seabed erosion) have not changed in the studied area. Taking into account that sedimentation in the study area is closely related to coastal erosion, one can note a certain correlation between the coastal dynamics of Baydaratskaya Bay and the sedimentation rates in the studied sediment cores. According to our data (Figure 6), the SARs have reached their maximum values in recent years, since the 2000s. The maximum coastal erosion of the bay also belongs to the same period [60]. In the 1990s, when coastal erosion of the bay slowed down, we noted a decrease in the sedimentation rate. At the same time, it is known that the dynamics of coastal erosion are closely related to the climate and depend on combinations of wind–wave and thermal factors within the region [60]. Consequently, the rates of marine sedimentation also depend on the climate.

*5.5. The Age of the Ice Scour*

The $^{210}$Pb method allowed us to date the upper part (0–0.4 m) of the sediments filling the ice scour, whereby thickness was estimated from seismic profiles to be about 0.6–0.8 m. We estimated the age of the ice scour for a filling sediment thickness of 0.7 m, assuming that age variations can occur both upwards and downwards due to variations in the actual thickness of the filling sediments. The age estimation of sediments in the range of 0.4–0.7 m is based on the extrapolation of the maximum and minimum possible mean sedimentation rates under certain sedimentation conditions.

It is known that marine sedimentation is determined by a few factors: climate, water salinity, basin depth, gas regime, the presence and nature of currents and biological activity [62]. At the same time, air temperature, as an important indicator of the climate, is a significant factor determining basin sedimentation. Often, there is a strong correlation between climatic components and sedimentation [63].

According to our calculations, the top 40 cm of sediment accumulated in the ice scour over about 105 ± 5 years, starting around 1917 ± 5 AD (mean SAR 0.38 cm/year). At the same time, more than 1/3 of the dated sediments (0–15 cm) accumulated over the past 20 ± 1 years (since around 2002 ± 1 AD) at a mean rate of about 0.75 cm/year. This period is characterized by an increase in air temperature in the Baydaratskaya Bay area, which accelerated the processes of coastal erosion [64], increasing the sediment supply in the accumulation basin. Conversely, a decrease in the rate of sedimentation is observed with a decrease in the sum of daily positive air temperatures in the period of 1975–1985. In the 1960s, the dependence of sedimentation rates on air temperatures persisted. The sum of daily positive air temperatures increased and sedimentation proceeded more dynamically than in the next decade.

Thus, based on the past temperature regime, we can indirectly assume the mean sedimentation rate in the ice scour before 1917. Due to the lack of meteorological observations in the work area, we decided to rely on data on climatic anomalies of the northern hemisphere from 1850 to the present [65–67]. According to these data, the value of the temperature anomaly in 1900–1930 was unchanged. Considering this, it can be assumed that the mean SAR in the ice scour in that time was not less than the mean SAR established for the period 1916.8–1934.3 (38–40 cm), i.e., around 0.11 cm/year. In this case, in 1900–1916.8, a layer of 40–41.8 cm was formed; however, we found that a 42–43 cm layer contained excess $^{210}$Pb. Therefore, its age cannot theoretically exceed 120 years (1901.8 AD) due to the dating limit of the method. Rounding the age of the beginning of the 42–43 cm layer formation to 1900 AD, it can be assumed that in 1900–1916.8, a layer of 40–43 cm was formed at a mean rate of 0.17 cm/year, rather than 40–41.8 cm. The values of temperature anomalies in 1850–1900 approximately corresponded to the temperature regime in 1930–1960. Therefore, we can assume that the mean SAR in the ice scour in 1850–1900 was not less than the mean SAR established for the period 1934.3–1959.2 (32–38 cm), i.e., around 0.24 cm/year. In this case, a layer of 43–55 cm was formed in 1850–1900. Thus, the mean sedimentation rate of the 15 cm stratum (40–55 cm) for 66.8 years (1850–1916.8) could be 0.22 cm/year. We consider this value as the minimum possible sedimentation rate before 1917. Considering that the warming since 2000 has been unprecedented, the mean rate of 0.38 cm/year determined for the thickness of 0–40 cm can be the maximum possible sedimentation rate for the period before 1917.

The extrapolation of the minimum (0.22 cm/year) and maximum (0.38 cm/year) possible sedimentation rates in the ice scour before 1917 allows us to estimate the time range at which the accumulation of the first sediments in the ice scour at a depth of 70 cm below the current seabed surface began. This time range is defined as 1780–1840. Thus, the age of the ice scour can be determined as 1810 ± 30 AD. The mean rate of ice scour filling with 70 cm thick sediments from the moment of its formation is around 0.33 cm/year.

## 6. Conclusions

- The studied ice scour, located in front of the entrance to Baydaratskaya Bay of the Kara Sea at a sea depth of about 28–32 m, has a serpentine-shaped plan configuration, changing its direction 2–3 times and reaching a length of at least 30–35 km. The maximum visible depth reaches 3.2 m, and the maximum width is up to 35 m. At present, it is the largest ice scour among those known in this region of the Kara Sea.
- Two sediment cores were studied, which were taken on 2 November 2021 (2021.8 A.D.) using a gravity corer directly in the ice scour and on the "background" seabed surface, which was not processed via ice scouring, and 140 m south of the first one. At the core sampling site, the apparent depth of the ice scour cutting into the background seabed surface was about 2.4 m. According to the seismic profiles, the thickness of sediments filling the ice scour was estimated at about 0.6–0.8 m, whereby the top 30 cm of which was presented exclusively by clayey silt. The pelite content varied from 52.9 to 68.4%, the silt content varied from 23.4 to 29.5% and the heterogeneous sand content varied from 4.2 to 16.4%. Some layers contained gravel grains. The highest content of fine gravel inclusions occurred in the upper part of the core. On the "background" seabed surface, the upper 50 cm of sediments was also represented by clayey silt. The pelite content varied from 56.6 to 72.7%, the silt content varied from 27.1 to 39.4% and the inclusion of heterogeneous sand and gravel was not significant. The low content of the sand and gravel in the sediments of the background surface is explained by

- higher hydrodynamics outside of the ice scour, which leads to the periodic erosion of sediments and the removal of rare sand and gravel particles.
- The excess $^{210}$Pb was found in all of the analyzed layers (up to 43 cm) of the ice scour sediments, reaching maximum values of specific activity (108 Bq/kg) in the upper horizons and decreasing to 48–55.9 Bq/kg toward the lower ones. The equilibrium of $^{210}$Pb with the initial radionuclide $^{226}$Ra was not revealed at the layer of 42–43 cm, but there was a tendency to approach it. The technogenic radionuclide $^{137}$Cs below 36 cm was not detected, while above its content it ranged from 47.3 Bq/kg to values less than the minimum detectable activity (<0.2 Bq/kg).
- In deposits on the "background" seabed surface, the excess $^{210}$Pb was only found in the upper 25.5 cm. Its specific activity decreased from top to bottom from 60 Bq/kg to 26.7 Bq/kg. The equilibrium of $^{210}$Pb with the initial radionuclide $^{226}$Ra was observed in the layer of 19.5–21.5 cm. The $^{137}$Cs specific activity was below 0.2 Bq/kg at all horizons, except for the 1–2 cm and 10–11 cm layers. The low content of the technogenic radionuclide $^{137}$Cs also indicates the periodic erosion of sediments.
- Based on $^{210}$Pb dating, the time of the beginning of sediment accumulation in the ice scour at a depth of 15 cm was estimated to be around 2002 AD; at a depth of 38 cm—around 1934 AD; at a depth of 40 cm—around 1917 AD and at a depth of 43 cm—around 1900 AD. Thus, over the past 120 years, there has been an increase in the mean SARs: 0.79 cm/year for the 0–15 cm horizon, 0.43 cm/year for the 0–38 cm horizon, 0.38 cm/year for the 0–40 cm thickness and not less than 0.35 cm/year for the 0–43 cm horizon.
- On the "background" seabed surface outside of the ice scour, the mean sedimentation rate over the past 110 years has been two times lower. The time of the beginning of sediment accumulation outside of the ice scour at a depth of 17.5 cm is estimated to be around 1910 AD (0.16 cm/year). Fluctuations in the mean SARs are not pronounced, which may be due to the periodic erosion of sediments outside of the ice scour.
- There is a close correlation between the marine sedimentation rates and air temperature fluctuations, as well as the coastal retreat rates of Baydaratskaya Bay, whereby the erosion products of which are the main source of seabed sediments due to the absence of large rivers in the area. Thus, since 2002, in the Baydaratskaya Bay area, there has been a sharp increase in air temperature, the rate of coastal retreat and the rate of sedimentation in the largest ice scour of this region.
- According to the results of $^{210}$Pb dating, the studied ice scour was formed no later than the end of the Little Ice Age (LIA) in the Arctic (turn of the 19th and 20th centuries). The age of the ice scour is estimated to be 1810 ± 30 AD based on the extrapolation of possible sedimentation rates prior to 1917 (0.22–0.38 cm/year). The mean rate of ice scour filling with 70 cm thick sediments from the moment of its formation is around 0.33 cm/year. Assuming that after the end of the LIA, the size of icebergs decreased, their penetration into Baydaratskaya Bay improved. Therefore, the ice scours of Baydaratskaya Bay were probably formed mainly after the end of the LIA, i.e., in the 20th century.
- Further study of the sedimentation chronology in ice scours will help to establish the periods of active ice scouring on the glaciated continental margins and to supplement knowledge about sedimentation on the Arctic shelf.

**Supplementary Materials:** The following supporting information can be downloaded at https://www.mdpi.com/article/10.3390/jmse11071404/s1: Table S1: Calculation of the sedimentary layers age (Year) and sedimentation rate (s(i)) in the "ice scour" core (ANS-52-16) using CF model (CF/43); Table S2: Calculation of the sedimentary layers age (Year) and sedimentation rate (s(i)) in the "background" core (ANS-52-17) using CF model (CF/19.5); Table S3: Calculation of the sedimentary layers age (Year') and sedimentation rate (s'(i)) in the "ice scour" core (ANS-52-16) using modified formulas of CF model (CFw/45).

**Author Contributions:** Conceptualization, O.K. and N.M.; methodology, O.K., I.U. and N.M.; field work, R.A., V.A., S.M. and N.S.; data analysis, O.K., I.U., N.M., R.A., V.A. and A.K.; writing—original draft preparation, O.K., I.U., N.M. and A.K.; writing—review and editing, O.K., A.K. and S.N. All authors have read and agreed to the published version of the manuscript.

**Funding:** The work was funded by the Russian Science Foundation, project no. 21-77-20038, GIN RAS, https://rscf.ru/en/project/21-77-20038/ (accessed on 10 July 2023). Dating results verification using Cs-137 was carried out within the framework of the MMBI RAS state assignment, topic no. 121091600105-4 (FMEE-2021-0029). Seismic profiling of previous cruises was obtained within the framework of the IO RAS state assignment, topic no. FMWE-2021-0005. A review of previous studies of the Baydaratskaya Bay ice scours was carried out within the framework of the MSU state assignment, topic no. 121051100167-1.

**Institutional Review Board Statement:** Not applicable.

**Informed Consent Statement:** Not applicable.

**Data Availability Statement:** Data available upon request from the authors.

**Acknowledgments:** We acknowledge the cruise administration, crew and participants of the R/V *Akademik Nikolaj Strakhov* (IO RAS) for their support and assistance in organizing the offshore activities and in obtaining and processing field data during the cruise. We are grateful to everyone who helped us greatly to improve the manuscript, and especially to the three reviewers, the JMSE issue editors for their helpful comments and Mitch Vowles for polishing our English.

**Conflicts of Interest:** The authors declare no conflict of interest.

# References

1. Lewis, C.F.M. Estimation of the frequency and magnitude of drift-ice groundings from the ice scour tracks in the Canadian Beaufort Sea. In Proceedings of the 4th International Conference on Port and Ocean Engineering under Arctic Conditions, St. John's, NL, Canada, 26–30 September 1977; Volume 1, pp. 568–579.
2. Barnes, P.W.; Rearic, D.M.; Reimnitz, E. Ice gouging characteristics and processes. In *The Alaskan Beaufort Sea: Ecosystems and Environments*; Barnes, P.W., Schell, D.M., Reimnitz, E., Eds.; Academic Press Inc.: Orlando, FL, USA, 1984; pp. 185–212.
3. Woodworth-Lynas, C.M.T. The Geology of Ice Scour. Ph.D. Thesis, The University of Wales, Cardiff, UK, November 1992.
4. Dowdeswell, J.A.; Villinger, H.; Whittington, R.J.; Marienfeld, P. Iceberg scouring in Scoresby Sund and on the East Greenland continental shelf. *Mar. Geol.* **1993**, *111*, 37–53. [CrossRef]
5. Ogorodov, S.; Arkhipov, V.; Kokin, O.; Marchenko, A.; Overduin, P.; Forbes, D. Ice Effect on Coast and Seabed in Baydaratskaya Bay, Kara Sea. *Geogr. Environ. Sustain.* **2013**, *6*, 21–37. [CrossRef]
6. Brown, C.S.; Newton, A.M.W.; Huuse, M.; Buckley, F. Iceberg scours, pits, and pockmarks in the North Falkland Basin. *Mar. Geol.* **2017**, *386*, 140–152. [CrossRef]
7. Ananiev, R.; Dmitrevskiy, N.; Jakobsson, M.; Lobkovsky, L.; Nikiforov, S.; Roslyakov, A.; Semiletov, I. Sea-ice ploughmarks in the eastern Laptev Sea, East Siberian Arctic shelf. In *Atlas of Submarine Glacial Landforms: Modern, Quaternary and Ancient*; Dowdeswell, J.A., Canals, M., Jakobsson, M., Todd, B.J., Dowdeswell, E.K., Hogan, K.A., Eds.; Geological Society: London, UK, 2016; pp. 301–302. [CrossRef]
8. Mironyuk, S.G.; Ivanova, A.A.; Kolyubakin, A.A. Extreme depths of modern ice gouging on the shelf of the northeastern part of the Barents Sea. *Ross. Polyarn. Issled.* **2018**, *1*, 12–14.
9. Maznev, S.; Ogorodov, S.; Baranskaya, A.; Vergun, A.; Arkhipov, V.; Bukharitsin, P. Ice-Gouging Topography of the Exposed Aral Sea Bed. *Remote Sens.* **2019**, *11*, 113. [CrossRef]
10. Bogoyavlensky, V.I.; Kishankov, A.V.; Kazanin, A.G. Heterogeneities in the Upper Part of the Section of the East Siberian Sea Sedimentary Cover: Gas Accumulations and Signs of Ice Gouging. *Dokl. Earth Sc.* **2022**, *505*, 411–415. [CrossRef]
11. Ottesen, D.; Dowdeswell, J.A. Distinctive iceberg ploughmarks on the mid-Norwegian margin: Tidally influenced chains of pits with implications for iceberg drift. *Arct. Antarct. Alp. Res.* **2022**, *54*, 163–175. [CrossRef]
12. Sokolov, S.Y.; Mazarovich, A.O.; Zakharov, V.G.; Zarayskaya, Y.A. Deep-Water Glacial Plow Marks in the Western Margin of the Barents Sea. *Dokl. Earth Sc.* **2022**, *503*, 75–80. [CrossRef]
13. Crane, K.; Vogt, P.R.; Sundvor, E. Deep Pleistocene Iceberg Plowmarks on the Yermak Plateau. In *Glaciated Continental Margins*; Springer: Dordrecht, The Netherlands, 1997; pp. 140–141. [CrossRef]
14. Arndt, J.E.; Niessen, F.; Jokat, W.; Dorschel, B. Deep Water Paleo-iceberg Scouring on Top of Hovgaard Ridge—Arctic Ocean. *Geophys. Res. Lett.* **2014**, *41*, 5068–5074. [CrossRef]
15. Condron, A.; Hill, J.C. Timing of iceberg scours and massive ice-rafting events in the subtropical North Atlantic. *Nat. Commun.* **2021**, *12*, 3668. [CrossRef]

16. Barnes, P.W.; Rearic, D.M. Rates of sediment disruption by sea ice as determined from characteristics of dated ice gouges created since 1975 on the inner shelf of the Beaufort Sea, Alaska. In *Open-File Report 85-463*; U.S. Geological Survey: Reston, VA, USA, 1985; pp. 1–35. [CrossRef]
17. Ogorodov, S.A.; Arkhipov, V.V.; Baranskaya, A.V.; Kokin, O.V.; Romanov, A.O. The Influence of Climate Change on the Intensity of Ice Gouging of the Bottom by Hummocky Formations. *Dokl. Earth Sci.* **2018**, *478*, 228–231. [CrossRef]
18. Aliyev, R.; Kalmykov, S. *Radioactivity: Tutorial*; Lan: St. Petersburg, Russia, 2013; 304p. (In Russian)
19. *Environmental Conditions of the Baydaratskaya Bay: The Main Results of Investigations for the Construction of the Yamal-Center Submarin Gas Pipeline System Crossing*; Baulin, V.V.; Dubikov, G.I.; Komarov, I.A.; Koreysha, M.M.; Parmuzin, S.Y.; Sovershaev, V.A.; Tuzhilkin, V.S. (Eds.) GEOS: Moscow, Russia, 1997; 432p. (In Russian)
20. *Ice Feartures of the Western Arctic Seas*; Zubakin, E.K. (Ed.) AANII: St. Petersburg, Russia, 2006; 272p. (In Russian)
21. Maznev, S.V.; Kokin, O.V.; Arkhipov, V.V.; Baranskaya, A.V. Modern and Relict Evidence of Iceberg Scouring at the Bottom of the Barents and Kara Seas. *Oceanology* **2023**, *63*, 84–94. [CrossRef]
22. Ogorodov, S.A.; Arkhipov, V.V.; Kokin, O.V. Climate Change Effect on the Intensity of Seabed Gouging by Hummocky Ice Floes. In *Arctic, Subarctic: Mosaic, Contrast, Variability of the Cryosphere: Proceedings of the International Conference*; Melnikov, V.P., Drozdov, D.S., Eds.; Epoha Publishing House: Tyumen, Russia, 2015; pp. 269–271. (In Russian)
23. Biryukov, V.Y.; Sovershaev, V.A. The relief of the bottom of the southwestern part of the Kara Sea and the history of its development in the Holocene. In *Geology and Geomorphology of Shelves and Continental Slopes*; Nauka: Moscow, Russia, 1985; pp. 89–95. (In Russian)
24. Ogorodov, S.A. *The Role of Sea Ice in Coastal Dynamics*; Moscow University Press: Moscow, Russia, 2011; 173p. (In Russian)
25. Ogorodov, S.A.; Arkhipov, V.V.; Kokin, O.V.; Marchenko, A.V. Comprehensive Monitoring of Ice Gouging Bottom Relief at Key Sites of Oil and Gas Development within the Coastal-Shelf Zone of the Yamal Peninsula, Kara Sea. In Proceedings of the International Conference on Port and Ocean Engineering under Arctic Conditions, POAC, Busan, Republic of Korea, 11–16 June 2017; pp. 123:1–123:12.
26. Arkhipov, V.V.; Kokin, O.V.; Ogorodov, S.A.; Godetyskiy, S.V.; Tsvetsinskiy, A.S.; Onishchenko, D.A. The Yamal coast fast ice edge of the Baidaratskaya Bay of the Kara Sea in 2012–2016: Dynamics and role in formation of modern ice gouges on the sea-bed. *Vesti Gazov. Nauk.* **2017**, *4*, 129–136. (In Russian)
27. Gurevich, V.I. *Modern Sedimentogenesis and Geoecology of the Western Arctic Shelf of Eurasia*; Nauchnyy Mir: Moscow, Russia, 2002; 135p. (In Russian)
28. Polyak, L.; Levitan, M.; Khusid, T.; Merklin, L.; Mukhina, V. Variations in the influence of riverine discharge on the Kara Sea during the last deglaciation and the Holocene. *Global Planet. Chang.* **2002**, *32*, 291–309. [CrossRef]
29. Stein, R.; Dittmers, K.; Niessen, F.; Fahl, K. Siberian river run-off and Late Quaternary glaciation in the southern Kara Sea, Arctic Ocean: Preliminary results. *Rep. Polar Mar. Res.* **2002**, *21*, 315–322. [CrossRef]
30. Galimov, E.M.; Kodina, L.A.; Stepanets, O.V.; Korobeinik, G.S. Biogeochemistry of the Russian Arctic. Kara Sea: Research Results under the SIRRO Project, 1995–2003. *Geochem. Int.* **2006**, *44*, 1139–1191. [CrossRef]
31. Stepanets, O.; Borisov, A.; Ligaev, A.; Galimov, E. The investigation of sedimentation rate of the Kara Sea modern sediments using radioactive tracer. *Rep. Polar Mar. Res.* **2001**, *393*, 205–212.
32. Stepanets, O.V.; Borisov, A.P.; Travkina, A.V.; Soloveva, G.Y.; Vladimirov, M.V.; Aliev, R.A. Application of the $^{210}$Pb and $^{137}$Cs radionuclides in the geochronology of modern sediments at the storage sites of solid radioactive wastes in the Arctic Basin. *Geochem. Int.* **2010**, *48*, 398–402. [CrossRef]
33. Rusakov, V.Y.; Borisov, A.P.; Solovieva, G.Y. Sedimentation rates in different facies–genetic types of bottom sediments in the Kara Sea: Evidence from the $^{210}$Pb and $^{137}$Cs radionuclides. *Geochem. Int.* **2019**, *57*, 1185–1200. [CrossRef]
34. Nikiforov, S.L.; Sorokhtin, N.O.; Ananiev, R.A.; Dmitrevskiy, N.N.; Moroz, E.A.; Kokin, O.V. Research in Barents and Kara Seas during cruise 52 of the R/V Akademik Nikolaj Strakhov. *Oceanology* **2022**, *62*, 433–434. [CrossRef]
35. Andreeva, I.A.; Lapina, N.N. *Method of Grain-Size Analysis of Bottom Sediments of the World Ocean and Geological Interpretation of the Results of Laboratory Study*; VNIIOkeangeologia: St. Petersburg, Russia, 1998. (In Russian)
36. Klenova, M.V. *Geology of the Seas*; Uchpedgiz: Moscow, Russia, 1948; 495p. (In Russian)
37. Appleby, P.G.; Nolan, P.J.; Gifford, D.W.; Godfrey, M.J.; Oldfield, F.; Anderson, N.J.; Battarbee, R.W. 210 Pb dating by low-background gamma. *Hydrobiologia* **1986**, *143*, 21–27. [CrossRef]
38. Schelske, C.L.; Peplow, A.; Brenner, M.; Spencer, C.N. Low-background gamma counting: Applications for 210 Pb dating of sediments. *J. Paleolimnol.* **1994**, *10*, 115–128. [CrossRef]
39. Appleby, P.G.; Piliposian, G.T. Efficiency corrections for variable sample height in well-type germanium gamma detectors. *Nucl. Instrum. Methods Phys. Res. Sect. B Beam Interact. Mater. At.* **2004**, *225*, 423–433. [CrossRef]
40. Aliev, R.A.; Bobrov, V.A.; Kalmykov, S.N.; Melgunov, M.S.; Vlasova, I.E.; Shevchenko, V.P.; Novigatsky, A.N.; Lisitzin, A.P. Natural and artificial radionuclides as a tool for sedimentation studies in the Arctic region. *J. Radioanal. Nucl. Chem.* **2007**, *274*, 315–321. [CrossRef]
41. Zaborska, A.; Carroll, J.; Papucci, C.; Torricelli, L.; Carroll, M.L.; Walkusz-Miotk, J.; Pempkowiak, J. Recent sediment accumulation rates for the Western margin of the Barents Sea. *Deep.-Sea Res.* **2008**, *55*, 2352–2360. [CrossRef]
42. Kuzyk, Z.; Gobeil, C.; Macdonald, R. $^{210}$Pb and $^{137}$Cs in margin sediments of the Arctic Ocean: Controls on boundary scavenging. *Glob. Biogeochem. Cycles* **2013**, *27*, 422–439. [CrossRef]

43. Goryachenkova, T.A.; Borisov, A.P.; Solov'eva, G.Y.; Lavrinovich, E.A.; Kazinskaya, I.E.; Ligaev, A.N.; Travkina, A.V.; Novikov, A.P. Content of Technogenic Radionuclides in Water, Bottom Sediments, and Benthos of the Kara Sea and Shallow Bays of the Novaya Zemlya Archipelago. *Geochem. Int.* **2019**, *57*, 1320–1326. [CrossRef]
44. Demina, L.; Dara, O.; Aliev, R.; Alekseeva, T.; Budko, D.; Novichkova, E.; Politova, N.; Solomatina, A.; Bulokhov, A. Elemental and Mineral Composition of the Barents Sea Recent and Late Pleistocene−Holocene Sediments: A Correlation with Environmental Conditions. *Minerals* **2020**, *10*, 593. [CrossRef]
45. Robbins, J.A.; Edgington, D.N. Determination of recent sedimentation rates in Lake Michigan using Pb-210 and Cs-137. *Geochim. Cosmochim. Acta* **1975**, *39*, 285–304. [CrossRef]
46. von Gunten, H.R.; Moser, R.N. How reliable is the $^{210}$Pb dating method? Old and new results from Switzerland. *J. Paleolimnotogy* **1993**, *9*, 161–178. [CrossRef]
47. Kirchner, G. $^{210}$Pb as a tool for establishing sediment chronologies: Examples of potentials and limitations of conventional dating models. *J. Environ. Radioact.* **2011**, *102*, 490–494. [CrossRef]
48. Pittauerova, D.; Hettwig, B.; Fischer, H.W. Pb-210 sediment chronology: Focused on supported lead. *Radioprotection* **2011**, *46*, S277–S282. [CrossRef]
49. Sanchez-Cabeza, J.A.; Ruiz-Fernández, A.C. $^{210}$Pb sediment radiochronology: An integrated formulation and classification of dating models. *Geochim. Cosmochim. Acta* **2012**, *82*, 183–200. [CrossRef]
50. Mabit, L.; Benmansour, M.; Abril, J.M.; Walling, D.E.; Meusburger, K.; Iurian, A.R. Fallout $^{210}$Pb as a soil and sediment tracer in catchment sediment budget investigations: A review. *Earth-Sci. Rev.* **2014**, *138*, 335–351. [CrossRef]
51. Gharibreza, M.; Zaman, M.; Arabkhedri, M.S.-Z. The off-site implications of deforestation on sedimentation rates and pollution in Abkenar open water (Anzali Lagoon, Caspian Sea) using radionuclide techniques and sediment quality indices. *Int. J. Sediment Res.* **2021**, *37*, 370–382. [CrossRef]
52. Appleby, P.G. Chronostratigraphic techniques in recent sediments. In *Tracking Environmental Change Using Lake Sediments, Volume I: Basin Analysis, Coring, and Chronological Techniques*; Last, W.S., Smol, J.P., Eds.; Springer: Dordrecht, The Netherlands, 2002; pp. 171–203. [CrossRef]
53. Longmore, M.E. The caesium-137 dating technique and associated applications in Australiaea review. In *Archaeometry: An Australasian Perspective*; Ambrose, B.W., Duerden, P., Eds.; Australian National University Press: Canberra, Australia, 1982; pp. 310–321.
54. Khalturin, V.I.; Rautian, T.G.; Richards, P.G.; Leith, W.S. A Review of Nuclear Testing by the Soviet Union at Novaya Zemlya, 1955–1990. *Sci. Glob. Secur.* **2005**, *13*, 1–42. [CrossRef]
55. Norris, R.S.; Arkin, W.M. Known Nuclear Tests Worldwide, 1945–1994. *Bull. At. Sci.* **1995**, *51*, 70–71. [CrossRef]
56. Bergqvist, N.-O.; Ferm, R. *Nuclear Explosions 1945–1998 (FOA-R-00-01572-180)*; Defence Research Establishment: Stocholm, Sweden, 2000; 43p.
57. Kautsky, H. Determination of distribution processes, transport routes and transport times in the North Sea and the northern Atlantic using artificial radionuclides as tracers. In *Radionuclides: A Tool for Oceanography*; Guary, B.J.C., Guegueniat, P., Pentreath, R.J., Eds.; Elsevier Applied Science: London, UK, 1988; pp. 271–280.
58. Kamalov, A.M.; Ogorodov, S.A.; Birukov, V.Y.; Sovershaeva, G.D.; Tsvetsinsky, A.S.; Arkhipov, V.V.; Belova, N.G.; Noskov, A.I.; Solomatin, V.I. Coastal and seabed morpholithodynamics of the Baydaratskaya bay at the route of gas pipeline crossing. *Earth's Cryosphere* **2006**, *10*, 3–14. (In Russian)
59. Romanenko, F.A.; Garankina, E.V.; Shilova, O.S. Loose deposits stratigraphy and the formation of the relief of Western Yamal in the late Pleistocene−Holocene. In *Fundamental Problems of the Quarter: Results of the Study and the Main Directions of Further Research, Proceedings of the VI All-Russian Meeting of the Quaternary Period Research, Novosibirsk, Russia, 19–23 October 2009*; SO RAN: Novosibirsk, Russia, 2009; pp. 505–508. (In Russian)
60. Kopa-Ovdienko, N.V.; Ogorodov, S.A. Peculiarities of dynamics of thermoabrasional coasts of the Baydaratskaya Bay (Kara Sea) today. *Geomorfologiya* **2016**, *3*, 12–21. (In Russian) [CrossRef]
61. Romanenko, F.A.; Belova, N.G.; Nikolaev, V.I.; Olyunina, O.S. Features of the loose deposits structure of the Yugra coast of the Baydaratskaya Bay, the Kara Sea. In *Fundamental Problems of the Quarter: Results of the Study and the Main Directions of Further Research, Proceedings of the V All-Russian Meeting of the Quaternary Period Research, Moscow, Russia, 7–9 November 2007*; GEOS: Moscow, Russia, 2007; pp. 348–351. (In Russian)
62. Strakhov, N.M. *Types of Lithogenesis and Their Evolution in the History of the Earth*; Gosgeoltekhizdat: Moscow, Russia, 1963; 535p. (In Russian)
63. Meshcheryakov, N.I.; Usyagina, I.S.; Sharin, V.V.; Dauvalter, V.A.; Dukhno, G.N. Chronology of sedimentation in Colesbukta, Spitsbergen (Svalbard Archipelago): The results of the 2018 expedition. *IOP Conf. Ser. Earth Environ. Sci.* **2021**, *937*, 042081. [CrossRef]
64. Shabanova, N.; Ogorodov, S.; Shabanov, P.; Baranskaya, A. Hydrometeorological forcing of Western Russian arctic coastal dynamics: XX-century history and current state. *Geogr. Environ. Sustain.* **2018**, *11*, 113–129. [CrossRef]
65. Kennedy, J.J.; Rayner, N.A.; Atkinson, C.P.; Killick, R.E. An ensemble data set of sea surface temperature change from 1850: The Met Office Hadley Centre HadSST. 4.0.0.0 data set. *J. Geophys. Res. Atmos.* **2019**, *124*, 7719–7763. [CrossRef]

66. Morice, C.P.; Kennedy, J.J.; Rayner, N.A.; Winn, J.P.; Hogan, E.; Killick, R.E. An updated assessment of near-surface temperature change from 1850: The HadCRUT5 data set. *J. Geophys. Res. Atmos.* **2021**, *126*, e2019JD032361. [CrossRef]
67. Osborn, T.J.; Jones, P.D.; Lister, D.H.; Morice, C.P.; Simpson, I.R.; Winn, J.P. Land surface air temperature variations across the globe updated to 2019: The CRUTEM5 data set. *J. Geophys. Res. Atmos.* **2021**, *126*, e2019JD032352. [CrossRef]

**Disclaimer/Publisher's Note:** The statements, opinions and data contained in all publications are solely those of the individual author(s) and contributor(s) and not of MDPI and/or the editor(s). MDPI and/or the editor(s) disclaim responsibility for any injury to people or property resulting from any ideas, methods, instructions or products referred to in the content.

MDPI AG
Grosspeteranlage 5
4052 Basel
Switzerland
Tel.: +41 61 683 77 34

*Journal of Marine Science and Engineering* Editorial Office
E-mail: jmse@mdpi.com
www.mdpi.com/journal/jmse

Disclaimer/Publisher's Note: The title and front matter of this reprint are at the discretion of the Guest Editors. The publisher is not responsible for their content or any associated concerns. The statements, opinions and data contained in all individual articles are solely those of the individual Editors and contributors and not of MDPI. MDPI disclaims responsibility for any injury to people or property resulting from any ideas, methods, instructions or products referred to in the content.

www.ingramcontent.com/pod-product-compliance
Lightning Source LLC
LaVergne TN
LVHW072332090526
838202LV00019B/2404